AF831087

General Quantum Numerical Analysis

This book is focused on the qualitative theory of general quantum calculus, the modern name for the investigation of calculus without limits. It centers on designing, analysing and applying computational techniques for general quantum differential equations.

The quantum calculus or q-calculus began with F.H. Jackson in the early twentieth century, but this kind of calculus had already been worked out by Euler and Jacobi. Recently, it has aroused interest due to the high demand of mathematics that models quantum computing and the connection between mathematics and physics.

Quantum calculus has many applications in different mathematical areas such as number theory, combinatorics, orthogonal polynomials, basic hyper-geometric functions and other sciences such as quantum theory, mechanics and the theory of relativity.

The authors summarize the most recent contributions in this area. *General Quantum Numerical Analysis* is intended for senior undergraduate students and beginning graduate students of engineering and science courses. The twelve chapters in this book are pedagogically organized, each concluding with a section of practical problems.

Svetlin G. Georgiev is a mathematician who has worked in various areas of mathematics. He currently focuses on harmonic analysis, functional analysis, partial differential equations, ordinary differential equations, Clifford and quaternion analysis, integral equations and dynamic calculus on time scales.

Khaled Zennir earned his PhD in mathematics in 2013 from Sidi Bel Abbès University, Algeria. In 2015, he received his highest diploma in Habilitation in mathematics from Constantine University, Algeria. He is currently an assistant professor at Qassim University in the Kingdom of Saudi Arabia. His research interests lie in the subjects of nonlinear hyperbolic partial differential equations: global existence, blowup and longtime behaviour.

Advances in Applied Mathematics
Series Editor: Daniel Zwillinger

Introduction to Quantum Control and Dynamics
Domenico D'Alessandro

Handbook of Radar Signal Analysis
Bassem R. Mahafza, Scott C. Winton, Atef Z. Elsherbeni

Separation of Variables and Exact Solutions to Nonlinear PDEs
Andrei D. Polyanin, Alexei I. Zhurov

Boundary Value Problems on Time Scales, Volume I
Svetlin G. Georgiev, Khaled Zennir

Boundary Value Problems on Time Scales, Volume II
Svetlin G. Georgiev, Khaled Zennir

Observability and Mathematics
Fluid Mechanics, Solutions of Navier-Stokes Equations, and Modeling
Boris Khots

Handbook of Differential Equations, Fourth Edition
Daniel Zwillinger, Vladimir Dobrushkin

Experimental Statistics and Data Analysis for Mechanical and Aerospace Engineers
James Middleton

Advanced Engineering Mathematics with MATLAB®, Fifth Edition
Dean G. Duffy

Handbook of Fractional Calculus for Engineering and Science
Harendra Singh, H. M. Srivastava, Juan J. Nieto

Advanced Engineering Mathematics
A Second Course with MATLAB®
Dean G. Duffy

Quantum Computation
Helmut Bez and Tony Croft

Computational Mathematics
An Introduction to Numerical Analysis and Scientific Computing with Python
Dimitrios Mitsotakis

Delay Ordinary and Partial Differential Equations
Andrei D. Polyanin, Vsevolod G. Sorkin, Alexi I. Zhurov

Clean Numerical Simulation
Shijun Liao

Multiplicative Partial Differential Equations
Svetlin G. Georgiev, Khaled Zennir

Engineering Statistics
A Matrix-Vector Approach with MATLAB®
Lester W. Schmerr Jr.

General Quantum Numerical Analysis
Svetlin G. Georgiev, Khaled Zennir

https://www.routledge.com/Advances-in-Applied-Mathematics/book-series/CRCADVAPPMTH?pd=published,forthcoming&pg=

General Quantum Numerical Analysis

Svetlin G. Georgiev
Khaled Zennir

CRC Press
Taylor & Francis Group
Boca Raton London New York

CRC Press is an imprint of the
Taylor & Francis Group, an **informa** business
A CHAPMAN & HALL BOOK

First edition published 2024
by CRC Press
2385 Executive Center Drive, Suite 320, Boca Raton, FL 33431

and by CRC Press
4 Park Square, Milton Park, Abingdon, Oxon, OX14 4RN

CRC Press is an imprint of Taylor & Francis Group, LLC

© 2024 Svetlin G. Georgiev and Khaled Zennir

Reasonable efforts have been made to publish reliable data and information, but the author and publisher cannot assume responsibility for the validity of all materials or the consequences of their use. The authors and publishers have attempted to trace the copyright holders of all material reproduced in this publication and apologize to copyright holders if permission to publish in this form has not been obtained. If any copyright material has not been acknowledged, please write and let us know so we may rectify in any future reprint.

Except as permitted under U.S. Copyright Law, no part of this book may be reprinted, reproduced, transmitted, or utilized in any form by any electronic, mechanical, or other means, now known or hereafter invented, including photocopying, microfilming, and recording, or in any information storage or retrieval system, without written permission from the publishers.

For permission to photocopy or use material electronically from this work, access www.copyright.com or contact the Copyright Clearance Center, Inc. (CCC), 222 Rosewood Drive, Danvers, MA 01923, 978-750-8400. For works that are not available on CCC, please contact mpkbookspermissions@tandf.co.uk

Trademark notice: Product or corporate names may be trademarks or registered trademarks and are used only for identification and explanation without intent to infringe.

Library of Congress Cataloging-in-Publication Data

Names: Georgiev, Svetlin, author. | Zennir, Khaled, author.
Title: General quantum numerical analysis / Svetlin G. Georgiev and Khaled Zennir.
Description: First edition. | Boca Raton : CRC Press, 2024. | Includes bibliographical references and index.
Identifiers: LCCN 2023052991 | ISBN 9781032741505 (hardback) | ISBN 9781032750507 (paperback) | ISBN 9781003472131 (ebook)
Subjects: LCSH: Differential equations--Qualitative theory | Quantum theory--Mathematics.
Classification: LCC QA371 .G425 2024 | DDC 515/.3--dc23/eng/20240310
LC record available at https://lccn.loc.gov/2023052991

ISBN: 978-1-032-74150-5 (hbk)
ISBN: 978-1-032-75050-7 (pbk)
ISBN: 978-1-003-47213-1 (ebk)

DOI: 10.1201/9781003472131

Typeset in Nimbus Roman font
by KnowledgeWorks Global Ltd.

Publisher's note: This book has been prepared from camera-ready copy provided by the authors.

Contents

Preface

Quantum calculus is the modern name for the investigation of calculus without limits. Quantum calculus, or q-calculus, began with F.H. Jackson in the early twentieth century, but this kind of calculus had already been worked out by Euler and Jacobi. Recently, interest in it has resurfaced due to the high demand of mathematics that models quantum computing. q-calculus appeared as a connection between mathematics and physics. It has a lot of applications in different mathematical areas such as number theory, combinatorics, orthogonal polynomials, basic hyper-geometric functions and other sciences quantum theory, mechanics and the theory of relativity. The book by Kac and Cheung [3] covers many of the fundamental aspects of quantum calculus. Recently they have defined general quantum difference operators that generalise quantum calculus. This book is devoted to the qualitative theory of general quantum calculus and designing, analysing and applying computational techniques for general quantum differential equations. The book is intended for senior undergraduate students and beginning graduate students of engineering and science. The twelve chapters in the book are pedagogically organised. Each chapter concludes with a section of practical problems.

In Chapter 1, the first and second kinds of general quantum operators are defined. The concept of general quantum differentiation is introduced. In the chapter the main properties and main rules for general quantum differentiation are deduced. They are given chain rules, and a mean values theorem, and the Leibniz formula is proven. Chapter 2 is devoted to the general quantum integration, and some of its properties are proved. In the chapter the basic rules for general quantum integration are deduced. The Taylor formula is deduced. In the chapter, improper integrals of the first and second kinds are investigated. In Chapter 3, some general quantum elementary functions are introduced. They are defined general quantum exponentials, and general quantum trigonometric and hyperbolic functions. Some of their basic properties are inferred. Chapter 4 deals with Lagrange, β-Lagrange, Hermite and β-Hermite interpolation polynomials. We show that in general Lagrange and β-Lagrange interpolation polynomials, as well as Hermite and β-Hermite interpolation polynomials, are different. From these interpolations, they are deducted approximations for the β-derivative of continuously β-differentiable functions. In Chapter 5, using the Lagrange, β-Lagrange, Hermite, and β-Hermite interpolation polynomials, we derive formulae for numerical β-integration. We deduce the trapezium and β-trapezium rules, Newton-Cotes and β-Newton-Cotes formulas, and we give some error estimates. In the chapter Gauss and β-Gauss quadrature rules are constructed. They are considered Euler-Maclauren and β-Euler-Maclauren expansions. In Chapter 6, piecewise approximations of a given function are introduced. They are defined linear interpolating splines, linear interpolating β-splines, cubic and Hermite splines. We deduce some of their properties. In Chapter 7, we investigate the Euler method for β-differential equations. We define local and global truncate errors and we evaluate them in both cases when $x_0 = s_0$ and $x_0 \neq s_0$. We provide the obtained results with two examples that are β-analogues of the classical logistic and Riccati equations. Chapter 8 is devoted on the order-two Taylor series method for first-order β-differential equations. We investigate its convergence. In the chapter, a β-analogue of the trapezoidal rule is deduced. In Chapter 9, we introduce the order-p Taylor series method. We investigate it for convergence and give estimations of the errors. As an application, we introduce the two-step Adams-Bashforth method. Chapter 10 deals with linear two- and k-step methods. We investigate these methods for consistency and give their constructions. In Chapter 11, we analyse quantum analogues of the Runge-Kutta method. We consider one-, two-, three- and s-stage methods. In Chapter 12, we derive a quantum analogue of the Adomian polynomial method and present its application to a β-differential equation of arbitrary order with a nonlinear term.

This book is addressed to a wide audience of specialists such as mathematicians, physicists, engineers and biologists. It can also be used as a textbook at the graduate level and as a reference book for several disciplines.

Paris, September 2023 *Svetlin G. Georgiev and Khaled Zennir*

1

General Quantum Differentiation

In this chapter, we define the first and second kinds of general quantum operators, and using them, we define the β-derivative and explore some of its properties. The chapter gives the main rules for β-differentiation. Some chain rules, a mean value theorem and the Leibniz formula are deduced and proved.

Suppose that $I \subset \mathbb{R}$ and $s_0 \in I$.

1.1 The β-Operator

Definition 1.1.1 *We say that $\beta : I \to \mathbb{R}$ is a first kind general quantum operator if $\beta : I \to \mathbb{R}$ is a strictly increasing continuous function on I with $\beta(t) \in I$ for any $t \in I$ and s_0 is the unique fixed point of β, and the following inequality holds.*

$$(t - s_0)(\beta(t) - t) \leq 0, \quad t \in I,$$

Definition 1.1.2 *We say that $\beta : I \to \mathbb{R}$ is a second kind general quantum operator if $\beta : I \to \mathbb{R}$ is a strictly increasing continuous function on I with $\beta(t) \in I$ for any $t \in I$ and s_0 is the unique fixed point of β, and the following inequality holds.*

$$(t - s_0)(\beta(t) - t) \geq 0, \quad t \in I,$$

Example 1.1.3 *Let $I = [3, \infty)$, $s_0 = 3$ and*

$$\beta(t) = 2\log\left(\frac{t}{3}\right) + 3, \quad t \in I.$$

Note that for any $t \in I$, we have

$$\log\left(\frac{t}{3}\right) \geq 0$$

DOI: 10.1201/9781003472131-1

and hence, $\beta(t) \in I$ for any $t \in I$. Also,

$$\begin{aligned} \beta'(t) &= \frac{2}{t} \\ &> 0, \quad t \in I, \end{aligned}$$

i.e., β is strictly increasing on I. Moreover,

$$\begin{aligned} \beta(3) &= 2\log\left(\frac{3}{3}\right)+3 \\ &= 2\log 1+3 \\ &= 3, \end{aligned}$$

i.e., 3 is the unique fixed point of β. Set

$$h(t) = 2\log\left(\frac{t}{3}\right)+3-t, \quad t \in I,$$

and

$$g(t) = (t-3)h(t), \quad t \in I.$$

We have

$$\begin{aligned} h'(t) &= \frac{2}{t}-1 \\ &= \frac{2-t}{t} \\ &< 0, \quad t \in I. \end{aligned}$$

Thus, h is strictly decreasing on I, and then

$$\begin{aligned} h(t) &\le h(3) \\ &= 2\log\left(\frac{3}{3}\right)+3-3 \\ &= 2\log 1 \\ &= 0. \end{aligned}$$

Hence, using that $t-3 \geq 0$, $t \in I$, we get

$$g(t) \le 0, \quad t \in I.$$

Therefore, β is a first kind general quantum operator.

Example 1.1.4 *Let $I = \mathbb{R}$, $q \in (0,1)$, $\omega \geq 0$, and*

$$\beta(t) = qt \pm \omega, \quad t \in I.$$

Then, $\beta(t) \in I$ for any $t \in I$, and

$$\begin{aligned} \beta'(t) &= q \\ &> 0, \quad t \in I. \end{aligned}$$

Thus, β is a strictly increasing continuous function on I. Consider the equation

$$q(s_0) = s_0,$$

or

$$s_0 = qs_0 \pm \omega,$$

or

$$(1-q)s_0 = \pm\omega,$$

whereupon

$$s_0 = \pm\frac{\omega}{1-q}$$

is the unique fixed point of β. Next, consider the function

$$g(t) = \left(t \mp \frac{\omega}{1-q}\right)((q-1)t \pm \omega), \quad t \in I.$$

1. *Let $t \geq \pm\frac{\omega}{1-q}$. Then*

$$t \mp \frac{\omega}{1-q} \geq 0,$$

and

$$\begin{aligned} (q-1)t \pm \omega &= -(1-q)t \pm \omega \\ &= -(1-q)\left(t \mp \frac{\omega}{1-q}\right) \\ &\leq 0. \end{aligned}$$

Hence, $g(t) \leq 0$.

2. *Let $t \leq \pm\frac{\omega}{1-q}$. Then*

$$t \mp \frac{\omega}{1-q} \leq 0,$$

and

$$\begin{aligned}(q-1)t \pm \omega &= -(1-q)\left(t \mp \frac{\omega}{1-q}\right)\\ &\geq 0.\end{aligned}$$

Hence, $g(t) \leq 0$.

Therefore, β *is a first kind general quantum operator.*

Example 1.1.5 *Let* $I = \mathbb{R}$, $q > 1$, $\omega \geq 0$ *and* β, g *and* s_0 *be as in Example 1.1.4. Then,* β *is a strictly increasing function on* I *with unique fixed point* s_0. *Moreover,*

1. *if* $t \geq \pm \frac{\omega}{1-q}$, *we have*

$$(q-1)t \pm \omega \geq 0$$

and $g(t) \geq 0$.

2. *if* $t \leq \pm \frac{\omega}{1-q}$, *then*

$$(q-1)t \pm \omega \leq 0$$

and $g(t) \geq 0$.

Thus, β *is a second kind general quantum operator.*

Example 1.1.6 *Let* $q \in (0,1)$, $n \in 2\mathbb{N}+1$, $I = \left(-q^{\frac{1}{1-n}}, q^{\frac{1}{1-n}}\right)$ *and*

$$\beta(t) = qt^n, \quad t \in I.$$

Then

$$\begin{aligned}\beta'(t) &= qnt^{n-1}\\ &> 0, \quad t \in I, \quad t \neq 0,\end{aligned}$$

and $\beta'(0) = 0$. *Thus,* β *is a strictly increasing continuous function on* I. *Hence, using that* $\beta(0) = 0$, *we conclude that* $s_0 = 0$ *is the unique fixed point of* β. *Next,*

$$\begin{aligned}t\left(qt^n - t\right) &= t^2\left(qt^{n-1} - 1\right)\\ &\leq t^2\left(q\left(q^{\frac{1}{1-n}}\right)^{n-1} - 1\right)\\ &= 0, \quad t \in I.\end{aligned}$$

Thus, β *is a first kind general quantum operator.*

Exercise 1.1.7 *Let* $I = \left[1, \frac{3}{2}\right]$,

$$\beta(t) = \frac{1}{2}t + \frac{3}{4}, \quad t \in I.$$

Prove that β *is a first kind general quantum operator.*

Definition 1.1.8 *Let* β *be a first kind or a second kind general quantum operator. For* $k \in \mathbb{N}_0$, *define*

$$\begin{aligned}
\beta^k(t) &= \underbrace{\beta \circ \beta \circ \cdots \circ \beta}_{k \quad times}(t), \\
\beta^{-k}(t) &= \underbrace{\beta^{-1} \circ \beta^{-1} \circ \cdots \circ \beta^{-1}}_{k \quad times}(t), \\
\beta^0(t) &= t, \quad t \in I.
\end{aligned}$$

Definition 1.1.9 *For* $q > 0$, $q \neq 1$, *and* $n \in \mathbb{N}_0$, *define*

$$[n]_q = \frac{1 - q^n}{1 - q}.$$

Example 1.1.10 *Let* I, β, q, ω *and* s_0 *be as in Example 1.1.4. Then*

$$qt = \beta(t) \mp \omega, \quad t \in I,$$

whereupon

$$t = \frac{\beta(t) \mp \omega}{q}, \quad t \in I.$$

Hence,

$$\beta^{-1}(t) = \frac{t \mp \omega}{q}, \quad t \in I.$$

Moreover, we have

$$\begin{aligned}
\beta^2(t) &= \beta \circ \beta(t) \\
&= q\beta(t) \pm \omega \\
&= q(qt \pm \omega) \pm \omega \\
&= q^2 t \pm q\omega \pm \omega \\
&= q^2 t \pm (q+1)\omega
\end{aligned}$$

$$\begin{aligned} &= q^2t \pm \frac{1-q^2}{1-q}\omega \\ &= q^2t \pm [2]_q\omega, \quad t \in I, \end{aligned}$$

and

$$\begin{aligned} \beta^3(t) &= \beta \circ \beta \circ \beta(t) \\ &= \beta \circ \beta^2(t) \\ &= \beta\left(\beta^2(t)\right) \\ &= q\beta^2(t) \pm \omega \\ &= q\left(q^2t \pm (1+q)\omega\right) \pm \omega \\ &= q^3t \pm (q+q^2)\omega \pm \omega \\ &= q^3t \pm (1+q+q^2)\omega \\ &= q^3t \pm \frac{1-q^3}{1-q}\omega \\ &= q^3t \pm [3]_q\omega, \quad t \in I. \end{aligned}$$

Assume that

$$\beta^l(t) = q^lt \pm [l]_q\omega, \quad t \in I,$$

for some $l \in \mathbb{N}$. We will prove that

$$\beta^{l+1}(t) = q^{l+1}t \pm [l+1]_q\omega, \quad t \in I.$$

Indeed, we have

$$\begin{aligned} \beta^{l+1}(t) &= \beta \circ \beta^l(t) \\ &= \beta\left(\beta^l(t)\right) \\ &= q\beta^l(t) \pm \omega \end{aligned}$$

$$\begin{aligned}
&= q\left(q^l t \pm (1+q+\cdots+q^{l-1})\omega\right) \pm \omega \\
&= q^{l+1}t \pm \left(q+q^2+\cdots+q^l\right)\omega \pm \omega \\
&= q^{l+1}t \pm \left(1+q+q^2+\cdots+q^l\right)\omega \\
&= q^{l+1}t \pm \frac{1-q^{l+1}}{1-q}\omega \\
&= q^{l+1}t \pm [l+1]_q\omega, \quad t \in I.
\end{aligned}$$

Hence, applying the principle of mathematical induction, we conclude that

$$\beta^n(t) = q^n t \pm [n]_q\omega, \quad t \in I,$$

for any $n \in \mathbb{N}_0$. Next,

$$\begin{aligned}
\beta^{-2}(t) &= \beta^{-1} \circ \beta^{-1}(t) \\
&= \beta^{-1}\left(\beta^{-1}(t)\right) \\
&= \frac{\beta^{-1}(t) \mp \omega}{q} \\
&= \frac{\frac{t\mp\omega}{q} \mp \omega}{q} \\
&= \frac{t \mp \omega \mp q\omega}{q^2} \\
&= \frac{t \mp (1+q)\omega}{q^2} \\
&= \frac{t \mp \frac{1-q^2}{1-q}\omega}{q^2} \\
&= \frac{t \mp [2]_q\omega}{q^2}, \quad t \in I,
\end{aligned}$$

and

$$\begin{aligned}
\beta^{-3}(t) &= \beta^{-1} \circ \beta^{-2}(t) \\
&= \beta^{-1}\left(\beta^{-2}(t)\right)
\end{aligned}$$

$$= \frac{\beta^{-2}(t) \mp \omega}{q}$$

$$= \frac{\frac{t \mp (1+q)\omega}{q^2} \mp \omega}{q}$$

$$= \frac{t \mp (1+q)\omega \mp q^2\omega}{q^3}$$

$$= \frac{t \mp (1+q+q^2)\omega}{q^3}$$

$$= \frac{t \mp \frac{1-q^3}{1-q}\omega}{q^3}$$

$$= \frac{t \mp [3]_q\omega}{q^3}, \quad t \in I.$$

Assume that

$$\beta^{-l}(t) = \frac{t \mp [l]_q\omega}{q^l}, \quad t \in I,$$

for some $l \in \mathbb{N}$. We will prove that

$$\beta^{-(l+1)}(t) = \frac{t \mp [l+1]_q\omega}{q^{l+1}}, \quad t \in I.$$

Indeed,

$$\begin{aligned}
\beta^{-(l+1)}(t) &= \beta^{-1} \circ \beta^{-l}(t) \\
&= \beta^{-1}\left(\beta^{-l}(t)\right) \\
&= \frac{\beta^{-l}(t) \mp \omega}{q} \\
&= \frac{\frac{t \mp (1+q+\cdots+q^{l-1})\omega}{q^l} \mp \omega}{q} \\
&= \frac{t \mp (1+q+\cdots+q^{l-1})\omega \mp q^l\omega}{q^{l+1}} \\
&= \frac{t \mp (1+q+\cdots+q^{l-1}+q^l)\omega}{q^{l+1}}
\end{aligned}$$

$$= \frac{t \mp \frac{1-q^{l+1}}{1-q}\omega}{q^{l+1}}$$

$$= \frac{t \mp [l+1]_q \omega}{q^{l+1}}, \quad t \in I.$$

Hence, applying the principle of mathematical induction, we conclude that

$$\beta^{-n}(t) = \frac{t \mp [n]_q \omega}{q^n}, \quad t \in I,$$

for any $n \in \mathbb{N}_0$.

Example 1.1.11 *Let* I, β, q *and* n *be as in Example 1.1.6. Then*

$$t^n = \frac{\beta(t)}{q}, \quad t \in I,$$

and

$$t = \left(\frac{\beta(t)}{q}\right)^{\frac{1}{n}}, \quad t \in I.$$

Therefore

$$\beta^{-1}(t) = \left(\frac{t}{q}\right)^{\frac{1}{n}}, \quad t \in I.$$

Moreover,

$$\begin{aligned} \beta^2(t) &= \beta \circ \beta(t) \\ &= \beta(\beta(t)) \\ &= q(\beta(t))^n \\ &= q(qt^n)^n \\ &= qq^n t^{n^2} \\ &= q^{1+n} t^{n^2} \\ &= q^{\frac{1-n^2}{1-n}} t^{n^2} \\ &= q^{[2]_n} t^{n^2}, \quad t \in I. \end{aligned}$$

Assume that

$$\beta^l(t) = q^{[l]}t^{n^l}, \quad t \in I,$$

for some $l \in \mathbb{N}$. *We will prove that*

$$\beta^{l+1}(t) = q^{[l+1]_n}t^{n^{l+1}}, \quad t \in I.$$

Really, we have

$$\begin{aligned}
\beta^{l+1}(t) &= \beta \circ \beta^l(t) \\
&= \beta\left(\beta^l(t)\right) \\
&= q\left(\beta^l(t)\right)^n \\
&= q\left(q^{1+n+\cdots+n^{l-1}}t^{n^l}\right)^n \\
&= qq^{n+n^2+\cdots+n^l}t^{n^{l+1}} \\
&= q^{1+n+\cdots+n^l}t^{n^{l+1}} \\
&= q^{\frac{1-n^{l+1}}{1-n}}t^{n^{l+1}} \\
&= q^{[l+1]_n}t^{n^{l+1}}, \quad t \in I.
\end{aligned}$$

Hence, applying the principle of mathematical induction, we get

$$\beta^k(t) = q^{[k]_n}t^{n^k}, \quad t \in I,$$

for any $k \in \mathbb{N}_0$. *Next,*

$$\begin{aligned}
\beta^{-2}(t) &= \beta^{-1} \circ \beta^{-1}(t) \\
&= \beta^{-1}\left(\beta^{-1}(t)\right) \\
&= \left(\frac{\beta^{-1}(t)}{q}\right)^{\frac{1}{n}} \\
&= \left(\frac{\left(\frac{t}{q}\right)^{\frac{1}{n}}}{q}\right)^{\frac{1}{n}}
\end{aligned}$$

$$
\begin{aligned}
&= \left(\frac{\frac{t^{\frac{1}{n}}}{q^{\frac{1}{n}}}}{q}\right)^{\frac{1}{n}} \\
&= \left(\frac{t^{\frac{1}{n}}}{q^{\frac{n+1}{n}}}\right)^{\frac{1}{n}} \\
&= \frac{t^{\frac{1}{n^2}}}{q^{\frac{n+1}{n^2}}} \\
&= \frac{t^{\frac{1}{n^2}}}{q^{\frac{1}{n^2}\frac{1-n^2}{1-n}}} \\
&= \frac{t^{\frac{1}{n^2}}}{q^{\frac{[2]_n}{n^2}}}, \quad t \in I.
\end{aligned}
$$

Assume that

$$
\beta^{-l}(t) = \frac{t^{\frac{1}{n^l}}}{q^{\frac{[l]_n}{n^l}}}, \quad t \in I,
$$

for some $l \in \mathbb{N}$. We will prove that

$$
\beta^{-(l+1)}(t) = \frac{t^{\frac{1}{n^{l+1}}}}{q^{\frac{[l+1]_n}{n^{l+1}}}}, \quad t \in I.
$$

Really,

$$
\begin{aligned}
\beta^{-(l+1)}(t) &= \beta^{-1} \circ \beta^{-l}(t) \\
&= \beta^{-1}\left(\beta^{-l}(t)\right) \\
&= \left(\frac{\beta^{-l}(t)}{q}\right)^{\frac{1}{n}} \\
&= \left(\frac{\frac{t^{\frac{1}{n^l}}}{q^{\frac{[l]_n}{n^l}}}}{q}\right)^{\frac{1}{n}}
\end{aligned}
$$

$$= \left(\frac{\frac{t^{\frac{1}{n^l}}}{q^{\frac{1+n+\cdots+n^{l-1}}{n^l}}}}{q} \right)^{\frac{1}{n}}$$

$$= \left(\frac{t^{\frac{1}{n^l}}}{q^{\frac{1+n+\cdots+n^{l-1}+n^l}{n^l}}} \right)^{\frac{1}{n}}$$

$$= \frac{t^{\frac{1}{n^{l+1}}}}{q^{\frac{\frac{1-n^{l+1}}{1-n}}{n^{l+1}}}}$$

$$= \frac{t^{\frac{1}{n^{l+1}}}}{q^{\frac{[l+1]_n}{n^{l+1}}}}, \quad t \in I.$$

Hence, applying the principle of mathematical induction, we get

$$\beta^{-k}(t) = \frac{t^{\frac{1}{n^k}}}{q^{\frac{[k]_n}{n^k}}}, \quad t \in I,$$

for any $k \in \mathbb{N}_0$.

Exercise 1.1.12 *Let I and β be as in Exercise 1.1.7 Find*

1. $\beta^{-1}(t)$, $t \in I$.
2. $\beta^k(t)$, $k \in \mathbb{N}_0$, $t \in I$.
3. $\beta^{-k}(t)$, $t \in I$, $k \in \mathbb{N}$.

Theorem 1.1.13 *Let β be a first or second kind general quantum operator. Then, we have the following.*

1. *The sequence $\{\beta^k\}_{k\in\mathbb{N}_0}$ is uniformly convergent to s_0 on any compact interval $J \subseteq I$ containing s_0.*
2. *The series*

$$\sum_{k=0}^{\infty} |\beta^k(t) - \beta^{k+1}(t)|$$

is uniformly convergent to $|t - s_0|$ on any compact interval $J \subset I$ containing s_0.

Proof 1.1.14 1. *Let $J = [a,b]$ and $s_0 \in [a,b]$.*

(a) *Suppose that* $t \in [s_0, b]$. *Then*

$$t - s_0 \geq 0$$

and

$$\beta(t) \leq t.$$

Hence, using that β *is strictly increasing on* I, *we get*

$$\begin{aligned} \beta^2(t) &\leq \beta(t), \\ \beta^3(t) &\leq \beta^2(t), \\ &\vdots \\ \beta^{k+1}(t) &\leq \beta^k(t), \quad k \in \mathbb{N}_0. \end{aligned}$$

Then $\{\beta^k\}_{k\in\mathbb{N}_0}$ *is decreasing to* s_0. *Now, applying the Dini test, we get that the sequence* $\{\beta^k\}_{k\in\mathbb{N}_0}$ *is uniformly convergent to* s_0 *on* $[s_0, b]$.

(b) *Suppose that* $t \in [a, s_0]$. *Then*

$$t - s_0 \leq 0$$

and

$$\beta(t) \geq t.$$

Hence, using that β *is strictly increasing on* I, *we get*

$$\begin{aligned} \beta^2(t) &\geq \beta(t), \\ \beta^3(t) &\geq \beta^2(t), \\ &\vdots \\ \beta^{k+1}(t) &\geq \beta^k(t), \quad k \in \mathbb{N}_0. \end{aligned}$$

Then $\{\beta^k\}_{k\in\mathbb{N}_0}$ *is increasing to* s_0. *Now, applying the Dini test, we get that the sequence* $\{\beta^k\}_{k\in\mathbb{N}_0}$ *is uniformly convergent to* s_0 *on* $[a, s_0]$.

2. *Consider the sequence* $\{S_n\}_{n\in\mathbb{N}}$, *where*

$$S_n(t) = \sum_{k=0}^{n} |\beta^k(t) - \beta^{k+1}(t)|, \quad t \in J, \quad n \in \mathbb{N}.$$

(a) *Let* $t \in [s_0, b]$. *Then*

$$\begin{aligned} S_n(t) &= t - \beta(t) + \beta(t) - \beta^2(t) + \cdots + \beta^n(t) - \beta^{n+1}(t) \\ &= t - \beta^{n+1}(t) \\ &\to t - s_0, \quad as \quad n \to \infty, \end{aligned}$$

uniformly.

(b) *Let* $t \in [a, s_0]$. *Then*

$$\begin{aligned} S_n(t) &= \sum_{k=0}^{n} \left(\beta^{k+1}(t) - \beta^k(t)\right) \\ &= \beta^{n+1}(t) - \beta^n(t) + \cdots + \beta(t) - t \\ &= \beta^{n+1}(t) - t \\ &\to s_0 - t, \quad as \quad n \to \infty, \end{aligned}$$

uniformly. This completes the proof.

Below, suppose that X is a Banach space.

Corollary 1.1.15 *Let* β *be a first kind with general quantum operator. If* $f : I \to X$ *is continuous at* s_0*, then the sequence* $\{f(\beta^k(t))\}_{k\in\mathbb{N}}$ *converges uniformly to* $f(s_0)$ *on any compact interval* $J \subseteq I$ *containing* s_0.

Theorem 1.1.16 *Let* β *be a first kind with general quantum operator. If* $f : I \to X$ *is continuous at* s_0*, then the series*

$$\sum_{k=0}^{\infty} \left|\beta^k(t) - \beta^{k+1}(t)\right| \left\|f\left(\beta^k(t)\right)\right\| \tag{1.1}$$

is uniformly convergent on any compact interval $J \subset I$ *containing* s_0.

Proof 1.1.17 *Let* $J \subset I$ *be any compact interval. By Corollary 1.1.15, we have that the sequence* $\{f\left(\beta^k(t)\right)\}_{k\in\mathbb{N}_0}$ *converges uniformly to* $f(s_0)$ *on* J*. Hence, there is a* $k_0 \in \mathbb{N}$ *such that*

$$\|f\left(\beta^k(t)\right) - f(s_0)\| < 1, \quad k \geq k_0, \quad t \in J.$$

Then

$$\|f\left(\beta^k(t)\right)\| < 1 + \|f(s_0)\|, \quad k \geq k_0, \quad t \in J.$$

Therefore

$$\left|\left(\beta^k(t)-\beta^{k+1}(t)\right)\right|\left\|f\left(\beta^k(t)\right)\right\| \le \left|\beta^k(t)-\beta^{k+1}(t)\right|(1+\|f(s_0)\|),$$

$k \ge k_0$, $t \in J$. *Set*

$$D_n(t) = \sum_{k=0}^{n}\left|\beta^k(t)-\beta^{k+1}(t)\right|\left\|f\left(\beta^k(t)\right)\right\|, \quad t \in J, \quad n \in \mathbb{N}_0,$$

and

$$C_n(t) = \sum_{k=0}^{n}\left|\beta^k(t)-\beta^{k+1}(t)\right|(1+\|f(s_0)\|), \quad t \in J, \quad n \in \mathbb{N}_0.$$

By Theorem 1.1.13, we have

$$C_n(t) \to_{n\to\infty} |t-s_0|(1+\|f(s_0)\|), \quad t \in J,$$

uniformly. Hence and the Cauchy test, it follows that for any $\varepsilon > 0$ *there is a* $m_0 \in \mathbb{N}_0$ *such that*

$$\|C_n(t)-C_m(t)\| < \varepsilon, \quad t \in J, \quad n,m \ge m_0.$$

Then

$$\begin{aligned}
\|D_n(t)-D_m(t)\| &= \Bigg\| \sum_{k=0}^{n}\left|\beta^k(t)-\beta^{k+1}(t)\right|\left\|f\left(\beta^k(t)\right)\right\| \\
&\quad -\sum_{k=0}^{m}\left|\beta^k(t)-\beta^{k+1}(t)\right|\left\|f\left(\beta^k(t)\right)\right\|\Bigg\| \\
&= \left\|\sum_{k=m+1}^{n}\left|\beta^k(t)-\beta^{k+1}(t)\right|\left\|f\left(\beta^k(t)\right)\right\|\right\| \\
&= \|C_n(t)-C_m(t)\| \\
&< \varepsilon, \quad t \in J, \quad n > m \ge \max\{k_0, m_0\}.
\end{aligned}$$

Thus, (1.1) *is uniformly convergent on* J. *This completes the proof.*

Example 1.1.18 *Let* I, β, q, ω *be as in Example 1.1.4. Then*

$$\begin{aligned}
\beta^k(t) &= q^k t \pm \frac{1-q^k}{1-q}\omega, \\
\beta^{-k}(t) &= \frac{t \mp \frac{1-q^k}{1-q}\omega}{q^k}, \quad t \in I, \quad k \in \mathbb{N}.
\end{aligned}$$

Then

$$\lim_{k\to\infty}\beta^k(t) = \lim_{k\to\infty}\left(q^k t \pm \frac{1-q^k}{1-q}\omega\right)$$

$$= \pm\frac{\omega}{1-q}$$

$$= s_0, \quad t \in I,$$

and

$$\begin{aligned}
\lim_{k\to\infty}\beta^{-k}(t) &= \lim_{k\to\infty}\left(\frac{t \mp \frac{1-q^k}{1-q}\omega}{q^k}\right)\\
&= \begin{cases} \infty & \text{if} \quad t > s_0 \\ -\infty & \text{if} \quad t < s_0. \end{cases}
\end{aligned}$$

Example 1.1.19 *Let I, β, q and n be as in Example 1.1.6. Then*

$$\begin{aligned}
\beta^k(t) &= q^{\frac{1-n^k}{1-n}} t^{n^k},\\
\beta^{-k}(t) &= \frac{t^{\frac{1}{n^k}}}{q^{\frac{1-n^k}{n^k(1-n)}}}, \quad t \in I, \quad k \in \mathbb{N}.
\end{aligned}$$

Then

$$\begin{aligned}
\lim_{k\to\infty}\beta^k(t) &= \lim_{k\to\infty}\left(q^{\frac{1-n^k}{1-n}} t^{n^k}\right)\\
&= 0, \quad t \in I,
\end{aligned}$$

and

$$\begin{aligned}
\lim_{k\to\infty}\beta^{-k}(t) &= \lim_{k\to\infty}\left(\frac{t^{\frac{1}{n^k}}}{q^{\frac{1-n^k}{n^k(1-n)}}}\right)\\
&= \begin{cases} q^{\frac{1}{1-n}} & \text{if} \quad t > 0 \\ 0 \quad \text{if} \quad t = 0 \\ -q^{\frac{1}{1-n}} & \text{if} \quad t < 0. \end{cases}
\end{aligned}$$

Exercise 1.1.20 *Let $n \in 2\mathbb{N}+1$, $I = (-1,1)$ and*

$$\beta(t) = t^n, \quad t \in I.$$

1. *Prove that* β *is a first kind general quantum operator.*
2. *Prove that*

$$\beta^k(t) = t^{n^k}, \quad t \in I, \quad k \in \mathbb{N}.$$

3. *Prove that*

$$\beta^{-k}(t) = t^{-n^k}, \quad t \in I, \quad k \in \mathbb{N}.$$

4. *Find*

$$\lim_{k \to \infty} \beta^k(t), \quad t \in I.$$

5. *Find*

$$\lim_{k \to \infty} \beta^{-k}(t), \quad t \in I.$$

Remark 1.1.21 *Note that Theorem 1.1.13, Corollary 1.1.15 and Theorem 1.1.16 are valid in the case when* β *is a a second kind with general quantum differential operator.*

1.2 Definition for β-Derivative

Below, throughout this book, suppose that $I \subset \mathbb{R}$, X is a Banach space and β is a first or second kind general quantum operator on I(shortly general quantum operator).

Definition 1.2.1 *For a function* $f : I \to X$*, define the* β*-difference operator of* f *as follows.*

$$D_\beta f(t) = \begin{cases} \dfrac{f(\beta(t)) - f(t)}{\beta(t) - t} & \text{if} \quad t \neq s_0 \\ \\ f'(s_0) \quad \text{if} \quad t = s_0, \end{cases}$$

provided that $f'(s_0)$ *exists. In this case, we say that* $D_\beta f(t)$ *is* β*-derivative of* f *at* t*. We say that* f *is* β*-differentiable on* I *if* $D_\beta f(t)$ *exists for any* $t \in I$*.*

Example 1.2.2 *Let* $I = \left[1, \frac{3}{2}\right]$ *and*

$$\beta(t) = \frac{1}{2}t + \frac{3}{4},$$

$$f(t) = 4t^2 - 9t, \quad t \in I.$$

We will find $D_\beta f(t)$*,* $t \in I$*. Here* $s_0 = \frac{3}{2}$*. Then*

$$f'(t) = 8t - 9, \quad t \in I,$$

and

$$D_\beta f(t) = \begin{cases} \dfrac{f(\beta(t)) - f(t)}{\beta(t) - t} & \textit{if} \quad t \neq \dfrac{3}{2} \\ f'\left(\dfrac{3}{2}\right) \quad \textit{if} \quad t = \dfrac{3}{2} \end{cases}$$

$$= \begin{cases} \dfrac{4\left(\frac{1}{2}t + \frac{3}{4}\right)^2 - \left(\frac{1}{2}t + \frac{3}{4}\right)}{\frac{1}{2}t + \frac{3}{4} - t} & \textit{if} \quad t \neq \dfrac{3}{2} \\ 8 \cdot \dfrac{3}{2} - 9 \quad \textit{if} \quad t = \dfrac{3}{2} \end{cases}$$

$$= \begin{cases} \dfrac{4\left(\frac{1}{4}t^2 + \frac{3}{4}t + \frac{9}{16}\right) - \frac{1}{2}t - \frac{3}{4}}{\frac{3}{4} - \frac{1}{2}t} & \textit{if} \quad t \neq \dfrac{3}{2} \\ 12 - 9 \quad \textit{if} \quad t = \dfrac{3}{2} \end{cases}$$

$$= \begin{cases} \dfrac{t^2 + 3t + \frac{9}{4} - \frac{1}{2}t - \frac{3}{4}}{\frac{3-2t}{4}} & \textit{if} \quad t \neq \dfrac{3}{2} \\ 3 \quad \textit{if} \quad t = \dfrac{3}{2} \end{cases}$$

$$= \begin{cases} \dfrac{t^2 + \frac{9}{2}t + \frac{3}{2}}{\frac{3-2t}{4}} & \textit{if} \quad t \neq \dfrac{3}{2} \\ 3 \quad \textit{if} \quad t = \dfrac{3}{2} \end{cases}$$

$$= \begin{cases} \dfrac{2(2t^2 + 9t + 3)}{3 - 2t} & \textit{if} \quad t \neq \dfrac{3}{2} \\ 3 \quad \textit{if} \quad t = \dfrac{3}{2}. \end{cases}$$

Example 1.2.3 *Let $I = [-1, 1]$ and*

$$\beta(t) = \frac{1}{4}t + \frac{1}{4},$$

$$f(t) = \begin{cases} t \quad \textit{if} \quad t \in (-1, 0) \\ -t \quad \textit{if} \quad t \in (0, 1) \\ 1 \quad \textit{if} \quad t = 0 \\ 0 \quad \textit{if} \quad t = 1, -1 \end{cases}$$

We will find $D_\beta f(t)$, $t \in I$. Here $s_0 = \frac{1}{3}$. Set

$$g(t) = -t, \quad t \in I.$$

Then

$$\begin{aligned} g'(t) &= -1, \\ \beta(0) &= \frac{1}{4}, \\ \beta(1) &= \frac{1}{2}, \\ \beta(-1) &= 0. \end{aligned}$$

Hence,

$$\begin{aligned} D_\beta f(t) &= \begin{cases} \dfrac{f(\beta(t)) - f(t)}{\beta(t) - t} & \text{if} \quad t \neq \dfrac{1}{3} \\ f'\left(\dfrac{1}{3}\right) & \text{if} \quad t = \dfrac{1}{3} \end{cases} \\ &= \begin{cases} \dfrac{\beta(t) - t}{\beta(t) - t} & \text{if} \quad t \in (-1, 0) \\ -\dfrac{\beta(t) - t}{\beta(t) - t} & \text{if} \quad t \in \left(0, \dfrac{1}{3}\right) \cup \left(\dfrac{1}{3}, 1\right) \\ g'\left(\dfrac{1}{3}\right) & \text{if} \quad t = \dfrac{1}{3} \\ \dfrac{f(\beta(0)) - f(0)}{\beta(0) - 0} & \text{if} \quad t = 0 \\ \dfrac{f(\beta(-1)) - f(-1)}{\beta(-1) + 1} & \text{if} \quad t = -1 \\ \dfrac{f(\beta(1)) - f(1)}{\beta(1) - 1} & \text{if} \quad t = 1 \end{cases} \end{aligned}$$

$$= \begin{cases} 1 \quad if \quad t \in (-1,0) \\ -1 \quad if \quad t \in (0,1) \\ \dfrac{f\left(\frac{1}{4}\right) - f(0)}{\frac{1}{4} - 0} \quad if \quad t = 0 \\ \dfrac{f(0) - f(-1)}{0+1} \quad if \quad t = -1 \\ \dfrac{f\left(\frac{1}{2}\right) - f(1)}{\frac{1}{2} - 1} \quad if \quad t = 1 \end{cases}$$

$$= \begin{cases} 1 \quad if \quad t \in (-1,0) \\ -1 \quad if \quad t \in (0,1) \\ \dfrac{-\frac{1}{4} - 1}{\frac{1}{4} - 0} \quad if \quad t = 0 \\ \dfrac{1-0}{1} \quad if \quad t = -1 \\ \dfrac{-\frac{1}{2} - 0}{-\frac{1}{2}} \quad if \quad t = 1 \end{cases}$$

$$= \begin{cases} 1 \quad if \quad t \in (-1,0) \\ -1 \quad if \quad t \in (0,1) \\ -5 \quad if \quad t = 0 \\ 1 \quad if \quad t = 1, -1. \end{cases}$$

Example 1.2.4 *Let $I = \mathbb{R}$ and*

$$\beta(t) = 2t + 3,$$

$$f(t) = t^{10}, \quad t \in I.$$

We will find $D_\beta f(t)$, $t \in I$. Here $s_0 = -3$ and

$$f'(t) = 10t^9, \quad t \in I.$$

Then

$$
\begin{aligned}
D_\beta f(t) &= \begin{cases} \dfrac{f(2t+3)-f(t)}{2t+3-t} & \text{if} \quad t \neq -3 \\ f'(-3) \quad \text{if} \quad t=-3 \end{cases} \\
&= \begin{cases} \dfrac{(2t+3)^{10}-t^{10}}{t+3} & \text{if} \quad t \neq -3 \\ 10\cdot(-3)^9 \quad \text{if} \quad t=-3 \end{cases} \\
&= \begin{cases} \dfrac{(2t+3-t)((2t+3)^9+t(2t+3)^8+\cdots+t^9)}{t+3} & \text{if} \\ -10\cdot 3^9 \quad \text{if} \quad t=-3 \end{cases} \\
&= \begin{cases} (2t+3)^9+t(2t+3)^8+\cdots+t^9 & \text{if} \quad t \neq -3 \\ -10\cdot 3^9 \quad \text{if} \quad t=-3. \end{cases}
\end{aligned}
$$

Example 1.2.5 *Let $I=\mathbb{R}$ and*

$$\beta(t)=\frac{1}{2}t+1, \quad t \in I,$$

and $f: I \to \mathbb{R}^2$ be defined as follows

$$f(t)=(t^2,2t), \quad t \in I.$$

We will find $D_\beta f(t)$, $t \in I$. Here $s_0=2$. Set

$$
\begin{aligned}
f_1(t) &= t^2, \\
f_2(t) &= 2t, \quad t \in I.
\end{aligned}
$$

Then

$$
\begin{aligned}
f_1'(t) &= 2t, \\
f_2'(t) &= 2, \quad t \in I,
\end{aligned}
$$

and

$$
D_\beta f_1(t) = \begin{cases} \dfrac{f_1\left(\frac{1}{2}t+1\right)-f_1(t)}{\frac{1}{2}t+1-t} & \text{if} \quad t \neq 2 \\ f_1'(2) \quad \text{if} \quad t=2 \end{cases}
$$

$$= \begin{cases} \dfrac{\left(\frac{1}{2}t+1\right)^2 - t^2}{1-\frac{1}{2}t} & \text{if} \quad t \neq 2 \\ 2\cdot 2 \quad \text{if} \quad t = 2 \end{cases}$$

$$= \begin{cases} \dfrac{\left(\frac{1}{2}t+1-t\right)\left(\frac{1}{2}t+1+t\right)}{1-\frac{1}{2}t} & \text{if} \quad t \neq 2 \\ 4 \quad \text{if} \quad t = 2 \end{cases}$$

$$= \begin{cases} 1+\dfrac{3}{2}t & \text{if} \quad t \neq 2 \\ 4 \quad \text{if} \quad t = 2 \end{cases}$$

and

$$D_\beta f_2(t) = \begin{cases} \dfrac{f_2\left(\frac{1}{2}t+1\right) - f_2(t)}{\frac{1}{2}t+1-t} & \text{if} \quad t \neq 2 \\ f_2'(2) \quad \text{if} \quad t = 2 \end{cases}$$

$$= \begin{cases} \dfrac{2\left(\frac{1}{2}t+1\right) - 2t}{1-\frac{1}{2}t} & \text{if} \quad t \neq 2 \\ 2 \quad \text{if} \quad t = 2 \end{cases}$$

$$= \begin{cases} \dfrac{2(t+2-2t)}{2-t} & \text{if} \quad t \neq 2 \\ 2 \quad \text{if} \quad t = 2 \end{cases}$$

$$= \begin{cases} 2 \quad \text{if} \quad t \neq 2 \\ 2 \quad \text{if} \quad t = 2. \end{cases}$$

Therefore

$$D_\beta f(t) = (D_\beta f_1(t), D_\beta f_2(t))$$

$$= \begin{cases} \left(1+\dfrac{3}{2}t, 2\right) & \text{if} \quad t \neq 2 \\ (4,2) \quad \text{if} \quad t = 2 \end{cases}$$

$$= \left(1+\frac{3}{2}t, 2\right), \quad t \in I.$$

Example 1.2.6 *Let $I = \mathbb{R}$,*

$$\beta(t) = \frac{1}{4}t, \quad t \in I,$$

and $f : I \to \mathscr{M}_{2\times 2}$ be defined ny

$$f(t) = \begin{pmatrix} t^3 & 1 \\ t & t^2 \end{pmatrix}, \quad t \in I.$$

We will find $D_\beta f(t)$, $t \in I$. Set

$$\begin{aligned} f_1(t) &= t^3, \\ f_2(t) &= 1, \\ f_3(t) &= t, \\ f_4(t) &= t^2, \quad t \in I. \end{aligned}$$

Here $s_0 = 0$ and

$$\begin{aligned} f_1'(t) &= 3t^2, \\ f_2'(t) &= 0, \\ f_3'(t) &= 1, \\ f_4'(t) &= 2t, \quad t \in I. \end{aligned}$$

Then

$$\begin{aligned} D_\beta f_1(t) &= \begin{cases} \dfrac{f_1\left(\frac{1}{4}t\right) - f_1(t)}{\frac{1}{4}t - t} & \text{if} \quad t \neq 0 \\ f_1'(0) \quad \text{if} \quad t = 0 \end{cases} \\ &= \begin{cases} \dfrac{\left(\frac{1}{4}t\right)^3 - t^3}{-\frac{3}{4}t} & \text{if} \quad t \neq 0 \\ 3 \cdot 0^2 \quad \text{if} \quad t = 0 \end{cases} \\ &= \begin{cases} \dfrac{\left(\frac{1}{4}t - t\right)\left(\frac{1}{16}t^2 + \frac{t^2}{4} + t^2\right)}{-\frac{3}{4}t} & \text{if} \quad t \neq 0 \\ 0 \quad \text{if} \quad t = 0 \end{cases} \end{aligned}$$

$$= \begin{cases} \frac{21}{16}t^2 & \textit{if} \quad t \neq 0 \\ 0 & \textit{if} \quad t = 0 \end{cases}$$

and

$$\begin{aligned} D_\beta f_2(t) &= \begin{cases} \dfrac{f_2\left(\frac{1}{4}t\right) - f_2(t)}{\frac{1}{4}t - t} & \textit{if} \quad t \neq 0 \\ f_2'(0) & \textit{if} \quad t = 0 \end{cases} \\ &= \begin{cases} \dfrac{1-1}{-\frac{3}{4}t} & \textit{if} \quad t \neq 0 \\ 0 & \textit{if} \quad t = 0 \end{cases} \\ &= \begin{cases} 0 & \textit{if} \quad t \neq 0 \\ 0 & \textit{if} \quad t = 0, \end{cases} \end{aligned}$$

and

$$\begin{aligned} D_\beta f_3(t) &= \begin{cases} \dfrac{f_3\left(\frac{1}{4}t\right) - f_3(t)}{\frac{1}{4}t - t} & \textit{if} \quad t \neq 0 \\ f_3'(0) & \textit{if} \quad t = 0 \end{cases} \\ &= \begin{cases} \dfrac{\frac{1}{4}t - t}{-\frac{3}{4}t} & \textit{if} \quad t \neq 0 \\ 1 & \textit{if} \quad t = 0 \end{cases} \\ &= \begin{cases} 1 & \textit{if} \quad t \neq 0 \\ 1 & \textit{if} \quad t = 0, \end{cases} \end{aligned}$$

and

$$\begin{aligned} D_\beta f_4(t) &= \begin{cases} \dfrac{f_4\left(\frac{1}{4}t\right) - f_4(t)}{\frac{1}{4}t - t} & \textit{if} \quad t \neq 0 \\ f_4'(0) & \textit{if} \quad t = 0 \end{cases} \\ &= \begin{cases} \dfrac{\left(\frac{1}{4}t\right)^2 - t^2}{\frac{1}{4}t - t} & \textit{if} \quad t \neq 0 \\ 2 \cdot 0 & \textit{if} \quad t = 0 \end{cases} \end{aligned}$$

$$= \begin{cases} \dfrac{\left(\frac{1}{4}t-t\right)\left(\frac{1}{4}t+t\right)}{-\frac{3}{4}t} & if \quad t \neq 0 \\ 0 \quad if \quad t=0 \end{cases}$$

$$= \begin{cases} \dfrac{5}{4}t \quad if \quad t \neq 0 \\ 0 \quad if \quad t=0. \end{cases}$$

Therefore

$$\begin{aligned} D_\beta f(t) &= \begin{pmatrix} D_\beta f_1(t) & D_\beta f_2(t) \\ D_\beta f_3(t) & D_\beta f_4(t) \end{pmatrix} \\ &= \begin{cases} \begin{pmatrix} \frac{21}{16}t^2 & 0 \\ 1 & \frac{5}{4}t \end{pmatrix} & if \quad t \neq 0 \\ \begin{pmatrix} 0 & 0 \\ 1 & 0 \end{pmatrix} \quad if \quad t=0 \end{cases} \\ &= \begin{pmatrix} \frac{21}{16}t^2 & 0 \\ 1 & \frac{5}{4}t \end{pmatrix}, \quad t \in I. \end{aligned}$$

Exercise 1.2.7 *Let $I = \mathbb{R}$ and*

$$\begin{aligned} \beta(t) &= \frac{1}{8}t + \frac{7}{8}, \\ f(t) &= t^2 - 2t + 2, \quad t \in I. \end{aligned}$$

1. *Prove that β is a first kind general quantum operator.*
2. *Find $D_\beta f(t)$, $t \in I$.*

1.3 Properties of the β-Derivative

In this section, we will deduct some basic properties of the β-derivative.

Theorem 1.3.1 *The β-difference operator is a linear operator.*

Proof 1.3.2 *Let $f, g : I \to X$ and $a \in \mathbb{C}$ be arbitrarily chosen. Then*

$$
\begin{aligned}
D_\beta(f+g)(t) &= \begin{cases} \dfrac{(f+g)(\beta(t))-(f+g)(t)}{\beta(t)-t} & \text{if} \quad t \neq s_0 \\[2ex] (f+g)'(s_0) \quad \text{if} \quad t = s_0 \end{cases} \\
&= \begin{cases} \dfrac{f(\beta(t))+g(\beta(t))-f(t)-g(t)}{\beta(t)-t} & \text{if} \quad t \neq s_0 \\[2ex] f'(s_0)+g'(s_0) \quad \text{if} \quad t = s_0 \end{cases} \\
&= \begin{cases} \dfrac{f(\beta(t))-f(t)}{\beta(t)-t} & \text{if} \quad t \neq s_0 \\[2ex] f'(s_0) \quad \text{if} \quad t = s_0 \end{cases} \\
&\quad + \begin{cases} \dfrac{g(\beta(t))-g(t)}{\beta(t)-t} & \text{if} \quad t \neq s_0 \\[2ex] g'(s_0) \quad \text{if} \quad t = s_0 \end{cases} \\
&= D_\beta f(t) + D_\beta g(t), \quad t \in I,
\end{aligned}
$$

and

$$
\begin{aligned}
D_\beta(af)(t) &= \begin{cases} \dfrac{(af)(\beta(t))-(af)(t)}{\beta(t)-t} & \text{if} \quad t \neq s_0 \\[2ex] (af)'(s_0) \quad \text{if} \quad t = s_0 \end{cases} \\
&= \begin{cases} \dfrac{af(\beta(t))-af(t)}{\beta(t)-t} & \text{if} \quad t \neq s_0 \\[2ex] af'(s_0) \quad \text{if} \quad t = s_0 \end{cases} \\
&= a \begin{cases} \dfrac{f(\beta(t))-f(t)}{\beta(t)-t} & \text{if} \quad t \neq s_0 \\[2ex] f'(s_0) \quad \text{if} \quad t = s_0 \end{cases} \\
&= aD_\beta f(t), \quad t \in I.
\end{aligned}
$$

Thus, D_β is a linear operator. This completes the proof.

Theorem 1.3.3 *If f is β-differentiable on I, then*

$$
f(\beta(t)) = f(t) + (\beta(t)-t)D_\beta f(t), \quad t \in I. \tag{1.2}
$$

Proof 1.3.4 *We will consider the following two cases.*

1. Let $t \neq s_0$. Then

$$D_\beta f(t) = \frac{f(\beta(t)) - f(t)}{\beta(t) - t},$$

whereupon we get (1.2).

2. Let $t = s_0$. Then

$$\beta(s_0) = s_0$$

and

$$\begin{aligned} f(\beta(s_0)) &= f(s_0) \\ &= f(s_0) + (\beta(s_0) - s_0) D_\beta f(s_0). \end{aligned}$$

This completes the proof.

Theorem 1.3.5 *Let f be β-differentiable on I. Then f is continuous at s_0.*

Proof 1.3.6 *Since $f'(s_0)$ exists, we have that f is continuous at s_0. This completes the proof.*

1.4 Rules for β-Differentiation

In this section, we will deduct the basic rules for β-differentiation.

Theorem 1.4.1 *Let $f : I \to X$ and $g : I \to \mathbb{R}$ be β-differentiable at $t \in I$. Then, we have the following.*

1. The product $fg : I \to X$ is β-differentiable at t and

$$\begin{aligned} D_\beta(fg)(t) &= (D_\beta f(t))g(t) + f(\beta(t)) D_\beta g(t) \\ &= (D_\beta f(t))g(\beta(t)) + f(t) D_\beta g(t), \quad t \in I. \end{aligned}$$

2. The quotient $\frac{f}{g}$ is β-differentiable at t and

$$D_\beta\left(\frac{f}{g}\right)(t) = \frac{(D_\beta f(t))g(t) - f(t) D_\beta g(t)}{g(t)g(\beta(t))}.$$

Proof 1.4.2 *1. We have the following two cases.*

(a) *Let $t = s_0$. Then*

$$\beta(s_0) = s_0$$

and

$$\begin{aligned}(fg)'(s_0) &= f'(s_0)g(s_0) + f(s_0)g'(s_0)\\ &= f'(s_0)g(\beta(s_0)) + f(s_0)g'(s_0)\\ &= D_\beta f(s_0)g(s_0) + f(\beta(s_0))D_\beta g(s_0).\end{aligned}$$

(b) *Let $t \neq s_0$. Then*

$$D_\beta(fg)(t) = \frac{(fg)(\beta(t)) - (fg)(t)}{\beta(t) - t}$$

$$\begin{aligned} &= \frac{f(\beta(t))g(\beta(t)) - f(t)g(t)}{\beta(t) - t}\\ &= \frac{f(\beta(t))(g(\beta(t)) - g(t)) + (f(\beta(t)) - f(t))g(t)}{\beta(t) - t}\\ &= f(\beta(t))\frac{g(\beta(t)) - g(t)}{\beta(t) - t} + \frac{f(\beta(t)) - f(t)}{\beta(t) - t}g(t)\\ &= f(\beta(t))D_\beta g(t) + (D_\beta f(t))g(t)\end{aligned}$$

and

$$D_\beta(fg)(t) = \frac{(fg)(\beta(t)) - (fg)(t)}{\beta(t) - t}$$

$$\begin{aligned} &= \frac{(f(\beta(t)) - f(t))g(\beta(t)) + f(t)(g(\beta(t)) - g(t))}{\beta(t) - t}\\ &= \frac{f(\beta(t)) - f(t)}{\beta(t) - t}g(\beta(t)) + f(t)\frac{g(\beta(t)) - g(t)}{\beta(t) - t}\\ &= (D_\beta f(t))g(\beta(t)) + f(t)D_\beta g(t).\end{aligned}$$

2. *We will consider the following two cases.*

(a) *Let $t = s_0$ and $g'(s_0) \neq 0$. Then*

$$g(s_0) = g(\beta(s_0))$$

and

$$\left(\frac{f}{g}\right)'(s_0) = \frac{f'(s_0)g(s_0)-f(s_0)g'(s_0)}{(g(s_0))^2}$$
$$= \frac{(D_\beta f(s_0))g(s_0)-f(s_0)D_\beta g(s_0)}{g(s_0)g(\beta(s_0))}.$$

(b) *Let* $t \neq s_0$. *Then*

$$D_\beta\left(\frac{f}{g}\right)(t) = \frac{\left(\frac{f}{g}\right)(\beta(t))-\left(\frac{f}{g}\right)(t)}{\beta(t)-t}$$

$$= \frac{\frac{f(\beta(t))}{g(\beta(t))}-\frac{f(t)}{g(t)}}{\beta(t)-t}$$
$$= \frac{f(\beta(t))g(t)-f(t)g(\beta(t))}{(\beta(t)-t)g(t)g(\beta(t))}$$
$$= \frac{(f(\beta(t))-f(t))g(t)-f(t)(g(\beta(t))-g(t))}{(\beta(t)-t)g(t)g(\beta(t))}$$
$$= \frac{\frac{f(\beta(t))-f(t)}{\beta(t)-t}g(t)-f(t)\frac{g(\beta(t))-g(t)}{\beta(t)-t}}{g(t)g(\beta(t))}$$
$$= \frac{(D_\beta f(t))g(t)-f(t)D_\beta g(t)}{g(t)g(\beta(t))}.$$

This completes the proof.

Example 1.4.3 *Let* $I = \mathbb{R}$ *and*

$$\beta(t) = \frac{1}{2}t+1,$$

$$h(t) = t^3+t^2+t, \quad t \in I.$$

Note that

$$\beta'(t) = \frac{1}{2}, \quad t \in I,$$

i.e., $\beta : I \to I$ *is a strictly increasing continuous function on* I. *Next,*

$$\beta(t) = t, \quad t \in I,$$

if and only if

$$\frac{1}{2}t+1 = t, \quad t \in I,$$

if and only if

$$\frac{1}{2}t = 1, \quad t \in I,$$

if and only if

$$t = 2,$$

which is the unique fixed point of β, *i.e., here* $s_0 = 2$. *Moreover,*

$$\begin{aligned}(t-s_0)(\beta(t)-t) &= (t-2)\left(\frac{1}{2}t+1-t\right)\\ &= \frac{(t-2)(2-t)}{2}\\ &= -\frac{(t-2)^2}{2}\\ &\leq 0, \quad t \in I.\end{aligned}$$

Therefore, β *is a first kind general quantum operator on* I. *We will find* $D_\beta h(t)$, $t \in I$. *We have*

$$h'(t) = 3t^2 + 2t + 1, \quad t \in I,$$

and

$$\begin{aligned}h'(2) &= 3 \cdot 2^2 + 2 \cdot 2 + 1\\ &= 3 \cdot 4 + 4 + 1\\ &= 12 + 5\\ &= 17\\ &= D_\beta h(2).\end{aligned}$$

Next,

$$\begin{aligned}D_\beta h(t) &= \frac{h(\beta(t)) - h(t)}{\beta(t) - t}\\ &= \frac{(\beta(t))^3 + (\beta(t))^2 + \beta(t) - t^3 - t^2 - t}{\beta(t) - t}\\ &= \frac{(\beta(t)-t)((\beta(t))^2 + t\beta(t) + t^2) + (\beta(t)-t)(\beta(t)+t) + \beta(t) - t}{\beta(t) - t}\end{aligned}$$

$$\begin{aligned}
&= (\beta(t))^2+t\beta(t)+t^2+\beta(t)+t+1\\
&= (\beta(t))^2+(t+1)\beta(t)+t^2+t+1\\
&= \left(\frac{1}{2}t+1\right)^2+(t+1)\left(\frac{1}{2}t+1\right)+t^2+t+1\\
&= \frac{1}{4}t^2+t+1+\frac{1}{2}t^2+t+\frac{1}{2}t+1+t^2+t+1\\
&= \frac{7}{4}t^2+\frac{7}{2}t+3, \quad t\neq 2.
\end{aligned}$$

Observe that

$$h(t)=t(t^2+t+1), \quad t\in I.$$

Set

$$\begin{aligned}
f(t) &= t,\\
g(t) &= t^2+t+1, \quad t\in I.
\end{aligned}$$

We have

$$h(t)=f(t)g(t), \quad t\in I,$$

and

$$\begin{aligned}
f'(t) &= 1,\\
g'(t) &= 2t+1, \quad t\in I,
\end{aligned}$$

and

$$\begin{aligned}
g(2) &= 2^2+2+1\\
&= 4+3\\
&= 7,\\
g'(2) &= 2\cdot 2+1\\
&= 5,
\end{aligned}$$

and

$$f(\beta(t)) = \beta(t)$$

$$= \frac{1}{2}t + 1,$$

$$g(\beta(t)) = (\beta(t))^2 + \beta(t) + 1$$

$$= \left(\frac{1}{2}t + 1\right)^2 + \frac{1}{2}t + 1 + 1$$

$$= \frac{1}{4}t^2 + t + 1 + \frac{1}{2}t + 2$$

$$= \frac{1}{4}t^2 + \frac{3}{2}t + 3, \quad t \in I.$$

Next,

$$D_\beta f(t) = 1, \quad t \in I,$$

and

$$D_\beta g(t) = \frac{g(\beta(t)) - g(t)}{\beta(t) - t}$$

$$= \frac{(\beta(t))^2 + \beta(t) + 1 - t^2 - t - 1}{\beta(t) - t}$$

$$= \frac{(\beta(t) - t)(\beta(t) + t) + \beta(t) - t}{\beta(t) - t}$$

$$= \beta(t) + t + 1$$

$$= \frac{1}{2}t + 1 + t + 1$$

$$= \frac{3}{2}t + 2, \quad t \in I, \quad t \neq 2,$$

and

$$D_\beta g(2) = g'(2)$$

$$= 5.$$

Therefore

$$D_\beta h(t) = D_\beta(fg)(t)$$

$$= (D_\beta f(t))g(t) + f(\beta(t))D_\beta g(t)$$

$$\begin{aligned}
&= 1\cdot(t^2+t+1)+\left(\frac{1}{2}t+1\right)\left(\frac{3}{2}t+2\right)\\
&= t^2+t+1+\frac{3}{4}t^2+t+\frac{3}{2}t+2\\
&= \frac{7}{4}t^2+\frac{7}{2}t+3, \quad t\neq 2,
\end{aligned}$$

and

$$\begin{aligned}
D_\beta h(t) &= D_\beta(fg)(t)\\
&= (D_\beta f(t))g(\beta(t))+f(t)D_\beta g(t)\\
&= 1\cdot\left(\frac{1}{4}t^2+\frac{3}{2}t+3\right)+t\left(\frac{3}{2}t+2\right)\\
&= \frac{1}{4}t^2+\frac{3}{2}t+3+\frac{3}{2}t^2+2t\\
&= \frac{7}{4}t^2+\frac{7}{2}t+3, \quad t\in I, \quad t\neq 2,
\end{aligned}$$

and

$$\begin{aligned}
D_\beta h(1) &= (D_\beta f(2))g(2)+f(2)(D_\beta g(2))\\
&= 1\cdot 7+2\cdot 5\\
&= 7+10\\
&= 17.
\end{aligned}$$

Example 1.4.4 *Let* $I=(-\sqrt{2}, \sqrt{2})$ *and*

$$\begin{aligned}
\beta(t) &= \frac{1}{2}t^3,\\
h(t) &= \frac{t^2-4t+1}{t^2+t+1}, \quad t\in I.
\end{aligned}$$

By Example 1.1.6, it follows that β *is a first-order general quantum operator. Here* $s_0=0$*. We will find* $D_\beta h(t)$, $t\in I$*. Let*

$$f(t) = t^2-4t+1,$$

$$g(t) = t^2+t+1, \quad t \in I.$$

Then

$$\begin{aligned} g(\beta(t)) &= g\left(\frac{1}{2}t^3\right) \\ &= \left(\frac{1}{2}t^3\right)^2 + \frac{1}{2}t^3+1 \\ &= \frac{1}{4}t^6 + \frac{1}{2}t^3+1, \\ f'(t) &= 2t-4, \\ g'(t) &= 2t+1, \quad t \in I, \end{aligned}$$

and

$$\begin{aligned} f(0) &= 1, \\ g(0) &= 1, \\ f'(0) &= -4, \\ g'(0) &= 1. \end{aligned}$$

Hence,

$$\begin{aligned} D_\beta h(0) &= \frac{f'(0)g(0)-f(0)g'(0)}{(g(0))^2} \\ &= \frac{-4\cdot 1 - 1\cdot 1}{1^2} \\ &= -4-1 \\ &= -5. \end{aligned}$$

Let $t \in (-\sqrt{2}, 0) \cup (0, \sqrt{2})$. *Then*

$$D_\beta f(t) = \frac{f(\beta(t))-f(t)}{\beta(t)-t}$$

$$\begin{aligned}
&= \frac{(\beta(t))^2 - 4\beta(t) + 1 - t^2 + 4t - 1}{\beta(t) - t} \\
&= \frac{(\beta(t) - t)(\beta(t) + t) - 4(\beta(t) - t)}{\beta(t) - t} \\
&= \beta(t) + t - 4 \\
&= \frac{1}{2}t^3 + t - 4
\end{aligned}$$

and

$$\begin{aligned}
D_\beta g(t) &= \frac{g(\beta(t)) - g(t)}{\beta(t) - t} \\
&= \frac{(\beta(t))^2 + \beta(t) + 1 - t^2 - t - 1}{\beta(t) - 1} \\
&= \frac{(\beta(t) - t)(\beta(t) + t) + \beta(t) - t}{\beta(t) - t} \\
&= \beta(t) + t + 1 \\
&= \frac{1}{2}t^3 + t + 1,
\end{aligned}$$

and

$$D_\beta h(t) = \frac{(D_\beta f(t))g(t) - f(t)D_\beta g(t)}{g(t)g(\beta(t))}$$

$$\begin{aligned}
&= \frac{\left(\frac{1}{2}t^3 + t - 4\right)(t^2 + t + 1) - (t^2 - 4t + 1)\left(\frac{1}{2}t^3 + t + 1\right)}{(t^2 + t + 1)\left(\frac{1}{4}t^6 + \frac{1}{2}t^3 + 1\right)} \\
&= \frac{1}{(t^2 + t + 1)\left(\frac{1}{4}t^6 + \frac{1}{2}t^3 + 1\right)}\left(\frac{1}{2}t^5 + \frac{1}{2}t^4 + \frac{1}{2}t^3\right. \\
&\quad + t^3 + t^2 + t - 4t^2 - 4t - 4 \\
&\quad \left. - \frac{1}{2}t^5 - t^3 - t^2 + 2t^4 + 4t^2 + 4t - \frac{1}{2}t^3 - t - 1\right) \\
&= \frac{\frac{5}{2}t^4 - 5}{(t^2 + t + 1)\left(\frac{1}{4}t^6 + \frac{1}{2}t^3 + 1\right)} \\
&= \frac{10(t^4 - 2)}{(t^2 + t + 1)(t^6 + 2t^3 + 4)}.
\end{aligned}$$

Example 1.4.5 *Let α be a constant and $m \in \mathbb{N}$, and*

$$\begin{aligned} f(t) &= (t-\alpha)^m, \\ g(t) &= \frac{1}{(t-\alpha)^m}, \quad t \in I. \end{aligned}$$

We have

$$D_\beta f(s_0) = m(s_0-\alpha)^{m-1}$$

and

$$\begin{aligned} D_\beta f(t) &= \frac{f(\beta(t))-f(t)}{\beta(t)-t} \\ &= \frac{(\beta(t)-\alpha)^m-(t-\alpha)^m}{\beta(t)-t} \\ &= \frac{1}{\beta(t)-t}(\beta(t)-t) \\ &\quad \times \left((\beta(t)-\alpha)^{m-1}+(t-\alpha)(\beta(t)-\alpha)^{m-2}+\cdots+(t-\alpha)^{m-1}\right) \\ &= \sum_{k=0}^{m-1}(\beta(t)-\alpha)^{m-1-k}(t-\alpha)^k, \quad t \in I, \quad t \neq s_0. \end{aligned}$$

Next,

$$D_\beta g(s_0) = -m(s_0-\alpha)^{m-1}$$

and

$$\begin{aligned} D_\beta g(t) &= D_\beta \frac{1}{f(t)} \\ &= \frac{(D_\beta 1)f(t)-D_\beta f(t)}{f(t)f(\beta(t))} \\ &= -\frac{\sum_{k=0}^{m-1}(\beta(t)-\alpha)^{m-1-k}(t-\alpha)^k}{(t-\alpha)^m(\beta(t)-\alpha)^m} \\ &= -\sum_{k=0}^{m-1}\frac{(\beta(t)-\alpha)^{m-1-k}(t-\alpha)^k}{(t-\alpha)^m(\beta(t)-\alpha)^m} \\ &= -\sum_{k=0}^{m-1}\frac{1}{(\beta(t)-\alpha)^{k+1}(t-\alpha)^{m-k}}, \quad t \in I, \quad t \neq s_0. \end{aligned}$$

Exercise 1.4.6 *Let $I = \mathbb{R}$ and*

$$\beta(t) = \frac{2}{3}t+1,$$

$$f(t) = \frac{t+1}{t^2+4}, \quad t \in I.$$

Find $D_\beta f(t)$, $t \in I$.

1.5 Properties of β-Differentiable Functions

In this section, we will deduct some of the properties of the β-differentiable functions.

Theorem 1.5.1 *Let $f : I \to X$ be β-differentiable function and*

$$D_\beta f(t) = 0, \quad t \in I. \tag{1.3}$$

Then

$$f(t) = f(s_0), \quad t \in I.$$

Proof 1.5.2 *Since* (1.3) *holds, we get that*

$$f(\beta(t)) = f(t), \quad t \in I, \quad t \neq s_0.$$

Hence, using that $\beta : I \to I$, we find

$$\begin{aligned} f(t) &= f(\beta(t)) \\ &= f(\beta^2(t)) \\ &\vdots \\ &= f(\beta^k(t)), \quad t \in I, \quad t \neq s_0, \quad k \in \mathbb{N}. \end{aligned} \tag{1.4}$$

By Corollary 1.1.15, we obtain

$$\lim_{k\to\infty} f\left(\beta^k(t)\right) = f(s_0), \quad t \in I.$$

Hence and (1.4)*, we arrive at*

$$\begin{aligned} f(t) &= \lim_{k\to\infty} f(t) \\ &= \lim_{k\to\infty} f\left(\beta^k(t)\right) \\ &= f(s_0), \quad t \in I. \end{aligned}$$

This completes the proof.

Corollary 1.5.3 *Let $f, g : I \to X$ be β-differentiable functions and*

$$D_\beta(f-g)(t) = 0, \quad t \in I. \tag{1.5}$$

Then

$$f(t) - f(t_0) = g(t) - g(t_0), \quad t \in I.$$

Hence, applying Theorem 1.5.1, we find

$$(f-g)(t) = (f-g)(s_0), \quad t \in I,$$

or

$$f(t) - g(t) = f(s_0) - g(s_0), \quad t \in I.$$

This completes the proof.

Definition 1.5.4 *Let $s_0 \in [a,b] \subseteq I$. We define the β-interval $[a,b]_\beta$ by*

$$[a,b]_\beta = \{\beta^k(a) : k \in \mathbb{N}_0\} \cup \{\beta^k(b) : k \in \mathbb{N}_0\} \cup \{s_0\}$$

and the class $[c]_\beta$ for $c \in I$ by

$$[c]_\beta = \{\beta^k(c) : k \in \mathbb{N}_0\} \cup \{s_0\}.$$

For any set $A \subseteq \mathbb{R}$, define

$$A^* = A \backslash \{s_0\}.$$

Example 1.5.5 *Let $I = (-\sqrt{3}, \sqrt{3})$ and*

$$\beta(t) = \frac{1}{3}t^3, \quad t \in I.$$

Take

$$[a,b] = [-1,1].$$

Then

$$[-1,1]^* = [-1,0) \cup (0,1].$$

Next,

$$\begin{aligned}
\beta(-1) &= -\frac{1}{3}, \\
\beta^2(-1) &= \beta(\beta(-1)) \\
&= \beta\left(-\frac{1}{3}\right)
\end{aligned}$$

$$\begin{aligned}
&= \frac{1}{3}\left(-\frac{1}{3}\right)^3 \\
&= -\frac{1}{3^4}, \\
\beta^3(-1) &= \beta(\beta^2(-1)) \\
&= \beta\left(-\frac{1}{3^4}\right) \\
&= \frac{1}{3}\left(-\frac{1}{3^4}\right)^3 \\
&= -\frac{1}{3^{13}}, \\
\beta^4(-1) &= \beta(\beta^3(-1)) \\
&= \beta\left(-\frac{1}{3^{13}}\right) \\
&= \frac{1}{3}\left(-\frac{1}{3^{13}}\right)^3 \\
&= -\frac{1}{3^{40}}, \\
&\vdots \quad ,
\end{aligned}$$

and

$$\begin{aligned}
\beta(1) &= \frac{1}{3}, \\
\beta^2(1) &= \frac{1}{3^4}, \\
\beta^3(1) &= \frac{1}{3^{13}}, \\
\beta^4(1) &= \frac{1}{3^{40}}, \\
&\vdots \quad .
\end{aligned}$$

Thus,

$$\begin{aligned}[-1,1]_{\frac{1}{3}t^3} &= \left\{-1,-\frac{1}{3},-\frac{1}{3^4},-\frac{1}{3^{13}},-\frac{1}{3^{40}},\ldots\right\} \\ &\cup\left\{1,\frac{1}{3},\frac{1}{3^4},\frac{1}{3^{13}},\frac{1}{3^{40}},\ldots\right\}\cup\{0\}.\end{aligned}$$

Moreover,

$$\begin{aligned}\beta\left(\frac{1}{2}\right) &= \frac{1}{3\cdot 2^3}, \\ \beta^2\left(\frac{1}{2}\right) &= \beta\left(\beta\left(\frac{1}{2}\right)\right) \\ &= \beta\left(\frac{1}{3\cdot 2^3}\right)^3 \\ &= \frac{1}{3^4\cdot 2^9}, \\ \beta^3\left(\frac{1}{2}\right) &= \beta\left(\beta^2\left(\frac{1}{2}\right)\right) \\ &= \beta\left(\frac{1}{3^4\cdot 2^9}\right) \\ &= \frac{1}{3}\left(\frac{1}{3^4\cdot 2^9}\right)^3 \\ &= \frac{1}{3^{13}\cdot 2^{27}}, \\ &\vdots\ .\end{aligned}$$

Therefore

$$\left[\frac{1}{2}\right]_{\frac{1}{3}t^3}=\left\{\frac{1}{2},\frac{1}{3\cdot 2^3},\frac{1}{3^4\cdot 2^9},\frac{1}{3^{13}\cdot 2^{27}},\ldots\right\}.$$

Exercise 1.5.6 *Let $I=\mathbb{R}$ and*

$$\beta(t)=\frac{1}{2}t+1,\quad t\in I.$$

Find

1. $[0,2]_\beta$.
2. $[1]_\beta$.

Theorem 1.5.7 *Let $[a,b] \subseteq I$ and $f : [a,b] \to \mathbb{R}$ be continuous at s_0. Then*

$$D_\beta f(t) > 0, \quad t \in [a,b]^*_\beta, \tag{1.6}$$

if and only if f is strictly increasing on $[a,b]_\beta$.

Proof 1.5.8

1. Suppose that (1.6) *holds and β is a first kind with general quantum operator. Without loss of generality, suppose that $s_0 \notin \{a,b\}$. Since*

$$(a-s_0)(\beta(a)-a) \leq a,$$

we get

$$a < \beta(a).$$

Now, using that β is strictly increasing on $[a,b]$, we find

$$a < \beta(a) < \beta^2(a) < \ldots < \beta^k(a) < \ldots < s_0, \quad k \in \mathbb{N}.$$

Next, we have

$$(b-s_0)(\beta(b)-b) \leq 0,$$

we obtain

$$\beta(b) < b.$$

Hence,

$$s_0 < \ldots < \beta^m(b) < \ldots < \beta^2(b) < \beta(b) < b, \quad m \in \mathbb{N}.$$

Now, applying Corollary 1.1.15, we arrive at

$$f(a) < f(\beta(a)) < \ldots < f(\beta^k(a)) < \ldots < f(s_0)$$

and

$$f(s_0) < \ldots < f(\beta^m(b)) < \ldots < f(\beta(b)) < f(b),$$

$k, m \in \mathbb{N}$, i.e., f is strictly increasing on $[a,b]_\beta$.

2. Let f be strictly increasing on $[a,b]_\beta$. Take $t \in [a,b]^*_\beta$ arbitrarily. If

$$\beta(t) > t,$$

then

$$f(\beta(t)) > f(t)$$

and (1.5.10) *holds. If*

$$\beta(t) < t,$$

then

$$f(\beta(t)) < f(t)$$

and (1.5.10) *holds. This completes the proof.*

Example 1.5.9 *Let* $I = \left[1, \frac{3}{2}\right]$ *and*

$$\begin{aligned} \beta(t) &= \frac{1}{2}t + \frac{3}{4}, \\ f(t) &= 4t^2 - 9t, \quad t \in I. \end{aligned}$$

Here $s_0 = \frac{3}{2}$ *and*

$$\begin{aligned} D_\beta f(t) &= \frac{f(\beta(t)) - f(t)}{\beta(t) - t} \\ &= \frac{4(\beta(t))^2 - 9\beta(t) - 4t^2 + 9t}{\beta(t) - t} \\ &= \frac{4(\beta(t) - t)(\beta(t) + t) - 9(\beta(t) - t)}{\beta(t) - t} \\ &= 4(\beta(t) + t) - 9 \\ &= 4\left(\frac{1}{2}t + \frac{3}{4} + t\right) - 9 \\ &= 6t + 3 - 9 \\ &= 6t - 6 \\ &\geq 0, \quad t \in I. \end{aligned}$$

On the other hand,

$$\begin{aligned} f(1) &= 4 \cdot 1^2 - 9 \cdot 1 \\ &= 4 - 9 \\ &= -5. \end{aligned}$$

Next,

$$\begin{aligned} f(1.1) &= 4 \cdot (1.1)^2 - 9 \cdot 1.1 \\ &= 4.84 - 9.9 \end{aligned}$$

$$= \quad -5.06.$$

Therefore, f is not strictly increasing on $\left[1, \frac{3}{2}\right]$.

As we have proved Theorem 1.5.7, one can prove the following result.

Theorem 1.5.10 *Let $[a,b] \subseteq I$ and $f : [a,b] \to \mathbb{R}$ be continuous at s_0. Then*

$$D_\beta f(t) < 0, \quad t \in [a,b]^*_\beta,$$

if and only if f is strictly decreasing on $[a,b]_\beta$.

1.6 Chain Rules

Theorem 1.6.1 *Let $g : I \to \mathbb{R}$ be a continuous function and β-differentiable and $f : I \to X$ be continuously differentiable. Then, there exists a point c between t and $\beta(t)$ so that*

$$D_\beta(f \circ g)(t) = f'(g(c))D_\beta g(t), \quad t \in I. \tag{1.7}$$

Proof 1.6.2 *If $t = s_0$, then the assertion is obvious. Let $t \neq s_0$. Then*

$$\begin{aligned} D_\beta(f \circ g)(t) &= \frac{f(g(\beta(t))) - f(g(t))}{\beta(t) - t} \\ &= \frac{f(g(\beta(t))) - f(g(t))}{g(\beta(t)) - g(t)} \frac{g(\beta(t)) - g(t)}{\beta(t) - t} \\ &= f'(\eta)D_\beta g(t), \end{aligned}$$

where η is between $g(\beta(t))$ and $g(t)$. Since g is a continuous function, there is a c between t and $\beta(t)$ such that

$$\eta = g(c)$$

Therefore, we get (1.7). This completes the proof.

Example 1.6.3 *Let $I = \mathbb{R}$ and*

$$\begin{aligned} \beta(t) &= \frac{4}{5}t + 1, \\ f(t) &= t^3, \\ g(t) &= t^2, \quad t \in I. \end{aligned}$$

We will search $c \in \mathbb{R}$ *such that*

$$D_\beta(f \circ g)(0) = f'(g(0))D_\beta g(0).$$

Here $s_0 = 5$. *Then*

$$\begin{aligned} f'(t) &= 3t^2, \\ f'(g(t)) &= 3(t^2)^2 \\ &= 3t^4, \quad t \in I, \\ f'(g(c)) &= 3c^4, \end{aligned}$$

and

$$\begin{aligned} D_\beta g(0) &= \frac{g(\beta(0)) - g(0)}{\beta(0) - 0} \\ &= \frac{(\beta(0))^2}{\beta(0)} \\ &= \beta(0) \\ &= 1, \end{aligned}$$

and

$$\begin{aligned} f(g(t)) &= (t^2)^3 \\ &= t^6, \quad t \in I, \\ D_\beta(f \circ g)(0) &= \frac{(\beta(0))^6 - 0^6}{\beta(0) - 0} \\ &= (\beta(0))^5 + (\beta(0))^4 + (\beta(0))^3 + (\beta(0))^2 + \beta(0) + 1 \\ &= 1 + 1 + 1 + 1 + 1 + 1 \\ &= 6. \end{aligned}$$

Therefore

$$6 = 3c^4$$

or

$$c^4 = 2,$$

or

$$c = \pm\sqrt[4]{2}.$$

Exercise 1.6.4 *Let $I = \mathbb{R}$ and*

$$\begin{aligned} \beta(t) &= \frac{1}{2}t + 1, \\ f(t) &= t^2, \\ g(t) &= t^2, \quad t \in I. \end{aligned}$$

We will find $c \in \mathbb{R}$ such that

$$D_\beta(f \circ g)(0) = f'(g(c))D_\beta g(0).$$

Theorem 1.6.5 *Let $f : \mathbb{R} \to \mathbb{R}$ be continuously differentiable and $g : I \to \mathbb{R}$ be β-differentiable. Then, $f \circ g : I \to \mathbb{R}$ is β-differentiable and*

$$D_\beta(f \circ g)(t) = \left(\int_0^1 f\left(g(t) + h(\beta(t) - t)D_\beta g(t)\right) dh \right) D_\beta g(t), \tag{1.8}$$

$t \in I$.

Proof 1.6.6 *For $t = s_0$ the assertion is obvious. Suppose that $t \neq s_0$. We have*

$$\begin{aligned} f(g(\beta(t))) - f(g(t)) &= \int_{g(t)}^{g(\beta(t))} f'(y)dt \\ &= \left(\int_0^1 f'\left(g(t) + h(g(\beta(t)) - g(t))\right) dh \right) (g(\beta(t)) - g(t)), \end{aligned}$$

whereupon

$$\begin{aligned} &\frac{f(g(\beta(t))) - f(g(t))}{\beta(t) - t} \\ &= \left(\int_0^1 f'\left(g(t) + h(\beta(t) - t)\frac{g(\beta(t)) - g(t)}{\beta(t) - t} \right) dh \right) \left(\frac{g(\beta(t)) - g(t)}{\beta(t) - t} \right) \end{aligned}$$

and we get (1.8). *This completes the proof.*

Example 1.6.7 *Let $I = \mathbb{R}$ and*

$$\beta(t) = \frac{7}{8}t + 1,$$

$$f(t) = \frac{1}{1+t^2},$$

$$g(t) = 1+t, \quad t \in I.$$

Then

$$f'(t) = -\frac{2t}{(1+t^2)^2},$$

$$D_\beta g(t) = 1, \quad t \in I, \quad t \neq 8,$$

and

$$\begin{aligned}
f(g(t)) &= \frac{1}{1+(t+1)^2}, \\
D_\beta(f \circ g)(t) &= \frac{f(g(\beta(t))) - f(g(t))}{\beta(t)-t} \\
&= \left(\frac{1}{1+(\beta(t)+1)^2} - \frac{1}{1+(t+1)^2}\right)\frac{1}{\beta(t)-t} \\
&= -\frac{(\beta(t)+1)^2-(t+1)^2}{\left(1+\left(\frac{7}{8}t+2\right)^2\right)(1+(t+1)^2)} \cdot \frac{1}{\beta(t)-t} \\
&= -\frac{\beta(t)+t+2}{\left(1+\frac{49}{64}t^2+\frac{7}{2}t+4\right)(1+t^2+2t+1)} \\
&= -\frac{\frac{7}{8}t+t+3}{\frac{1}{64}(49t^2+224t+320)(t^2+2t+2)} \\
&= -\frac{\frac{15}{8}t+3}{\frac{1}{64}(49t^2+224t+320)(t^2+2t+2)} \\
&= -\frac{8(15t+24)}{(49t^2+224t+320)(t^2+2t+2)}, \quad t \in I, \quad t \neq 8,
\end{aligned}$$

and

$$\begin{aligned}
g(t)+h(\beta(t)-t)D_\beta g(t) &= t+1+h\left(\frac{7}{8}t+1-t\right) \\
&= t+1+h\left(1-\frac{1}{8}t\right), \\
f'(g(t)+h(\beta(t)-t)D_\beta g(t)) &= -\frac{2\left(t+1+h\left(1-\frac{1}{8}t\right)\right)}{\left(1+\left(t+1+h\left(1-\frac{1}{8}t\right)\right)^2\right)^2},
\end{aligned}$$

$t \in I$, $h \in [0,1]$, *and*

$$\begin{aligned}
&\left(\int_0^1 f'(g(t)+h(\beta(t)-t)D_\beta g(t))dh\right) D_\beta g(t) \\
&= -\int_0^1 \frac{2\left(t+1+h\left(1-\frac{1}{8}t\right)\right)}{\left(1+\left(t+1+h\left(1-\frac{1}{8}t\right)\right)^2\right)^2} dh \\
&= -\frac{1}{1-\frac{1}{8}t}\int_0^1 \frac{d\left(t+1+h\left(1-\frac{1}{8}t\right)\right)^2}{\left(1+\left(t+1+h\left(1-\frac{1}{8}t\right)\right)^2\right)^2} \\
&= \frac{1}{1-\frac{1}{8}t}\frac{1}{1+\left(t+1+h\left(1-\frac{1}{8}t\right)\right)^2}\Bigg|_{h=0}^{h=1} \\
&= \frac{1}{1-\frac{1}{8}t}\left(\frac{1}{1+\left(t+1+1-\frac{1}{8}t\right)^2} - \frac{1}{1+(t+1)^2}\right) \\
&= \frac{1}{1-\frac{1}{8}t}\cdot\frac{1+(t+1)^2-1-\left(\frac{7}{8}t+2\right)^2}{\left(1+\left(2+\frac{7}{8}t\right)^2\right)(1+(t+1)^2)} \\
&= \frac{1}{1-\frac{1}{8}t}\cdot\frac{\left(t+1-\frac{7}{8}t-2\right)\left(t+1+\frac{7}{8}t+2\right)}{\left(1+\left(2+\frac{7}{8}t\right)^2\right)(1+(t+1)^2)} \\
&= -\frac{\frac{15}{8}t+3}{\left(\frac{49}{64}t^2+\frac{7}{2}t+4+1\right)(t^2+2t+1+1)} \\
&= -\frac{\frac{1}{8}(15t+24)}{\frac{1}{64}(49t^2+224t+320)(t^2+2t+2)} \\
&= -\frac{8(15t+24)}{(49t^2+224t+320)(t^2+2t+2)} \\
&= D_\beta(f\circ g)(t), \quad t\in I, \quad t\neq 8.
\end{aligned}$$

Exercise 1.6.8 *Let* $I=\mathbb{R}$ *and*

$$\begin{aligned}
\beta(t) &= \frac{1}{3}t+1, \\
f(t) &= t^3+2t^2+2, \\
g(t) &= 2+3t, \quad t\in I.
\end{aligned}$$

Check (1.8).

Theorem 1.6.9 *Let $g : I \to I$ be β-differentiable, $\gamma : I \to I$ be a general quantum operator such that*

$$g(\beta(t)) = \gamma(g(t)), \quad t \in I,$$

and $f : \mathbb{R} \to \mathbb{R}$ be γ-differentiable on I. Then

$$D_\beta(f \circ g)(t) = D_\gamma f(g(t)) D_\beta g(t), \quad t \in I. \tag{1.9}$$

Proof 1.6.10 *Let $s_1 \in I$ be the unique fixed point of γ. Then*

$$g(s_0) = \gamma(g(s_0)).$$

Hence,

$$g(s_0) = s_1$$

and (1.9) holds. Let $t \in I$, $t \neq s_0$. Then

$$\begin{aligned}
D_\beta(f \circ g)(t) &= \frac{f(g(\beta(t))) - f(g(t))}{\beta(t) - t} \\
&= \frac{f(\gamma(g(t))) - f(g(t))}{\beta(t) - t} \\
&= \frac{f(\gamma(g(t))) - f(g(t))}{\gamma(g(t)) - g(t)} \cdot \frac{\gamma(g(t)) - g(t)}{\beta(t) - t} \\
&= \frac{f(\gamma(g(t))) - f(g(t))}{\gamma(g(t)) - g(t)} \cdot \frac{g(\beta(t)) - g(t)}{\beta(t) - t} \\
&= D_\gamma f(g(t)) D_\beta g(t).
\end{aligned}$$

This completes the proof.

Example 1.6.11 *Let $I = \mathbb{R}$ and*

$$\begin{aligned}
\beta(t) &= \frac{1}{2}t + 2, \\
\gamma(t) &= \frac{1}{2}t + 4, \\
f(t) &= t^2, \\
g(t) &= t + 4, \quad t \in I.
\end{aligned}$$

Then

$$f(g(t)) = (g(t))^2$$

$$= (t+4)^2$$

$$= t^2+8t+16, \quad t \in I.$$

Here $s_0 = 4$ *is the unique fixed point of* β *and* $s_1 = 8$ *is the unique fixed point of* γ. *We have*

$$\begin{aligned} D_\beta f(g(t)) &= \frac{(\beta(t))^2+8\beta(t)+16-t^2-8t-16}{\beta(t)-t} \\ &= \frac{(\beta(t)-t)(\beta(t)+t)+8(\beta(t)-t)}{\beta(t)-t} \\ &= \beta(t)+t+8 \\ &= \frac{1}{2}t+2+t+8 \\ &= \frac{3}{2}t+10, \quad t \in I, \quad t \neq 4, \end{aligned}$$

and

$$\begin{aligned} D_\beta f(g(s_0)) &= 2s_0+8 \\ &= 2\cdot 4+8 \\ &= 8+8 \\ &= 16. \end{aligned}$$

Next,

$$\begin{aligned} g(\beta(t)) &= \beta(t)+4 \\ &= \frac{1}{2}t+2+4 \\ &= \frac{1}{2}t+6, \\ \gamma(g(t)) &= \frac{1}{2}g(t)+4 \\ &= \frac{1}{2}(t+4)+4 \end{aligned}$$

$$= \frac{1}{2}t + 2 + 4$$

$$= \frac{1}{2}t + 6, \quad t \in I.$$

Therefore

$$g(\beta(t)) = \gamma(g(t)), \quad t \in I.$$

Moreover,

$$\begin{aligned}
D_\gamma f(g(t)) &= \frac{f(\gamma(t)) - f(t)}{\gamma(t) - t} \\
&= \frac{(\gamma(t))^2 - t^2}{\gamma(t) - t} \\
&= \frac{(\gamma(t) - t)(\gamma(t) + t)}{\gamma(t) - t} \\
&= \gamma(t) + t \\
&= \frac{1}{2}t + 4 + t \\
&= \frac{3}{2}t + 4, \\
D_\gamma f(g(t)) &= \frac{3}{2}g(t) + 4 \\
&= \frac{3}{2}(t + 4) + 4 \\
&= \frac{3}{2}t + 6 + 4 \\
&= \frac{3}{2}t + 10, \quad t \in I, \quad t \neq 4.
\end{aligned}$$

Note that

$$g(s_0) = s_1$$

and

$$\begin{aligned}
D_\gamma f(g(s_0)) &= D_\gamma f(s_1) \\
&= \frac{3}{2}s_1 + 4
\end{aligned}$$

$$\begin{aligned}
&= \frac{3}{2}\cdot 8+4\\
&= 12+4\\
&= 16,
\end{aligned}$$

$$\begin{aligned}
D_\beta g(t) &= \frac{g(\beta(t))-g(t)}{\beta(t)-t}\\
&= \frac{\beta(t)+4-t-4}{\beta(t)-t}\\
&= 1,\\
D_\beta g(s_0) &= 1.
\end{aligned}$$

Thus,

$$\begin{aligned}
D_\gamma f(g(t))D_\beta g(t) &= \left(\frac{3}{2}t+10\right)\cdot 1\\
&= \frac{3}{2}t+10\\
&= D_\beta(f\circ g)(t), \quad t\in I, \quad t\neq s_0,
\end{aligned}$$

and

$$\begin{aligned}
D_\gamma f(g(s_0))D_\beta g(s_0) &= 16\cdot 1\\
&= 16\\
&= D_\beta(f\circ g)(s_0).
\end{aligned}$$

Thus, (1.9) *holds.*

Exercise 1.6.12 *Let* $I=\mathbb{R}$ *and*

$$\begin{aligned}
\beta(t) &= \frac{1}{4}t+1,\\
\gamma(t) &= \frac{1}{4}t+5,
\end{aligned}$$

$$g(t) = 2t+6,$$

$$f(t) = t^3, \quad t \in I.$$

Check (1.9).

1.7 A Mean Value Theorem

In this section, we will deduct some mean value theorems in the area of the β-differentiable calculus. We will start with the following useful result.

Theorem 1.7.1 *Let f be β-differentiable at $t \in I$. Then, there exists a function g defined in a neighbourhood U of t with*

$$\lim_{s \to t} g(s) = g(t) = 0,$$

such that

$$f(\beta(t)) = f(s) + (D_\beta f(t) + g(s))(\beta(t) - s)$$

for any $s \in I$.

Proof 1.7.2 *Define*

$$g(s) = \begin{cases} \dfrac{f(\beta(t)) - f(s)}{\beta(t) - s} - D_\beta f(t) & if \quad s \neq \beta(t) \\ 0 \quad if \quad s = \beta(t). \end{cases}$$

For $s = \beta(t) \in U$, we get the desired result. For $s \neq \beta(t)$, $s \in U$, we have

$$\frac{f(\beta(t)) - f(s)}{\beta(t) - s} = D_\beta f(t) + g(s),$$

whereupon we get the desired result. This completes the proof.

1.8 Higher Order β-Derivatives

Let $n \in \mathbb{N}$. With $S_k^{(n)}$ we denote the set of all possible strings of length n containing k times β and $n-k$ times D_β. We also denote

$$f^{D_\beta \beta}(t) = (D_\beta f)(\beta(t)), \quad t \in I,$$

and

$$f^{\beta D_\beta}(t) = D_\beta(f \circ \beta)(t), \quad t \in I,$$

and f^Λ is defined accordingly for $\Lambda \in S_k^{(n)}$.

Definition 1.8.1 *If f is β-differentiable n-times over I, then the higher order derivatives of f are defined by*

$$D_\beta^n f = D_\beta(D_\beta^{n-1} f), \quad n \in \mathbb{N},$$

where

$$D_\beta^0 f = f.$$

For $k \in \mathbb{N}_0$, with $\mathscr{C}_\beta^k(I)$ we will denote the set of all functions $f : I \to \mathbb{R}$ that are continuous and $D_\beta^l f$ exist and are continuous for any $l \in \{0, \ldots, k\}$.

Example 1.8.2 *Let $I = \mathbb{R}$ and*

$$\begin{aligned} \beta(t) &= \frac{2}{3}t + 1, \\ f(t) &= t^2 + t, \quad t \in I. \end{aligned}$$

Then $s_0 = 3$. For $t \neq 3$, we have

$$\begin{aligned} D_\beta f(t) &= \frac{f(\beta(t)) - f(t)}{\beta(t) - t} \\ &= \frac{(\beta(t))^2 + \beta(t) - t^2 - t}{\beta(t) - t} \\ &= \frac{(\beta(t) - t)(\beta(t) + t) + (\beta(t) - t)}{\beta(t) - t} \\ &= \beta(t) + t + 1 \\ &= \frac{2}{3}t + 1 + t + 1 \\ &= \frac{5}{3}t + 2 \end{aligned}$$

and hence,

$$D_\beta^2 f(t) = \frac{5}{3}.$$

Exercise 1.8.3 *Let $I = \mathbb{R}$ and*

$$\beta(t) = \frac{3}{4}t + 1,$$

$$f(t) = t^2 - t + 2, \quad t \in I.$$

Find $D^2_\beta f(t)$, $t \in I$, $t \neq 4$.

Theorem 1.8.4 (Leibniz Formula) *If*

$$f^\Lambda \quad \text{exists} \quad \text{for} \quad \text{all} \quad \Lambda \in S_k^{(n)},$$

then

$$D^n_\beta (fg) = \sum_{k=0}^{n} \left(\sum_{\Lambda \in S_k^{(n)}} f^\Lambda \right) D^k_\beta g. \tag{1.10}$$

Proof 1.8.5 *We will use induction.*

1. $n = 1$. *Then*

$$S_0^{(1)} = 0, \quad S_1^{(1)} = \beta.$$

Hence,

$$\begin{aligned} \sum_{k=0}^{1} \left(\sum_{\Lambda \in S_k^{(1)}} f^\Lambda \right) D^2_\beta g &= \sum_{\Lambda \in S_0^{(1)}} f^\Lambda g + \sum_{\Lambda \in S_1^{(1)}} f^\Lambda D_\beta g \\ &= D_\beta f g + f^\beta D_\beta g \\ &= D_\beta (fg), \end{aligned}$$

i.e., the assertion holds for $n = 1$.

2. *Assume that the assertion is valid for some* $n \in \mathbf{N}$.

We will prove that

$$D^{n+1}_\beta (fg) = \left(\sum_{\Lambda \in S_k^{(n+1)}} f^\Lambda \right) D^k_\beta g.$$

Using (1.10), we have

$$D^{n+1}_\beta (fg) = D_\beta \left(D^n_\beta (fg) \right)$$

$$= D_\beta \left(\sum_{k=0}^{n} \left(\sum_{\Lambda \in S_k^{(n)}} f^\Lambda \right) D^k_\beta g \right)$$

$$
\begin{aligned}
&= \sum_{k=0}^{n} D_\beta \left(\left(\sum_{\Lambda \in S_k^{(n)}} f^{\Lambda} \right) D_\beta^k g \right) \\
&= \sum_{k=0}^{n} \left(\left(\sum_{\Lambda \in S_k^{(n)}} f^{\Lambda\beta} \right) D_\beta^{k+1} g + \left(\sum_{\Lambda \in S_k^{(n)}} D_\beta f^{\Lambda} \right) D_\beta^k g \right) \\
&= \sum_{k=0}^{n} \left(\sum_{\Lambda \in S_k^{(n)}} f^{\Lambda\beta} \right) D_\beta^{k+1} g + \sum_{k=0}^{n} \left(\sum_{\Lambda \in S_k^{(n)}} f^{\Lambda D_\beta} \right) D_\beta^k g \\
&= \sum_{k=1}^{n+1} \left(\sum_{\Lambda \in S_{k-1}^{(n)}} f^{\Lambda\beta} \right) D_\beta^k g + \sum_{k=0}^{n} \left(\sum_{\Lambda \in S_k^{(n)}} f^{\Lambda\beta} \right) D_\beta^k g \\
&= \sum_{k=1}^{n} \left(\sum_{\Lambda \in S_{k-1}^{(n)}} f^{\Lambda\beta} \right) D_\beta^k g + \left(\sum_{\Lambda \in S_k^{(n)}} f^{\Lambda\beta} \right) D_\beta^{n+1} g \\
&\quad + \sum_{k=1}^{n} \left(\sum_{\Lambda \in S_k^{(n)}} f^{\Lambda\beta} \right) D_\beta^k g + \left(\sum_{\Lambda \in S_0^{(n)}} f^{\Lambda\beta} \right) g \\
&= \sum_{k=1}^{n} \left(\sum_{\Lambda \in S_{k-1}^{(n)}} f^{\Lambda\beta} + \sum_{\Lambda \in S_k^{(n)}} f^{\Lambda D_\beta} \right) D_\beta^k g \\
&\quad + \left(\sum_{\Lambda \in S_n^{(n)}} f^{\Lambda D_\beta} + \sum_{\Lambda \in S_0^{(n)}} f^{\Lambda D_\beta} \right) g \\
&= \sum_{k=1}^{n} \left(\sum_{\Lambda \in S_k^{(n+1)}} f^{\Lambda} \right) D_\beta^k g + \left(\sum_{\Lambda \in S_{n+1}^{(n+1)}} f^{\Lambda} \right) D_\beta^{n+1} g \\
&\quad + \left(\sum_{\Lambda \in S_0^{(n+1)}} f^{\Lambda} \right) g
\end{aligned}
$$

$$= \sum_{k=0}^{n+1} \left(\sum_{\Lambda \in S_k^{(n+1)}} f^{\Lambda} \right) D_\beta^k g.$$

This completes the proof.

1.9 The β-Rolle Theorem

Definition 1.9.1 *Let $y : I \to \mathbb{R}$ be $(k-1)$-times differentiable, $k \in \mathbb{N}$. We say that y has a general zero(GZ) of order greater than or equal to k at $t \in I$ provided*

$$D_\beta^i y(t) = 0, \quad i \in \{0, \ldots, k-1\}, \tag{1.11}$$

or

$$D_\beta^i y(t) = 0 \quad \text{for} \quad i \in \{0, \ldots, k-2\} \tag{1.12}$$

and

$$D_\beta^{k-1} y(\beta^{-1}(t)) D_\beta^{k-1} y(t) < 0$$

holds.

Remark 1.9.2 *Note that in the case* (1.12) *t must be left-scattered. Otherwise $\beta^{-1}(t) = t$ and*

$$\begin{aligned} 0 &> D_\beta^{k-1} y(\beta^{-1}(t)) D_\beta^{k-1} y(t) \\ &= \left(D_\beta^{k-1} y(t) \right)^2 \\ &\geq 0, \end{aligned}$$

which is a contradiction.

Theorem 1.9.3 *The condition* (1.12) *holds if and only if*

$$D_\beta^j y(t) = 0, \quad j \in \{0, \ldots, k-2\}, \tag{1.13}$$

and

$$(-1)^{k-1} y(\beta^{-1}(t)) D_\beta^{k-1} y(t) < 0.$$

Proof 1.9.4 *1. Let* (1.12) *holds. Then,* $t < \beta(t)$, $\beta(\beta^{-1}(t)) = t$ *and*

$$\begin{aligned} D_\beta^{k-1}y(\beta^{-1}(t)) &= \frac{D_\beta^{k-2}y(\beta(\beta^{-1}(t))) - D_\beta^{k-2}y(\beta^{-1}(t))}{t-\beta^{-1}(t)} \\ &= -\frac{D_\beta^{k-2}y(\beta^{-1}(t))}{t-\beta^{-1}(t)} \qquad (1.14) \\ &= \cdots \\ &= (-1)^{k-1}\frac{y(\beta^{-1}(t))}{(t-\beta^{-1}(t))^{k-1}}. \end{aligned}$$

Hence,

$$\begin{aligned} 0 &> D_\beta^{k-1}y(\beta^{-1}(t))D_\beta^{k-1}y(t) \\ &= (-1)^{k-1}\frac{y(\beta^{-1}(t))}{(t-\beta^{-1}(t))^{k-1}}D_\beta^{k-1}y(t) \end{aligned}$$

and (1.13) *holds.*

2. Assume (1.13). *Let* $\beta(t) = t = s_0$. *Then,* $\beta^{-1}(t) = t$ *and we have the following cases.*

(a) Let $k = 1$. *Then,* $(y(t))^2 < 0$, *which is a contradiction.*

(b) Let $k \geq 2$. *Then,* $y(t) = 0$ *and*

$$(-1)^{k-1}y(t)D_\beta^{k-1}y(t) < 0.$$

This is a contradiction.

Consequently t is left-scattered. Hence, using (1.14), *we obtain* (1.12). *This completes the proof.*

Theorem 1.9.5 *Let* $j \in \mathbb{N}_0$ *and* $t \in I^{\kappa^j}$. *Then*

$$D_\beta^i y(t) = 0, \quad 0 \leq i \leq j, \tag{1.15}$$

if and only if

$$D_\beta^i y(\beta^l(t)) = 0, \quad 0 \leq i \leq j-l, \quad 0 \leq l \leq j. \tag{1.16}$$

In this case,

$$D_\beta^{j+1-l}y(\beta^l(t)) = \prod_{s=0}^{l-1}(\beta^{s+1}(t)\beta^s(t))D_\beta^{j+1}y(t). \tag{1.17}$$

Proof 1.9.6 *1. Let* (1.15) *holds.*

(a) *Suppose that* $j = 0$. *Then,* $l = i = 0$, $y(t) = 0$ *and*

$$D^i_\beta y(\beta^l(t)) = y(t) = 0.$$

(b) *Suppose that* $j > 0$. *Then*

$$y(t) = D_\beta y(t) = \ldots = D^j_\beta y(t) = 0.$$

i. *Let* $j = 1$. *Then* $i \in \{0,1\}$, $l \in \{0,1\}$. *Let* $l = 0$. *Then* $i \in \{0,1\}$ *and* (1.16) *holds. Let* $l = 1$. *Then* $i = 0$ *and*

$$\begin{aligned} y(\beta(t)) &= y(t) + (\beta(t) - t) D_\beta y(t) \\ &= 0. \end{aligned}$$

ii. *Assume that* (1.16) *is true for some* $j \in \mathbb{N}$, *i.e., assume*

$$\begin{aligned} y(t) &= D_\beta y(t) = \ldots = D^j_\beta y(t) = 0, \\ y(\beta(t)) &= D_\beta y(\beta(t)) = \ldots = D^{j-1}_\beta y(\beta(t)) = 0, \\ y(\beta^2(t)) &= D_\beta y(\beta^2(t)) = \ldots = D^{j-2}_\beta y(\beta^2(t)) = 0, \\ &\vdots \\ y(\beta^l(t)) &= D_\beta y(\beta^l(t)) = \ldots = D^{j-l}_\beta y(\beta^l(t)) = 0, \\ &\vdots \\ y(\beta^j(t)) &= 0. \end{aligned} \tag{1.18}$$

iii. *We will prove* (1.16) *for* $j + 1$, *i.e., we will prove that*

$$\begin{aligned} y(t) &= D_\beta y(t) = \ldots = D^{j+1}_\beta y(t) = 0, \\ y(\beta(t)) &= D_\beta y(\beta(t)) = \ldots = D^j_\beta y(\beta(t)) = 0, \\ y(\beta^2(t)) &= D_\beta y(\beta^2(t)) = \ldots = D^{j-1}_\beta y(\beta^2(t)) = 0, \\ &\vdots \\ y(\beta^l(t)) &= D_\beta y(\beta^l(t)) = \ldots = D^{j-l+1}_\beta y(\beta^l(t)) = 0, \end{aligned}$$

$$\vdots$$

$$y(\beta^{j+1}(t)) = 0.$$

By (1.18), *it follows that we have to prove*

$$\begin{aligned} D_\beta^{j+1}y(t) &= D_\beta^j y(\beta(t)) = \ldots = D_\beta^{j=l+1}y(\beta^l(t)) \\ &= \ldots = y(\beta^{j+1}(t)) = 0. \end{aligned}$$

By (1.15), *we have* $D_\beta^{j+1}y(t) = 0$. *Then*

$$\begin{aligned} D_\beta^j y(\beta(t)) &= D_\beta^j y(t) + (\beta(t) - t)D_\beta^{j+1}y(t) \\ &= 0, \\ D_\beta^{j-1}y(\beta^2(t)) &= D_\beta^{j-1}y(\beta(t)) + (\beta(t) - t)D_\beta^j y(\beta(t)) \\ &= 0, \\ &\vdots \\ D_\beta y(\beta^j(t)) &= 0, \\ y(\beta^{j+1}(t)) &= y(\beta^j(t)) + (\beta(t) - t)D_\beta y(\beta^j(t)) \\ &= 0. \end{aligned}$$

Hence and the principle of mathematical induction, we conclude that (1.16) *holds for any* $j \in \mathbb{N}_0$.

2. *Suppose that* (1.16) *holds.*

(a) *Let* $j = 0$. *Then* $i = l = 0$ *and*

$$D_\beta^i y(\beta^l(t)) = y(t) = 0.$$

(b) *Let* $j > 0$

i. *Assume that* $j = 1$. *Then, by* (1.16), *we have*

$$y(t) = D_\beta y(t) = y(\beta(t)) = 0.$$

ii. *Assume that* (1.15) *holds for some* $j \in \mathbb{N}_0$.

iii. *We will prove that* (1.15) *holds for* $j+1$. *Since* (1.16) *holds for* $j+1$, *by* (1.18) *we obtain*

$$y(t) = D_\beta y(t) = \ldots = D_\beta^{j+1} y(t) = 0.$$

Hence and the principle of mathematical induction, it follows that (1.15) *holds for any* $j \in \mathbb{N}_0$.

Suppose (1.16), *i.e., we assume* (1.18). *Then, we will prove* (1.17). *We have*

$$\begin{aligned}
D_\beta^j y(\beta(t)) &= D_\beta^j y(t) + (\beta(t) - t) D_\beta^{j+1} y(t) \\
&= (\beta(t) - t) D_\beta^{j+1} y(t), \\
D_\beta^{j-1} y(\beta^2(t)) &= D_\beta^{j-1} y(\beta(t)) + (\beta^2(t) - \beta(t)) D_\beta^j y(\beta(t)) \\
&= (\beta^2(t) - \beta(t)) D_\beta^j y(\beta(t)) \\
&= (\beta^2(t) - \beta(t))(\beta(t) - t) D_\beta^{j+1} y(t), \\
&\vdots \\
D_\beta^{j+1-l} y(\beta^l(t)) &= \prod_{s=0}^{l-1} (\beta^{s+1}(t) - \beta^s(t)) D_\beta^{j+1} y(t).
\end{aligned}$$

This completes the proof.

Definition 1.9.7 *If* y *has a GZ of order greater than or equal to* k *at* t *we will say that* y *has at least* k *GZs, counting multiplicities. By Theorem 1.9.5, it follows that if* y *has a GZ of order greater than or equal to* k *at* t, *then* y *has a GZ of order greater than or equal to* $k-1$ *at* $\beta(t)$. *Therefore, if* y *has a GZ of order greater than or equal to* k *at* t_1 *and* y *has a GZ of order greater than or equal to* k_2 *at* t_2 *and* $\beta^{k-1}(t_1) < t_2$, *then we will say that* y *has ate least* $k_1 + k_2$ *GZs, counting multiplicities.*

Theorem 1.9.8 (Rolle's Theorem) *If* y *has at least* $k \in \mathbb{N}$ *GZs on* $[a,b]$, *counting multiplicities, then* $D_\beta y$ *has at least* $k-1$ *GZs on* $[a,b]$, *counting multiplicities.*

Proof 1.9.9 1. *Firstly, we will prove that if* y *has a GZ of order* $\leq m$ *at* t, *then* $D_\beta y$ *has a GZ of order* $\leq m-1$ *at* t.

Since y *has a GZ of order* $\leq m$ *at* t, *then*

$$D_\beta^i y(t) = 0, \quad i \in \{0, \ldots, m-1\},$$

or

$$D^i_\beta y(t) = 0, \quad i \in \{0, \dots, m-2\},$$

and

$$D^{m-1}_\beta y(\beta^{-1}(t)) D^{m-1}_\beta y(t) < 0.$$

If $m = 1$, then $y(t_0$ or $y(\beta^{-1}(t))y(t) < 0$. If $D_\beta y(t) = 0$ or $D_\beta y(\beta^{-1}(t)) D_\beta y(t) < 0$, then $m > 1$, which is a contradiction. Therefore, $D_\beta y$ has no GZ at t. Let $m \geq 2$. Then

$$D^{i-1}_\beta (D_\beta y)(t) = 0, \quad i \in \{0, \dots, m-2\},$$

or

$$D^i_\beta (D_\beta y)(t) = 0, \quad i \in \{0, \dots, m-3\},$$

and

$$D^{m-2}_\beta (D_\beta y)(\beta^{-1}(t)) D^{m-2}_\beta (D_\beta y)(t) < 0.$$

Thus, $D_\beta y$ has a GZ of order $\leq m-1$ at t.

2. Now, we will prove that if y has a GZ of order $\leq m \in \mathbb{N}$ at t and y has a GZ of order ≤ 1 at s with $\beta^{m-1}(t) < s$, then $D_\beta y$ has at least m GZs in $[t, s)$ which is equivalent to

a)*If $y(r) = 0$ and $D_\beta y$ has no GZ in $[r, s)$, where $r < \beta^{-1}(s)$, then y has no GZ at s.*

b)*If $y(\beta^{-1}(r))y(r) < 0$ and $D_\beta y$ has no GZ in $[r, s)$, where $r < \beta^{-1}(s)$, then ty has no GZ at s.*

If the assumptions of a) hold, then $D_\beta y(\tau) > 0$, $\tau \in [r, s)$, or $D_\beta y(\tau) < 0$, $\tau \in [r, s)$. Thus,

$$\begin{aligned} y(\beta^{-1}(s))y(s) &= \left(\int_r^{\beta^{-1}(s)} D_\beta y(\tau) d_\beta \tau \right) \left(\int_r^s D_\beta y(\tau) d_\beta \tau \right) \\ &> 0. \end{aligned}$$

Therefore, y has no GZ at s. If the assumptions of b) hold, then $\beta^{-1}(r) < r$ and

$$\begin{aligned} y(\beta^{-1}(r)) D_\beta y(\beta^{-1}(r)) &= y(\beta^{-1}(r)) \frac{y(r) - y(\beta^{-1}(r))}{r - \beta^{-1}(r)} \\ &< 0. \end{aligned}$$

Since $D_\beta y$ has a constant sign on $[\beta^{-1}(r), s)$, then

$$y(\beta^{-1}(r)) D_\beta y(\tau) < 0, \quad \tau[\beta^{-1}(r), s).$$

Then

$$y(\beta^{-1}(r))y(t) = y(\beta^{-1}(r))\left(y(r)+\int_r^t D_\beta y(\tau)d_\beta\tau\right) < 0, \quad t\in\{\beta^{-1}(s),s\}.$$

Hence,

$$y(\beta^{-1}(r))y(\beta^{-1}(s))<0, \quad y(\beta^{-1}(r))y(s)<0,$$

and

$$(y(\beta^{-1}(r)))^2y(\beta^{-1}(s))y(s)>0.$$

Therefore, $y(\beta^{-1}(s))y(s)>0$ *and* y *has no GZ at* s. *This completes the proof.*

1.10 Advanced Practical Problems

Problem 1.10.1 *Let* $I=\mathbb{R}$,

$$\beta(t)=3t, \quad t\in I.$$

Prove that β *is a first kind general quantum operator.*

Problem 1.10.2 *Let* $I=[-1,1]$,

$$\beta(t)=\frac{1}{4}t+\frac{1}{4}, \quad t\in I.$$

Prove that β *is a first-order general quantum operator.*

Problem 1.10.3 *Let* I *and* β *be as in Problem 1.10.1. Find*

1. $\beta^{-1}(t)$, $t\in I$.
2. $\beta^k(t)$, $t\in I$, $k\in\mathbb{N}_0$.
3. $\beta^{-k}(t)$, $t\in I$, $k\in\mathbb{N}$.

Problem 1.10.4 *Let* I *and* β *be as in Problem 1.10.2. Find*

1. $\beta^{-1}(t)$, $t\in I$.
2. $\beta^k(t)$, $t\in I$, $k\in\mathbb{N}_0$.
3. $\beta^{-k}(t)$, $t\in I$, $k\in\mathbb{N}$.

Problem 1.10.5 *Let* $I=[1,\infty)$ *and*

$$\beta(t)=\log t+1, \quad t\in I.$$

1. *Prove that β is a first kind general quantum operator.*
2. *Prove that*

$$\beta^{-1}(t) = e^{t-1}, \quad t \in I.$$

3. *Find*

$$\lim_{k\to\infty} \beta^k(t), \quad t \in I.$$

4. *Find*

$$\lim_{k\to\infty} \beta^{-k}(t), \quad t \in I.$$

Problem 1.10.6 *Let $I = \mathbb{R}$ and*

$$\begin{aligned} \beta(t) &= \frac{1}{2}t, \\ f(t) &= t^2 + 4, \quad t \in I. \end{aligned}$$

1. *Prove that β is a first kind general quantum operator.*
2. *Find $D_\beta f(t)$, $t \in I$.*

Problem 1.10.7 *Let $I = \mathbb{R}$ and*

$$\begin{aligned} \beta(t) &= \frac{3}{4}t + 1, \\ f(t) &= \frac{t-1}{t^2+1}, \quad t \in I. \end{aligned}$$

Find $D_\beta f(t)$, $t \in I$.

Problem 1.10.8 *Let $I = \mathbb{R}$ and*

$$\beta(t) = \frac{2}{3}t + \frac{1}{3}, \quad t \in I.$$

Find

1. $[0,2]_\beta$.
2. $[3]_\beta$.

Problem 1.10.9 *Let $I = \mathbb{R}$ and*

$$\begin{aligned} \beta(t) &= \frac{2}{3}t + 1, \\ f(t) &= t^2, \\ g(t) &= t + 2, \quad t \in I. \end{aligned}$$

Find $c \in \mathbb{R}$ such that

$$D_\beta(f \circ g)(0) = f'(g(c)) D_\beta g(0).$$

Problem 1.10.10 *Let $I = \mathbb{R}$ and*

$$\begin{aligned} \beta(t) &= \frac{1}{2}t + 1, \\ f(t) &= \frac{t+1}{t^2+1}, \\ g(t) &= 1 + t + t^2, \quad t \in I. \end{aligned}$$

Check (1.8).

Problem 1.10.11 *Let $I = \mathbb{R}$ and*

$$\begin{aligned} \beta(t) &= \frac{1}{8}t + 1, \\ \gamma(t) &= \frac{1}{8}t + 6, \\ g(t) &= 5t + 2, \\ f(t) &= t^3 + t, \quad t \in I. \end{aligned}$$

Check (1.9).

Problem 1.10.12 *Let $I = \mathbb{R}$ and*

$$\begin{aligned} \beta(t) &= \frac{7}{8}t + 1, \\ f(t) &= 2t^2 - 2t + 7, \quad t \in I. \end{aligned}$$

Find $D_\beta^2 f(t)$, $t \in I$, $t \neq 4$.

1.11 Notes and References

In this chapter, we define first and second kind general quantum operators. We introduce the concept for β-differentiation. In the chapter are deduced the main properties and main rules for β-differentiation. They are given chain rules, a mean values theorem and it is proved the Leibniz formula. Some of the results in this chapter can be found in [1].

2

General Quantum Integration

In this chapter we define β-integral and deduce some of its basic properties, and they are given rules for β-integration. In the chapter is deduced the Taylor formula. They are introduced improper integrals of the first and second kind. Let $I \subseteq \mathbb{R}$ and β be first kind or second kind general quantum operator on I. Let also, X be a Banach space.

2.1 β-Antiderivatives

Definition 2.1.1 *We say that a function $F : I \to X$ is a β-antiderivative of the function $f : I \to X$ if F is β-differentiable on I and*

$$D_\beta F(t) = f(t), \quad t \in I.$$

β-Antiderivative

Example 2.1.2 *Let $I = (-\sqrt{2}, \sqrt{2})$ and*

$$\begin{aligned} \beta(t) &= \frac{1}{2}t^3, \\ f(t) &= \frac{1}{4}t^6 + \frac{1}{2}t^4 - \frac{1}{2}t^3 + t^2 - t + 1, \\ F(t) &= t^3 - t^2 + t, \quad t \in I. \end{aligned}$$

Note that $s_0 = 0$. For $t \neq 0$, we have

$$D_\beta F(t) = \frac{F(\beta(t)) - F(t)}{\beta(t) - t}$$

$$= \frac{(\beta(t))^3 - (\beta(t))^2 + \beta(t) - t^3 + t^2 - t}{\beta(t) - t}$$

$$= \frac{(\beta(t) - t)((\beta(t))^2 + t\beta(t) + t^2) - (\beta(t) - t)(\beta(t) + t) + \beta(t) - t}{\beta(t) - t}$$

DOI: 10.1201/9781003472131-2

$$
\begin{aligned}
&= (\beta(t))^2 + t\beta(t) + t^2 - \beta(t) - t + 1 \\
&= \left(\frac{1}{2}t^3\right)^2 + t\left(\frac{1}{2}t^3\right) + t^2 - \frac{1}{2}t^3 - t + 1 \\
&= \frac{1}{4}t^6 + \frac{1}{2}t^4 - \frac{1}{2}t^3 + t^2 - t + 1 \\
&= f(t).
\end{aligned}
$$

Next,

$$
\begin{aligned}
F'(0) &= (3t^2 - 2t + 1)\Big|_{t=0} \\
&= 1 \\
&= f(0).
\end{aligned}
$$

Thus, F is a β-antiderivative of f.

Exercise 2.1.3 *Let $I = \mathbb{R}$ and*

$$
\begin{aligned}
\beta(t) &= \frac{1}{2}t + 2, \\
f(t) &= \frac{3}{2}t + 1, \\
F(t) &= t^2 - t + 1, \quad t \in I.
\end{aligned}
$$

Prove that F is a β-antiderivative of f.

Definition 2.1.4 *With Ω we denote the vector space of all functions $g : I \to X$ which are continuous at s_0 and $g(s_0) = 0$.*

Define the operator $T_\beta : \Omega \to \Omega$ as follows

$$T_\beta(g)(t) = g(\beta(t)), \quad t \in I, \quad g \in \Omega.$$

Denote

$$Y = \text{range}(\mathscr{I} - T_\beta),$$

where $\mathscr{I}$ is the identity operator.

Theorem 2.1.5 *For any $g \in Y$, the series*

$$\sum_{k=0}^{\infty} g(\beta^k(t)), \quad t \in I, \tag{2.1}$$

is uniformly convergent on I.

Proof 2.1.6 *Let $\varepsilon > 0$ be arbitrarily chosen. By Corollary 1.1.15 and the definitions for Ω and Y, we get*

$$\begin{aligned}\lim_{k\to\infty} g(\beta^k(t)) &= g(s_0)\\ &= 0.\end{aligned}$$

Thus, there is a $K \in \mathbb{N}$ such that

$$|g(\beta^k(t))| < \frac{\varepsilon}{(k+1)^2}, \quad t \in I,$$

for any $k \geq K$. Note that the series

$$\sum_{k=0}^{\infty} \frac{1}{(k+1)^2}$$

is convergent. Now, applying the Weierstrass criteria, we conclude that the series (2.1) is uniformly convergent on I. This completes the proof.

Theorem 2.1.7 *The operator $\mathscr{I} - T_\beta : \Omega \to \Omega$ is one-to-one.*

Proof 2.1.8 *Let $f, g \in \Omega$ be such that*

$$(\mathscr{I} - T_\beta)(f)(t) = (\mathscr{I} - T_\beta)(g)(t), \quad t \in I,$$

or

$$f(t) - f(\beta(t)) = g(t) - g(\beta(t)), \quad t \in I.$$

We have the following cases.

1. *Let $t = s_0$. Then*

$$f(s_0) = g(s_0).$$

2. *Let $t \neq s_0$. Then*

$$\frac{f(\beta(t)) - f(t)}{\beta(t) - t} = \frac{g(\beta(t)) - g(t)}{\beta(t) - t},$$

or

$$D_\beta f(t) = D_\beta g(t),$$

or

$$D_\beta (f - g)(t).$$

Now, applying Corollary 1.5.3, we get

$$\begin{aligned}f(t) - g(t) &= f(s_0) - g(s_0)\\ &= 0.\end{aligned}$$

Thus, $\mathscr{I} - T_\beta : \Omega \to \Omega$ is injection.

3. *Let $h \in \Omega$ be arbitrarily chosen. Set*

$$l(t) = \sum_{k=0}^{\infty} h(\beta^k(t)), \quad t \in I.$$

By Theorem 2.1.5, it follows that $l \in \Omega$. Moreover, we have

$$\begin{aligned}
(\mathscr{I} - T_\beta)(l)(t) &= l(t) - T_\beta l(t) \\
&= \sum_{k=0}^{\infty} h(\beta^k(t)) - T_\beta \sum_{k=0}^{\infty} h(\beta^k(t)) \\
&= \sum_{k=0}^{\infty} h(\beta^k(t)) - \sum_{k=0}^{\infty} h(\beta^{k+1}(t)) \\
&= \sum_{k=0}^{\infty} h(\beta^k(t)) - \sum_{k=1}^{\infty} h(\beta^k(t)) \\
&= h(\beta^0(t)) \\
&= h(t), \quad t \in I.
\end{aligned}$$

Therefore, $\mathscr{I} - T_\beta : \Omega \to \Omega$ is a surjection and hence, it is one-to-one. This completes the proof.

Corollary 2.1.9 *The operator $A : Y \to \Omega$ defined by*

$$A(g)(t) = \sum_{k=0}^{\infty} g(\beta^k(t)), \quad t \in I,$$

is the inverse operator of the operator $\mathscr{I} - T_\beta$.

Theorem 2.1.10 *Let $f : I \to X$ be continuous at s_0. Then the function*

$$F(t) = \sum_{k=0}^{\infty} (\beta^k(t) - \beta^{k+1}(t)) f(\beta^k(t)), \quad t \in I, \tag{2.2}$$

is a β-antiderivative of f with $F(s_0) = 0$. Conversely, if F is a β-antiderivative of f that vanishes at s_0, then F is given by (2.3).

Proof 2.1.11 1. *Let F be given by (2.2). Then*

$$\begin{aligned}
F(s_0) &= \sum_{k=0}^{\infty} (\beta^k(s_0) - \beta^{k+1}(s_0)) f(\beta^k(t)) \\
&= 0
\end{aligned}$$

and

$$\begin{aligned}
\frac{F(\beta(t))-F(t)}{\beta(t)-t} &= \frac{1}{\beta(t)-t}\left(\sum_{k=0}^{\infty}(\beta^{k+1}(t)-\beta^{k+2}(t))f(\beta^k(t))\right.\\
&\quad \left.-\sum_{k=0}^{\infty}(\beta^k(t)-\beta^{k+1}(t))f(\beta^k(t))\right)\\
&= \frac{1}{\beta(t)-t}\left(\sum_{k=1}^{\infty}(\beta^k(t)-\beta^{k+1}(t))f(\beta^k(t))\right.\\
&\quad \left.-\sum_{k=0}^{\infty}(\beta^k(t)-\beta^{k+1}(t))f(\beta^k(t))\right)\\
&= \frac{1}{\beta(t)-t}(\beta(t)-t)f(t)\\
&= f(t), \quad t \neq s_0, \quad t \in I.
\end{aligned}$$

Let $\varepsilon > 0$ be arbitrarily chosen. Since f is continuous at s_0, there is a $\delta > 0$ such that $0 < h < \delta$, $k \geq 0$, imply

$$|f(\beta^k(s_0+h)) - f(s_0)| < \varepsilon.$$

Hence, for $0 < h < \delta$, $k \geq 0$, we have

$$\left|\frac{1}{h}F(s_0+h) - f(s_0)\right|$$

$$\begin{aligned}
&= \left|\frac{1}{h}\sum_{k=0}^{\infty}(\beta^k(s_0+h)-\beta^{k+1}(s_0+h))f(\beta^k(s_0+h)) - f(s_0)\right|\\
&= \left|\frac{1}{h}\sum_{k=0}^{\infty}(\beta^k(s_0+h)-\beta^{k+1}(s_0+h))(f(\beta^k(s_0+h)) - f(s_0))]\right|\\
&\leq \frac{1}{h}\sum_{k=0}^{\infty}(\beta^k(s_0+h)-\beta^{k+1}(s_0+h))|f(\beta^k(s_0+h)) - f(s_0)|\\
&< \varepsilon\frac{1}{h}\sum_{k=0}^{\infty}(\beta^k(s_0+h)-\beta^{k+1}(s_0+h))f(\beta^k(s_0+h))\\
&= \varepsilon.
\end{aligned}$$

Thus,

$$F'(s_0) = f(s_0)$$

and F is a β-antiderivative of f.

2. *Let F be a β-antiderivative of f with $F(s_0)=0$. Then*

$$\begin{aligned} f(t) &= D_\beta F(t) \\ &= \frac{F(\beta(t))-F(t)}{\beta(t)-t} \\ &= \frac{1}{t-\beta(t)}(\mathscr{I}-T_\beta)F(t), \quad t\neq s_0, \quad t\in I, \end{aligned}$$

whereupon

$$(\mathscr{I}-T_\beta)F(t)=f(t)(t-\beta(t)), \quad t\in I.$$

Hence,

$$\begin{aligned} F(t) &= (\mathscr{I}-T_\beta)^{-1}(t-\beta(t))f(t) \\ &= \sum_{k=0}^{\infty}(\beta^k(t)-\beta^{k+1}(t))f(\beta^k(t)), \quad t\in I. \end{aligned}$$

This completes the proof.

2.2 Definition for β-Integral

Definition 2.2.1 *Let $f:I\to X$ and $a,b\in I$. Define*

$$\int_{s_0}^{a} f(t)d_\beta t=\sum_{k=0}^{\infty}(\beta^k(a)-\beta^{k+1}(a))f(\beta^k(a))$$

and

$$\int_{a}^{b} f(t)d_\beta t=\int_{s_0}^{b} f(t)d_\beta t-\int_{s_0}^{a} f(t)d_\beta t. \tag{2.3}$$

The equation (2.3) can be rewritten in the form

$$\begin{aligned} \int_{a}^{b} f(t)d_\beta t &= \sum_{k=0}^{\infty}(\beta^k(b)-\beta^{k+1}(b))f(\beta^k(b)) \\ &\quad -\sum_{k=0}^{\infty}(\beta^k(a)-\beta^{k+1}(a))f(\beta^k(a)). \end{aligned}$$

Example 2.2.2 *Let $I=(-\sqrt{2},\sqrt{2})$ and*

$$\beta(t)=\frac{1}{2}t^3, \quad t\in I.$$

By Example 1.1.10, we have

$$\begin{aligned}
\beta^k(t) &= \left(\frac{1}{2}\right)^{[k]_3} t^{3^k} \\
&= \left(\frac{1}{2}\right)^{\frac{3^k-1}{2}} t^{3^k} \\
&= \sqrt{2}\left(\frac{1}{2}\right)^{\frac{3^k}{2}} t^{3^k} \\
&= \sqrt{2}\left(\frac{1}{\sqrt{2}}\right)^{3^k} t^{3^k} \\
&= \sqrt{2}\left(\frac{t}{\sqrt{2}}\right)^{3^k}, \quad t \in I.
\end{aligned}$$

Hence,

$$\begin{aligned}
\int_0^x t d_\beta t &= \sum_{k=0}^{\infty} (\beta^k(x) - \beta^{k+1}(x))\beta^k(x) \\
&= \sqrt{2}\sum_{k=0}^{\infty}\left(\left(\frac{x}{\sqrt{2}}\right)^{3^k} - \left(\frac{x}{\sqrt{2}}\right)^{3^{k+1}}\right)\left(\frac{x}{\sqrt{2}}\right)^{3^k} \\
&= \sqrt{2}\sum_{k=0}^{\infty}\left(\left(\frac{x}{\sqrt{2}}\right)^{2\cdot 3^k} - \left(\frac{x}{\sqrt{2}}\right)^{4\cdot 3^k}\right) \\
&= \sqrt{2}\sum_{k=0}^{\infty}\left(\left(\frac{x^2}{2}\right)^{3^k} - \left(\frac{x^4}{4}\right)^{3^k}\right), \quad x \in I.
\end{aligned}$$

Example 2.2.3 *Let $I = \mathbb{R}$ and*

$$\beta(t) = \frac{1}{2}t + 1, \quad t \in I.$$

Here $s_0 = 2$. We will compute

$$\begin{aligned}
I_1 &= \int_2^3 t d_\beta t, \\
I_2 &= \int_2^4 t d_\beta t, \\
I_3 &= \int_3^4 t d_\beta t.
\end{aligned}$$

By Example 1.1.10, we have

$$\beta^k(t) = \left(\frac{1}{2}\right)^k t + [k]_{\frac{1}{2}}, \quad t \in I, \quad k \in \mathbb{N}.$$

Moreover,

$$\begin{aligned} [k]_{\frac{1}{2}} &= \frac{1-\left(\frac{1}{2}\right)^k}{1-\frac{1}{2}} \\ &= 2\left(1-\left(\frac{1}{2}\right)^k\right) \end{aligned}$$

and

$$\begin{aligned} [k]_{\frac{1}{2}} - [k+1]_{\frac{1}{2}} &= \frac{1-\left(\frac{1}{2}\right)^k}{1-\frac{1}{2}} - \frac{1-\left(\frac{1}{2}\right)^{k+1}}{1-\frac{1}{2}} \\ &= 2\left(1-\left(\frac{1}{2}\right)^k - 1 + \left(\frac{1}{2}\right)^{k+1}\right) \\ &= 2\left(-\left(\frac{1}{2}\right)^k + \left(\frac{1}{2}\right)^{k+1}\right), \quad k \in \mathbb{N}. \end{aligned}$$

Hence,

$$\begin{aligned} I_1 &= \int_2^3 t d_\beta t \\ &= \sum_{k=0}^{\infty} \left(\beta^k(3) - \beta^{k+1}(3)\right) \beta^k(3) \\ &= \sum_{k=0}^{\infty} \left(3 \cdot \left(\frac{1}{2}\right)^k + [k]_{\frac{1}{2}} - 3 \cdot \left(\frac{1}{2}\right)^{k+1} - [k+1]_{\frac{1}{2}}\right) \\ &\quad \times \left(3 \cdot \left(\frac{1}{2}\right)^k + [k]_{\frac{1}{2}}\right) \\ &= \sum_{k=0}^{\infty} \left(3 \cdot \left(\frac{1}{2}\right)^k - 3 \cdot \left(\frac{1}{2}\right)^{k+1} - 2 \cdot \left(\frac{1}{2}\right)^k + 2 \cdot \left(\frac{1}{2}\right)^{k+1}\right) \\ &\quad \times \left(3 \cdot \left(\frac{1}{2}\right)^k + [k]_{\frac{1}{2}}\right) \\ &= \sum_{k=0}^{\infty} \left(\left(\frac{1}{2}\right)^k - \left(\frac{1}{2}\right)^{k+1}\right) \left(3 \cdot \left(\frac{1}{2}\right)^k + [k]_{\frac{1}{2}}\right) \end{aligned}$$

$$\begin{aligned}
&= \sum_{k=0}^{\infty}\left(1-\frac{1}{2}\right)\left(\frac{1}{2}\right)^k\left(3\cdot\left(\frac{1}{2}\right)^k+[k]_{\frac{1}{2}}\right)\\
&= \frac{1}{2}\sum_{k=0}^{\infty}\left(3\cdot\left(\frac{1}{4}\right)^k+[k]_{\frac{1}{2}}\left(\frac{1}{2}\right)^k\right)\\
&= \frac{3}{2}\sum_{k=0}^{\infty}\left(\frac{1}{4}\right)^k+\frac{1}{2}\sum_{k=0}^{\infty}\left(\frac{1-(\frac{1}{2})^k}{1-\frac{1}{2}}\right)\left(\frac{1}{2}\right)^k\\
&= \frac{3}{2}\cdot\frac{1}{1-\frac{1}{4}}+\sum_{k=0}^{\infty}\left(\frac{1}{2}\right)^k-\sum_{k=0}^{\infty}\left(\frac{1}{4}\right)^k\\
&= 2+\frac{1}{1-\frac{1}{2}}-\frac{1}{1-\frac{1}{4}}\\
&= 2+2-\frac{4}{3}\\
&= 4-\frac{4}{3}\\
&= \frac{8}{3}
\end{aligned}$$

and

$$\begin{aligned}
I_2 &= \int_2^4 t d_\beta t\\
&= \sum_{k=0}^{\infty}\left(\beta^k(4)-\beta^{k+1}(4)\right)\beta^k(4)\\
&= \sum_{k=0}^{\infty}\left(4\cdot\left(\frac{1}{2}\right)^k+[k]_{\frac{1}{2}}-4\cdot\left(\frac{1}{2}\right)^{k+1}-[k+1]-\frac{1}{2}\right)\\
&\quad\times\left(4\cdot\left(\frac{1}{2}\right)^k+[k]_{\frac{1}{2}}\right)\\
&= \sum_{k=0}^{\infty}\left(4\cdot\left(\frac{1}{2}\right)^k-4\cdot\left(\frac{1}{2}\right)^{k+1}-2\cdot\left(\frac{1}{2}\right)^k+2\cdot\left(\frac{1}{2}\right)^{k+1}\right)\\
&\quad\times\left(4\cdot\left(\frac{1}{2}\right)^k+[k]_{\frac{1}{2}}\right)
\end{aligned}$$

$$
\begin{aligned}
&= \sum_{k=0}^{\infty}\left(2\cdot\left(\frac{1}{2}\right)^k - 2\cdot\left(\frac{1}{2}\right)^{k+1}\right)\left(4\cdot\left(\frac{1}{2}\right)^k + \frac{1-\left(\frac{1}{2}\right)^k}{1-\frac{1}{2}}\right) \\
&= 2\sum_{k=0}^{\infty}\left(\frac{1}{2}\right)^k\left(1-\frac{1}{2}\right)\left(4\cdot\left(\frac{1}{2}\right)^k + 2 - 2\cdot\left(\frac{1}{2}\right)^k\right) \\
&= \sum_{k=0}^{\infty}\left(\frac{1}{2}\right)^k\left(2+2\cdot\left(\frac{1}{2}\right)^k\right) \\
&= 2\sum_{k=0}^{\infty}\left(\frac{1}{2}\right)^k + 2\sum_{k=0}^{\infty}\left(\frac{1}{4}\right)^k \\
&= 2\cdot\frac{1}{1-\frac{1}{2}} + 2\cdot\frac{1}{1-\frac{1}{4}} \\
&= 1+\frac{8}{3} \\
&= \frac{11}{3},
\end{aligned}
$$

and

$$
\begin{aligned}
I_3 &= \int_3^4 t d_\beta t \\
&= \int_2^4 t d_\beta t - \int_2^3 t d_\beta t \\
&= I_2 - I_1 \\
&= \frac{11}{3} - \frac{8}{3} \\
&= 1.
\end{aligned}
$$

Remark 2.2.4 *Note that in the classical case, we have*

$$
\begin{aligned}
\int_3^4 t dt &= \left.\frac{t^2}{2}\right|_{t=3}^{t=4} \\
&= \frac{16}{2} - \frac{9}{2} \\
&= \frac{7}{2}.
\end{aligned}
$$

Example 2.2.5 *Let $I = \mathbb{R}$ and*

$$\beta(t) = \frac{3}{4}t - 1, \quad t \in I.$$

Here $s_0 = -4$. We will find

$$I = \int_1^3 (t^2 - t + 1) d_\beta t.$$

For this aim, we will compute

$$\begin{aligned} I_1 &= \int -4^1 (t^2 - t + 1) d_\beta t, \\ I_2 &= \int_{-4}^3 (t^2 - t + 1) d_\beta t. \end{aligned}$$

By Example 1.1.10, we have

$$\begin{aligned} \beta^k(t) &= \left(\frac{3}{4}\right)^k t - [k]_{\frac{3}{4}} \\ &= \left(\frac{3}{4}\right)^k - \frac{1 - \left(\frac{3}{4}\right)^k}{1 - \frac{3}{4}} \\ &= \left(\frac{3}{4}\right)^k - 4\left(1 - \left(\frac{3}{4}\right)^k\right), \quad t \in I. \end{aligned}$$

Then

$$\begin{aligned} I_1 &= \int_{-4}^1 (t^2 - t + 1) d_\beta t \\ &= \sum_{k=0}^{\infty} \left(\beta^k(1) - \beta^{k+1}(1)\right)\left((\beta^k(1))^2 - \beta^k(1) + 1\right) \\ &= \sum_{k=0}^{\infty} \left(\left(\left(\frac{3}{4}\right)^k - 4 + 4\left(\frac{3}{4}\right)^k - \left(\frac{3}{4}\right)^{k+1} + 4 - 4\left(\frac{3}{4}\right)^{k+1}\right)\right. \\ &\quad \times \left.\left(\left(\left(\frac{3}{4}\right)^k - 4 + 4\left(\frac{3}{4}\right)^k\right)^2 - \left(\frac{3}{4}\right)^k + 4 - 4\left(\frac{3}{4}\right)^k + 1\right)\right) \\ &= \sum_{k=0}^{\infty} \left(\left(5\left(\frac{3}{4}\right)^k - \frac{3}{4}\left(\frac{3}{4}\right)^k - 3\left(\frac{3}{4}\right)^k\right)\right. \\ &\quad \times \left.\left(\left(5\left(\frac{3}{4}\right)^k - 4\right)^2 - 5\left(\frac{3}{4}\right)^k + 5\right)\right) \end{aligned}$$

$$\begin{aligned}
&= \sum_{k=0}^{\infty}\left(\frac{5}{4}\left(\frac{3}{4}\right)^k\left(25\left(\frac{9}{16}\right)^k-40\left(\frac{3}{4}\right)^k+16-5\left(\frac{3}{4}\right)^k+5\right)\right)\\
&= \frac{5}{4}\sum_{k=0}^{\infty}\left(\frac{3}{4}\right)^k\left(25\left(\frac{9}{16}\right)^k-45\left(\frac{3}{4}\right)^k+21\right)\\
&= \frac{125}{4}\sum_{k=0}^{\infty}\left(\frac{27}{64}\right)^k-\frac{225}{4}\sum_{k=0}^{\infty}\left(\frac{9}{16}\right)^k+\frac{105}{4}\sum_{k=0}^{\infty}\left(\frac{3}{4}\right)^k\\
&= \frac{125}{4}\cdot\frac{1}{1-\frac{27}{64}}-\frac{225}{4}\cdot\frac{1}{1-\frac{9}{16}}+\frac{105}{4}\cdot\frac{1}{1-\frac{3}{4}}\\
&= \frac{125\cdot 16}{37}-\frac{225\cdot 4}{7}+105\\
&= \frac{2000}{37}-\frac{900}{7}+105\\
&= \frac{2000\cdot 7-900\cdot 37+105\cdot 37\cdot 7}{259}\\
&= \frac{14000-33300+27195}{259}\\
&= \frac{7895}{259}.
\end{aligned}$$

Next,

$$\begin{aligned}
I_2 &= \int_{-4}^{3}(t^2-t+1)d_\beta t\\
&= \sum_{k=0}^{\infty}\left(\beta^k(3)-\beta^{k+1}(3)\right)\left((\beta^k(3))^2-\beta^k(3)+1\right)\\
&= \sum_{k=0}^{\infty}\left(\left(3\left(\frac{3}{4}\right)^k-4+4\left(\frac{3}{4}\right)^k-3\left(\frac{3}{4}\right)^{k+1}+4-4\left(\frac{3}{4}\right)^{k+1}\right)\right.\\
&\quad\left.\times\left(\left(3\left(\frac{3}{4}\right)^k-4+4\left(\frac{3}{4}\right)^k\right)^2-3\left(\frac{3}{4}\right)^k+4-4\left(\frac{3}{4}\right)^k+1\right)\right)\\
&= \sum_{k=0}^{\infty}\left(\left(7\left(\frac{3}{4}\right)^k-\frac{21}{4}\left(\frac{3}{4}\right)^k\right)\right.\\
&\quad\left.\times\left(\left(7\left(\frac{3}{4}\right)^k-4\right)^2-7\left(\frac{3}{4}\right)^k+5\right)\right)
\end{aligned}$$

$$
\begin{aligned}
&= \sum_{k=0}^{\infty}\left(\frac{7}{4}\left(\frac{3}{4}\right)^k\left(49\left(\frac{9}{16}\right)^k-56\left(\frac{3}{4}\right)^k+16-7\left(\frac{3}{4}\right)^k+5\right)\right)\\
&= \frac{7}{4}\sum_{k=0}^{\infty}\left(\frac{3}{4}\right)^k\left(49\left(\frac{9}{16}\right)^k-63\left(\frac{3}{4}\right)^k+21\right)\\
&= \frac{343}{4}\sum_{k=0}^{\infty}\left(\frac{27}{64}\right)^k-\frac{441}{4}\sum_{k=0}^{\infty}\left(\frac{9}{16}\right)^k+\frac{147}{4}\sum_{k=0}^{\infty}\left(\frac{3}{4}\right)^k\\
&= \frac{343}{4}\cdot\frac{1}{1-\frac{27}{64}}-\frac{441}{4}\cdot\frac{1}{1-\frac{9}{16}}+\frac{147}{4}\cdot\frac{1}{1-\frac{3}{4}}\\
&= \frac{343\cdot 16}{37}-\frac{441\cdot 4}{7}+147\\
&= \frac{5488}{37}-\frac{1764}{7}+147\\
&= \frac{5488\cdot 7-1764\cdot 37+147\cdot 37\cdot 7}{259}\\
&= \frac{38416-65268+38073}{259}\\
&= \frac{11221}{259}.
\end{aligned}
$$

Therefore

$$
\begin{aligned}
I &= \int_1^3 (t^2-t+1)d_\beta t\\
&= \int_{-4}^3 (t^2-t+1)d_\beta t-\int_{-4}^1 (t^2-t+1)d_\beta t\\
&= I_2-I_1\\
&= \frac{11221}{259}-\frac{7895}{259}\\
&= \frac{3326}{259}.
\end{aligned}
$$

Exercise 2.2.6 *Let* $I=\mathbb{R}$ *and*

$$\beta(t)=\frac{1}{2}t-1,\quad t\in I.$$

Compute

1.

$$\int_{-2}^{-1}(t-1)d_\beta t.$$

2.

$$\int_{-2}^{1}(t-1)d_\beta t.$$

3.

$$\int_{-1}^{1}(t-1)d_\beta t.$$

2.3 Properties of β-Integrals

In this section, we will deduct some important properties of β-integrals. Suppose that $[a,b] \subseteq I$.

Theorem 2.3.1 *The β-integral is a linear operator.*

Proof 2.3.2 *Let $f,g : I \to X$ be β-integrable and $\alpha \in \mathbb{C}$ be arbitrarily chosen. Then, we have the following.*

$$\begin{aligned}
\int_a^b (f(t)+g(t))d_\beta t &= \int_a^b (f+g)(t)d_\beta t \\
&= \sum_{k=0}^{\infty}(\beta^k(b)-\beta^{k+1}(b))(f+g)(\beta^k(b)) \\
&\quad -\sum_{k=0}^{\infty}(\beta^k(a)-\beta^{k+1}(a))(f+g)(\beta^k(a)) \\
&= \sum_{k=0}^{\infty}(\beta^k(b)-\beta^{k+1}(b))f(\beta^k(b)) \\
&\quad -\sum_{k=0}^{\infty}(\beta^k(a)-\beta^{k+1}(a))f(\beta^k(a)) \\
&\quad +\sum_{k=0}^{\infty}(\beta^k(b)-\beta^{k+1}(b))g(\beta^k(b)) \\
&\quad -\sum_{k=0}^{\infty}(\beta^k(a)-\beta^{k+1}(a))g(\beta^k(a)) \\
&= \int_a^b f(t)d_\beta t + \int_a^b g(t)d_\beta t
\end{aligned}$$

and

$$\begin{aligned}
\int_a^b \alpha f(t)d_\beta t &= \int_a^b (\alpha f)(t)d_\beta t \\
&= \sum_{k=0}^{\infty}(\beta^k(b)-\beta^{k+1}(b))(\alpha f)(\beta^k(b)) \\
&\quad -\sum_{k=0}^{\infty}(\beta^k(a)-\beta^{k+1}(a))(\alpha f)(\beta^k(a)) \\
&= \alpha\sum_{k=0}^{\infty}(\beta^k(b)-\beta^{k+1}(b))f(\beta^k(b)) \\
&\quad -\alpha\sum_{k=0}^{\infty}(\beta^k(a)-\beta^{k+1}(a))f(\beta^k(a)) \\
&= \alpha\left(\sum_{k=0}^{\infty}(\beta^k(b)-\beta^{k+1}(b))f(\beta^k(b))\right. \\
&\quad \left.-\sum_{k=0}^{\infty}(\beta^k(a)-\beta^{k+1}(a))f(\beta^k(a))\right) \\
&= \alpha\int_a^b f(t)d_\beta t.
\end{aligned}$$

This completes the proof.

Theorem 2.3.3 *Let $f: I \to X$ be β-integrable over $[a,b]$. Then*

$$\int_a^a f(t)d_\beta t = 0.$$

Proof 2.3.4 *We have*

$$\begin{aligned}
\int_a^a f(t)d_\beta t &= \sum_{k=0}^{\infty}(\beta^k(a)-\beta^{k+1}(a))f(\beta^k(a)) \\
&\quad -\sum_{k=0}^{\infty}(\beta^k(a)-\beta^{k+1}(a))f(\beta^k(a)) \\
&= 0.
\end{aligned}$$

This completes the proof.

Theorem 2.3.5 *Let $f: I \to X$ be β-integrable over $[a,b]$. Then*

$$\int_a^b f(t)d_\beta t = -\int_b^a f(t)d_\beta t.$$

Proof 2.3.6 *We have*

$$\begin{aligned}\int_a^b f(t)d_\beta t &= \sum_{k=0}^{\infty}(\beta^k(b)-\beta^{k+1}(b))f(\beta^k(b))\\ &\quad -\sum_{k=0}^{\infty}(\beta^k(a)-\beta^{k+1}(a))f(\beta^k(a))\\ &= -\Big(-\sum_{k=0}^{\infty}(\beta^k(b)-\beta^{k+1}(b))f(\beta^k(b))\\ &\quad +\sum_{k=0}^{\infty}(\beta^k(a)-\beta^{k+1}(a))f(\beta^k(a))\Big)\\ &= -\int_b^a f(t)d_\beta t.\end{aligned}$$

This completes the proof.

Theorem 2.3.7 *Let $f: I \to X$ be β-integrable over $[a,b]$. Then, for any $c \in [a,b]$, we have*

$$\int_a^b f(t)d_\beta t = \int_a^c f(t)d_\beta t + \int_c^b f(t)d_\beta t.$$

Proof 2.3.8 *We have*

$$\begin{aligned}\int_a^b f(t)d_\beta t &= \sum_{k=0}^{\infty}(\beta^k(b)-\beta^{k+1}(b))f(\beta^k(b))\\ &\quad -\sum_{k=0}^{\infty}(\beta^k(a)-\beta^{k+1}(a))f(\beta^k(a))\\ &= \sum_{k=0}^{\infty}(\beta^k(c)-\beta^{k+1}(c))f(\beta^k(c))\\ &\quad -\sum_{k=0}^{\infty}(\beta^k(a)-\beta^{k+1}(a))f(\beta^k(a))\\ &\quad +\sum_{k=0}^{\infty}(\beta^k(b)-\beta^{k+1}(b))f(\beta^k(b))\\ &\quad -\sum_{k=0}^{\infty}(\beta^k(c)-\beta^{k+1}(c))f(\beta^k(c))\\ &= \int_a^c f(t)d_\beta t + \int_c^b f(t)d_\beta t.\end{aligned}$$

This completes the proof.

Theorem 2.3.9 (Fundamental Theorem of β-Differential Calculus) *Let $f : I \to X$ be continuous at s_0. Then*

$$F(x) = \int_{s_0}^{x} f(t) d_\beta t, \quad x \in I,$$

is continuous at s_0, $D_\beta F(x)$ exists for any $x \in I$ and

$$D_\beta F(x) = f(x), \quad x \in I.$$

Proof 2.3.10 *We have*

$$F(x) = \sum_{k=0}^{\infty} (\beta^k(x) - \beta^{k+1}(x)) f(\beta^k(x)), \quad x \in I.$$

Hence and Theorem 2.1.10, it follows that F is antiderivative of f. This completes the proof.

Corollary 2.3.11 *Let $f : I \to X$ is continuous at x_0. Then*

$$\int_{t}^{\beta(t)} f(x) d_\beta x = (\beta(t) - t) f(t), \quad t \in I.$$

Proof 2.3.12 *Set*

$$F(t) = \int_{s_0}^{t} f(x) d_\beta x, \quad t \in I.$$

Then, applying Theorem 2.3.9, we get

$$\begin{aligned}
\int_{t}^{\beta(t)} f(x) d_\beta x &= \int_{s_0}^{\beta(t)} f(x) d_\beta x - \int_{s_0}^{t} f(x) d_\beta x \\
&= F(\beta(t)) - F(t) \\
&= (\beta(t) - t) D_\beta F(t) \\
&= (\beta(t) - t) f(t), \quad t \in I.
\end{aligned}$$

This completes the proof.

Corollary 2.3.13 *If $f : I \to X$ is continuous at s_0 and $\Phi : I \to X$ is a β-antiderivative of f on I, then*

$$\int_{a}^{b} f(t) d_\beta t = \Phi(b) - \Phi(a).$$

Proof 2.3.14 *We have*

$$\Phi(x) = \int_{s_0}^{x} f(t)d_\beta t, \quad x \in I.$$

Hence,

$$\begin{aligned}\int_a^b f(t)d_\beta t &= \int_{s_0}^b f(t)d_\beta t - \int_{s_0}^a f(t)d_\beta t \\ &= \Phi(b) - \Phi(a).\end{aligned}$$

This completes the proof.

Theorem 2.3.15 *Assume that $f, g : I \to X$ are β-differentiable functions on I and $D_\beta f$, $D_\beta g$ both continuous at s_0. Then*

$$\int_a^b f(t)D_\beta g(t)d_\beta t = (f(t)g(t))\Big|_{t=a}^{t=b} - \int_a^b D_\beta f(t)g(\beta(t))d_\beta t.$$

Proof 2.3.16 *We have that fg is β-antiderivative of $fD_\beta g + g^\beta D_\beta f$. Hence, the result follows from Corollary 2.3.13. This completes the proof.*

2.4 Inequalities and β-Integrals

We will start the investigations in this section with the following useful lemmas. Suppose that $[c,d] \subseteq I$. With D we will denote the set of all real-valued functions defined on $[c,d]_\beta$.

Lemma 2.4.1 *Let $f \in D$. Then*

$$\int_c^d f(t)h(\beta(t))d_\beta t = 0 \tag{2.4}$$

for any $h \in D$ with $h(c) = h(d) = 0$ if and only if

$$f(t) = 0, \quad t \in [c,d]_\beta. \tag{2.5}$$

Proof 2.4.2 *1. Let (2.5) holds. Then*

$$f(\beta^k(t)) = 0, \quad t \in [c,d], \quad k \in \mathbb{N}.$$

Hence,

$$\int_c^d f(t)h(\beta(t))d_\beta t = \sum_{k=0}^{\infty} (\beta^k(d) - \beta^{k+1}(d))f(\beta^k(d))h(\beta^{k+1}(d))$$

$$-\sum_{k=0}^{\infty}(\beta^k(c)-\beta^{k+1}(c))f(\beta^k(c))h(\beta^{k+1}(c))$$

$$= \quad 0,$$

i.e., (2.4) *holds.*

2. *Let* (2.4) *holds. Assume that there is a* $l \in [c,d]_\beta$ *so that* $f(l) \neq 0$.

(*a*) *Let* $l \neq s_0$. *Then* $l = \beta^k(c)$ *or* $l = \beta^k(d)$ *for some* $k \in \mathbb{N}_0$.

i. *Let* $l = \beta^k(c)$ *for some* $k \in \mathbb{N}_0$. *Define*

$$h(t) = \begin{cases} f(l) & \text{if} \quad t = \beta(l) \\ 0 & \text{otherwise.} \end{cases}$$

We have that $h \in D$ *and*

$$h(c) = h(d) = 0.$$

Hence,

$$\begin{aligned} 0 &= \int_c^d f(t)h(\beta(t))d_\beta t \\ &= \int_c^{\beta(c)} f(t)h(\beta(t))d_\beta t \\ &= (\beta(c)-c)f(c)h(\beta(c)) \\ &= (\beta(c)-c)(f(c))^2 \\ &\neq 0, \end{aligned}$$

which is a contradiction.

ii. *The case* $c = \beta^k(d)$ *for some* $k \in \mathbb{N}_0$ *is treated similarly.*

(*b*) *Let* $c = s_0$. *Without loss of generality, assume that* $f(s_0) > 0$. *Since* f *is continuous at* s_0, *we have*

$$\begin{aligned} \lim_{k\to\infty} f(\beta^k(d)) &= \lim_{k\to\infty} f(\beta^k(c)) \\ &= f(s_0). \end{aligned}$$

Thus, there is a $k_0 \in \mathbb{N}_0$ *such that*

$$\begin{aligned} f(\beta^k(d)) &> 0, \\ f(\beta^k(c)) &> 0, \quad k > k_0. \end{aligned}$$

i. Let $s_0 \notin [c,d]$. Define

$$h(t) = \begin{cases} f(\beta^k(c)) & \text{if} \quad t = \beta^{k+1}(c), \quad k > k_0, \\ f(\beta^k(d)) & \text{if} \quad t = \beta^{k+1}(d), \quad k > k_0, \\ 0 & \text{otherwise}. \end{cases}$$

Hence, $h \in D$ and

$$\begin{aligned} & \int_c^d f(t)h(\beta(t))d_\beta t \\ = & \sum_{k=k_0}^{\infty} (\beta^k(d) - \beta^{k+1}(d)) f(\beta^k(d)) h(\beta^{k+1}(d)) \\ & - \sum_{k=k_0}^{\infty} (\beta^k(c) - \beta^{k+1}(c)) f(\beta^k(c)) h(\beta^{k+1}(c)) \\ = & \sum_{k=k_0}^{\infty} (\beta^k(d) - \beta^{k+1}(d)) (f(\beta^k(d)))^2 h \\ & - \sum_{k=k_0}^{\infty} (\beta^k(c) - \beta^{k+1}(c)) (f(\beta^k(c)))^2 \\ \neq & \ 0, \end{aligned}$$

which is a contradiction.

ii. Let $s_0 = c$. Define

$$h(t) = \begin{cases} f(\beta^k(d)) & \text{if} \quad t = \beta^{k+1}(d), \quad k > k_0, \\ 0 & \text{if} \quad \text{otherwise}. \end{cases}$$

Therefore

$$\begin{aligned} & \int_c^d f(t)h(\beta(t))d_\beta t \\ = & \sum_{k=k_0}^{\infty} (\beta^k(d) - \beta^{k+1}(d)) f(\beta^k(d)) h(\beta^{k+1}(d)) \\ = & \sum_{k=k_0}^{\infty} (\beta^k(d) - \beta^{k+1}(d)) (f(\beta^k(d)))^2 \\ \neq & \ 0, \end{aligned}$$

which is a contradiction.

iii. *The case $s_0 = d$ is treated similarly. This completes the proof.*

Definition 2.4.3 *Let $g : [r]_\beta \times (-\theta, \theta) \to \mathbb{R}$, $\theta_0 \in (\theta, \theta)$. We say that $g(t, \cdot)$ is continuous at θ_0 uniformly in $t \in [r]_\beta$ if for any $\varepsilon > 0$ there exists a $\delta = \delta(\varepsilon) > 0$ such that $|\theta - \theta_0| < \delta$ implies*

$$|g(t,\theta) - g(t,\theta_0)| < \varepsilon, \quad t \in [r]_\beta.$$

Definition 2.4.4 *Let $g : [r]_\beta \times (-\theta, \theta) \to \mathbb{R}$, $\theta_0 \in (\theta, \theta)$. We say that $g(t, \cdot)$ is differentiable at θ_0 uniformly in $t \in [r]_\beta$ if for any $\varepsilon > 0$ there exists a $\delta = \delta(\varepsilon) > 0$ such that $|\theta - \theta_0| < \delta$ implies*

$$\left| \frac{g(t,\theta) - g(t,\theta_0)}{\theta - \theta_0} - g_\theta(t,\theta_0) \right| < \varepsilon, \quad t \in [r]_\beta.$$

Lemma 2.4.5 *Assume that $g(t, \cdot)$ is differentiable at θ_0 uniformly in $t \in [r]_\beta$. Let also,*

$$G(\theta) \quad = \quad \int_{s_0}^{r} g(t,\theta) d_\beta t$$

and

$$\int_{s_0}^{r} g_\theta(t,\theta_0) d_\beta t$$

exist for $\theta \in \mathbb{R}$. Then G is differentiable at θ_0 and

$$G'(\theta_0) = \int_{s_0}^{r} g_\theta(t,\theta_0) d_\beta t.$$

Proof 2.4.6 *Let $\varepsilon > 0$ be arbitrarily chosen. Since $g(t, \cdot)$ is differentiable at θ_0 uniformly in $t \in [r]_\beta$, there is a $\delta = \delta(\varepsilon) > 0$ such that*

$$\left| \frac{g(t,\theta) - g(t,\theta_0)}{\theta - \theta_0} - g_\theta(t,\theta_0) \right| < \frac{\varepsilon}{r - s_0}, \quad t \in [r]_\beta.$$

Hence,

$$\begin{aligned}
&\left| \frac{G(\theta) - G(\theta_0)}{\theta - \theta_0} - \int_{s_0}^{r} g_\theta(t,\theta_0) d_\beta t \right| \\
&= \left| \int_{s_0}^{r} \left(\frac{g(t,\theta) - g(t,\theta_0)}{\theta - \theta_0} - g_\theta(t,\theta_0) \right) d_\beta t \right| \\
&\leq \int_{s_0}^{r} \left| \frac{g(t,\theta) - g(t,\theta_0)}{\theta - \theta_0} - g_\theta(t,\theta_0) \right| d_\beta t \\
&< \frac{\varepsilon}{r - s_0} (r - s_0) \\
&= \varepsilon.
\end{aligned}$$

This completes the proof.

Lemma 2.4.7 *Let $f : I \to X$, $g : I \to \mathbb{R}$ be β-integrable on I. If*

$$\|f(t)\| \le g(t), \quad t \in [a,b]_\beta, \quad a, b \in I, \quad a < b,$$

then for any $x, y \in [a,b]$, $x < s_0 < y$, we have

$$\begin{aligned}
\left\| \int_{s_0}^{y} f(t) d_\beta t \right\| &\le \int_{s_0}^{y} g(t) d_\beta t, \\
\left\| \int_{s_0}^{x} f(t) d_\beta t \right\| &\le -\int_{s_0}^{x} g(t) d_\beta t, \\
\left\| \int_{x}^{y} f(t) d_\beta t \right\| &\le \int_{x}^{y} g(t) d_\beta t.
\end{aligned}$$

Proof 2.4.8 *Since $y > s_0$, we have*

$$\beta^{k+1}(y) < \beta^k(y), \quad k \in \mathbb{N}.$$

Then

$$\begin{aligned}
\left\| \int_{s_0}^{y} f(t) d_\beta t \right\| &= \left\| \sum_{k=0}^{\infty} (\beta^k(y) - \beta^{k+1}(y)) f(\beta^k(y)) \right\| \\
&\le \sum_{k=0}^{\infty} (\beta^k(y) - \beta^{k+1}(y)) \left\| f(\beta^k(y)) \right\| \\
&\le \sum_{k=0}^{\infty} (\beta^k(y) - \beta^{k+1}(y)) g(\beta^k(y)) \\
&= \int_{s_0}^{y} g(t) d_\beta t.
\end{aligned}$$

Next, using that $x < s_0$, we have

$$\beta^k(x) < \beta^{k+1}(x), \quad k \in \mathbb{N}.$$

Then

$$\begin{aligned}
\left\| \int_{s_0}^{x} f(t) d_\beta t \right\| &= \left\| \sum_{k=0}^{\infty} (\beta^k(x) - \beta^{k+1}(x)) f(\beta^k(x)) \right\| \\
&\le \sum_{k=0}^{\infty} (\beta^{k+1}(x) - \beta^k(x)) \left\| f(\beta^k(x)) \right\| \\
&\le -\sum_{k=0}^{\infty} (\beta^k(x) - \beta^{k+1}(x)) g(\beta^k(x)) \\
&= -\int_{s_0}^{x} g(t) d_\beta t.
\end{aligned}$$

Then

$$\begin{aligned}\left\|\int_x^y f(t)d_\beta t\right\| &= \left\|\int_{s_0}^y f(t)d_\beta t - \int_{s_0}^x f(t)d_\beta t\right\| \\ &\leq \left\|\int_{s_0}^y f(t)d_\beta t\right\| + \left\|\int_{s_0}^x f(t)d_\beta t\right\| \\ &\leq \int_{s_0}^y g(t)d_\beta t - \int_{s_0}^x g(t)d_\beta t \\ &= \int_x^y g(t)d_\beta t.\end{aligned}$$

This completes the proof.

Lemma 2.4.9 *Let $f : I \to X$ and $g : I \to \mathbb{R}$ be β-differentiable on I and*

$$\|D_\beta f(t)\| \leq D_\beta g(t), \quad t \in [a,b]_\beta, \quad a,b \in I, \quad a < b.$$

Then

$$\|f(y) - f(x)\| \leq g(y) - g(x),$$

for any $x,y \in [a,b]_\beta$, $x < s_0 < y$.

Proof 2.4.10 *Applying Lemma 2.4.7, we get*

$$\begin{aligned}\|f(y) - f(x)\| &= \left\|\int_x^y D_\beta f(t)d_\beta t\right\| \\ &\leq \int_x^y \|D_\beta f(t)\| d_\beta t \\ &\leq \int_x^y D_\beta g(t)d_\beta t \\ &= g(y) - g(x).\end{aligned}$$

This completes the proof.

Theorem 2.4.11 (The Mean Value Theorem) *Suppose that $f,g : I \to X$ are β-differentiable on I. Then*

$$\|f(y) - f(x)\| \leq \sup_{t\in I} \|D_\beta f(t)\|(y - x)$$

for any $x,y \in [a,b]_\beta$, $x < s_0 < y$.

Proof 2.4.12 *We have*

$$\|f(y) - f(x)\| = \left\|\int_x^y D_\beta f(t)d_\beta t\right\|$$

$$\leq \int_x^y \|D_\beta f(t)\| d_\beta t$$

$$\leq \sup_{t\in I} \|D_\beta f(t)\|(y-x), \quad x,y \in [a,b]_\beta, x < s_0 < y.$$

This completes the proof.

Theorem 2.4.13 *Let $k : I \times I \to \mathbb{R}$ be continuous and β-differentiable with respect to its first argument. Let also,*

$$g(x) = \int_{s_0}^x k(x,s) d_\beta s, \quad x \in I.$$

Then

$$D_\beta g(x) = k(\beta(x),x) + \int_{s_0}^x D_{\beta x} k(x,s) d_\beta s, \quad x \in I.$$

Proof 2.4.14 *1. Let $x \neq s_0$. Then*

$$\begin{aligned} g(\beta(x)) - g(x) &= \int_{s_0}^{\beta(x)} k(\beta(x),s) d_\beta s - \int_{s_0}^x k(x,s) d_\beta s \\ &= \int_{s_0}^x (k(\beta(x),s) - k(x,s)) d_\beta s + \int_x^{\beta(x)} k(\beta(x),s) d_\beta s \\ &= \int_{s_0}^k (k(\beta(x),s) - k(x,s)) d_\beta s + (\beta(x) - x) k(\beta(x),x), \end{aligned}$$

whereupon

$$\begin{aligned} D_\beta g(x) &= \frac{g(\beta(x)) - g(x)}{\beta(x) - x} \\ &= \int_{x_0}^x \frac{k(\beta(x),s) - k(x,s)}{\beta(x) - x} d_\beta s + k(\beta(x),x) \\ &= k(\beta(x),x) + \int_{s_0}^x D_{\beta x} k(x,s) d_\beta s. \end{aligned}$$

2. Let $x = s_0$. For any $h \neq 0$ so that $s_0 + h \in I$, we have

$$\begin{aligned} g(s_0+h) &= \sum_{k=0}^{\infty} (\beta^k(s_0+h) - \beta^{k+1}(s_0+h)) k(s_0+h, \beta^k(s_0+h)) \\ &= \sum_{k=0}^{\infty} (\beta^k(s_0+h) - \beta^{k+1}(s_0+h))(k(s_0+h, \beta^k(s_0+h)) - k(s_0,s_0)) \end{aligned}$$

$$+\sum_{k=0}^{\infty}(\beta^k(s_0+h)-\beta^{k+1}(s_0+h))k(s_0,s_0).$$

Therefore

$$\begin{aligned}\frac{g(s_0+h)}{h} &= \sum_{k=0}^{\infty}\frac{\beta^k(s_0+h)-\beta^{k+1}(s_0+h)}{h}\\ &\quad\times(k(s_0+h,\beta^k(s_0+h))-k(s_0,s_0))\\ &\quad+\sum_{k=0}^{\infty}\frac{\beta^k(s_0+h)-\beta^{k+1}(s_0+h)}{h}k(s_0,s_0).\end{aligned}$$

Hence, we get

$$\begin{aligned}\lim_{h\to 0}\frac{g(s_0+h)}{h} &= \lim_{h\to 0}\sum_{k=0}^{\infty}\frac{\beta^k(s_0+h)-\beta^{k+1}(s_0+h)}{h}\\ &\quad\times(k(s_0+h,\beta^k(s_0+h))-k(s_0,s_0))\\ &\quad+\lim_{h\to 0}\sum_{k=0}^{\infty}\frac{\beta^k(s_0+h)-\beta^{k+1}(s_0+h)}{h}k(s_0,s_0)\\ &= 0+k(s_0,s_0)\\ &= k(s_0,s_0).\end{aligned}$$

This completes the proof.

2.5 General Quantum Monomials

We introduce the general quantum monomials $h_k : I\times I\to\mathbb{R}$, $k\in\mathbb{N}_0$, by

$$\begin{aligned}h_0(x,y) &= 1,\\ h_k(x,y) &= \int_y^x h_{k-1}(z,y)d_\beta z,\quad k\in\mathbb{N},\quad x,y\in I.\end{aligned}$$

Note that

$$D_{\beta x}h_k(x,y)=h_{k-1}(x,y),\quad k\in\mathbb{N},\quad x,y\in I.$$

Example 2.5.1 *Let $I = \mathbb{R}$ and*

$$\beta(t) = \frac{1}{2}t + 1, \quad t \in I.$$

Then

$$\begin{aligned}
h_0(x,y) &= 1,\\
h_1(x,y) &= \int_y^x h_0(z,y)d_\beta z\\
&= \int_y^x d_\beta z\\
&= x-y,\\
h_2(x,y) &= \int_y^x h_1(z,y)d_\beta z\\
&= \int_y^x (z-y)d_\beta z\\
&= \int_y^x zd_\beta z - y\int_y^x d_\beta z\\
&= \int_2^x zd_\beta z - \int_2^y zd_\beta z - y(x-y)\\
&= \left(\left(1-\frac{1}{2}\right)x-1\right)\sum_{k=0}^{\infty}\left(\frac{1}{2}\right)^k\left(\left(\frac{1}{2}\right)^k x + \frac{1-\left(\frac{1}{2}\right)^k}{1-\frac{1}{2}}\right)\\
&\quad -\left(\left(1-\frac{1}{2}\right)y-1\right)\sum_{k=0}^{\infty}\left(\frac{1}{2}\right)^k\left(\left(\frac{1}{2}\right)^k y + \frac{1-\left(\frac{1}{2}\right)^k}{1-\frac{1}{2}}\right)\\
&\quad -y(x-y)\\
&= \left(\frac{x}{2}-1\right)\sum_{k=0}^{\infty}\left(\frac{1}{2}\right)^k\left(\left(\frac{1}{2}\right)^k x + 2 - 2\left(\frac{1}{2}\right)^k\right)\\
&\quad -\left(\frac{y}{2}-1\right)\sum_{k=0}^{\infty}\left(\frac{1}{2}\right)^k\left(\left(\frac{1}{2}\right)^k y + 2 - 2\left(\frac{1}{2}\right)^k\right)\\
&\quad -y(x-y)
\end{aligned}$$

$$
\begin{aligned}
&= \left(\frac{x}{2}-1\right)\left(\sum_{k=0}^{\infty}\left(\frac{1}{4}\right)^k x+2\sum_{k=0}^{\infty}\left(\frac{1}{2}\right)^k-2\sum_{k=0}^{\infty}\left(\frac{1}{4}\right)^k\right)\\
&\quad -\left(\frac{y}{2}-1\right)\left(\sum_{k=0}^{\infty}\left(\frac{1}{4}\right)^k y+2\sum_{k=0}^{\infty}\left(\frac{1}{2}\right)^k-2\sum_{k=0}^{\infty}\left(\frac{1}{4}\right)^k\right)\\
&\quad -y(x-y)\\
&= \left(\frac{x}{2}-1\right)\left(\frac{1}{1-\frac{1}{4}}x+\frac{2}{1-\frac{1}{2}}-\frac{2}{1-\frac{1}{4}}\right)\\
&\quad -\left(\frac{y}{2}-1\right)\left(\frac{1}{1-\frac{1}{4}}y+\frac{2}{1-\frac{1}{2}}-\frac{2}{1-\frac{1}{4}}\right)\\
&\quad -y(x-y)\\
&= \left(\frac{x}{2}-1\right)\left(\frac{4}{3}x+4-\frac{8}{3}\right)-\left(\frac{y}{2}-1\right)\left(\frac{4}{3}y+4-\frac{8}{3}\right)\\
&\quad -y(x-y)\\
&= \left(\frac{x}{2}-1\right)\left(\frac{4}{3}x+\frac{4}{3}\right)-\left(\frac{y}{2}-1\right)\left(\frac{4}{3}y+\frac{4}{3}\right)-y(x-y)\\
&= \frac{2}{3}((x-2)(x+1)-(y-2)(y+1))-y(x-y)\\
&= \frac{2}{3}(x^2+x-2x-2-y^2-y+2y+2)-y(x-y)\\
&= \frac{2}{3}(x^2-y^2-x+y)-y(x-y)\\
&= \frac{2}{3}(x-y)(x+y-1)-y(x-y)\\
&= \frac{2}{3}(x-y)(x+y-1-3y)\\
&= \frac{2}{3}(x-y)(x-2y-1), \quad x,y\in I.
\end{aligned}
$$

Theorem 2.5.2 *We have*

$$
h_{k+m+1}(x,s_0)=\int_{s_0}^{x} h_k(x,\beta(y))h_m(y,s_0)d_\beta y, \quad x\in I.
$$

Proof 2.5.3 *Let*

$$g(x) = \int_{s_0}^{x} h_k(x, \beta(y)) h_m(y, s_0) d_\beta y, \quad x \in I.$$

Then

$$\begin{aligned} D_\beta g(x) &= h_k(\beta(x), \beta(x)) h_k(x, s_0) \\ &\quad + \int_{s_0}^{x} D_{\beta x} h_k(x, \beta(y)) h_m(y, s_0) d_\beta y \\ &= \int_{s_0}^{x} h_{k-1}(x, \beta(y)) h_m(y, s_0) d_\beta y \\ &\vdots \\ D_\beta^k g(x) &= \int_{s_0}^{x} h_m(y, s_0) d_\beta y \\ &= h_{m+1}(x, s_0), \quad x \in I. \end{aligned}$$

Hence,

$$\begin{aligned} D_\beta^{k-1} g(x) - D_\beta^{k-1} g(s_0) &= \int_{s_0}^{x} h_{m+1}(y, s_0) d_\beta y \\ &= h_{m+2}(x, s_0), \quad x \in I, \end{aligned}$$

or

$$D_\beta^{k-1} g(x) = h_{m+2}(x, s_0), \quad x \in I.$$

An so on,

$$D_\beta g(x) = h_{k+m}(x, s_0), \quad x \in I,$$

whereupon

$$\begin{aligned} g(x) - g(s_0) &= \int_{s_0}^{x} h_{k+m}(y, s_0) d_\beta y \\ &= h_{k+m+1}(x, s_0), \quad x \in I, \end{aligned}$$

or

$$g(x) = h_{k+m+1}(x, s_0), \quad x \in I.$$

This completes the proof.

Theorem 2.5.4 *For any $k \in \mathbb{N}_0$, we have*

$$h_k(x, s_0) \geq \frac{(x - s_0)^k}{k!}, \quad k \in \mathbb{N}, \quad x \geq s_0, \quad x \in I. \tag{2.6}$$

Proof 2.5.5 *Let*

$$g(x) = (x - s_0)^{k+1}, \quad x \in I, \quad x \geq s_0.$$

We have

$$D_\beta g(x) = \sum_{l=0}^{k} (\beta(x) - s_0)^{k-l}(x - s_0)^l, \quad x \in I, \quad x \geq s_0.$$

Note that the inequality (2.6) holds for $k = 0$. Assume that the inequality (2.6) holds for some $k \in \mathbb{N}$. We will prove the inequality for $k+1$. We have

$$\begin{aligned}
h_{k+1}(x, s_0) &= \int_{s_0}^{x} h_k(z, s_0) d_\beta z \\
&\geq \frac{1}{k!} \int_{s_0}^{x} (z - s_0)^k d_\beta z \\
&= \frac{1}{(k+1)!} \int_{s_0}^{x} \sum_{j=0}^{k} (z - s_0)^k d_\beta z \\
&= \frac{1}{(k+1)!} \int_{s_0}^{x} \sum_{j=0}^{k} (z - s_0)^{k-j}(z - s_0)^j d_\beta z \\
&\geq \frac{1}{(k+1)!} \int_{s_0}^{x} \sum_{j=0}^{k} (\beta(z) - s_0)^{k-j}(z - s_0)^j d_\beta z \\
&= \frac{1}{(k+1)!} \int_{s_0}^{x} D_\beta g(z) d_\beta z \\
&= \frac{1}{(k+1)!} g(z)\Big|_{z=s_0}^{z=x} \\
&= \frac{(x - s_0)^{k+1}}{(k+1)!}, \quad x \geq s_0, \quad x \in I.
\end{aligned}$$

By the principle of mathematical induction, it follows that the inequality (2.6) holds for any $k \in \mathbb{N}$. This completes the proof.

Theorem 2.5.6 *The functions $h_n(x, y)$ satisfy for all $t \in I$ the relationship*

$$h_n(\beta^k(t), t) = 0$$

for all $n \in \mathbb{N}$ and all $0 \leq k \leq n - 1$.

Proof 2.5.7 *Let $n \in \mathbb{N}$ be arbitrarily chosen. Then*

$$\begin{aligned}
h_n(\beta^0(t), t) &= h_n(t, t) \\
&= \int_{t}^{t} h_{n-1}(z, t) d_\beta z
\end{aligned}$$

$$= \quad 0.$$

Assume that

$$h_{n-1}(\beta^k(t),t) = 0 \quad \text{and} \quad h_n(\beta^k(t),t) = 0$$

for some $k \in \{0, \ldots, n-2\}$. *We will prove that*

$$h_n(\beta^{k+1}(t),t) = 0.$$

We have

$$\begin{aligned} h_n(\beta^{k+1}(t),t) &= h_n(\beta^k(t),t) + (\beta(t)-t)D_\beta h_n(\beta^k(t),t) \\ &= h_n(\beta^k(t),t) + (\beta(t)-t)h_{n-1}(\beta^k(t),t) \\ &= 0. \end{aligned}$$

This completes the proof.

Define, for $t,s \in I$, the monomials

$$\begin{aligned} g_0(t,s) &= 1, \\ g_n(t,s) &= \int_s^t g_{n-1}(\beta(z),s)d_\beta z, \quad n \in \mathbb{N}. \end{aligned}$$

We have

$$D_\beta g_n(t,s) = g_{n-1}(\beta(t),s), \quad t,s \in I, \quad n \in \mathbb{N}.$$

2.6 The Taylor Formula

Theorem 2.6.1 *Let* $m \in \mathbb{N}$ *and suppose that* f *is* m*-times* β*-differentiable at* $\beta^{m-1}(t)$*. Then*

$$f(t) = \sum_{k=0}^{m-1} (-1)^k D_\beta^k f(\beta^{m-1-k}(t))h_k(\beta^{m-1}(t),t), \quad t \in I.$$

Proof 2.6.2 *1. For* $m = 1$*, we have*

$$\begin{aligned} \sum_{k=0}^{0} (-1)^k D_\beta^k f(\beta^{0-k}(t))h_k(\beta^0(t),t) &= (-1)^0 f(t)h_0(t,t) \\ &= f(t), \quad t \in I. \end{aligned}$$

2. *Assume that*

$$f(t)=\sum_{k=0}^{m-1}(-1)^k D_\beta^k f(\beta^{m-1-k}(t))h_k(\beta^{m-1}(t),t), \quad t\in I.$$

for some $m\in\mathbb{N}$.

3. *We will prove that*

$$f(t)=\sum_{k=0}^{m}(-1)^k D_\beta^k f(\beta^{m-k}(t))h_k(\beta^{m}(t),t), \quad t\in I.$$

Using that

$$h_k(\beta^m(t),t)=h_k(\beta^{m-1}(t),t)+(\beta(t)-t)h_{k-1}(\beta^{m-1}(t),t), \quad t\in I,$$

and

$$h_m(\beta^{m-1}(t),t)=0, \quad t\in I,$$

we get

$$\begin{aligned}
&\sum_{k=0}^{m}(-1)^k D_\beta^k f(\beta^{m-k}(t))h_k(\beta^m(t),t)\\
&= f(\beta^m(t))+\sum_{k=1}^{m}(-1)^k D_\beta^k f(\beta^{m-k}(t))h_k(\beta^m(t),t)\\
&= f(\beta^m(t))+\sum_{k=1}^{m}(-1)^k D_\beta^k f(\beta^{m-k}(t))h_k(\beta^{m-1}(t),t)\\
&\quad +(\beta(t)-t)\sum_{k=1}^{m}(-1)^k D_\beta^k f(\beta^{m-k}(t))h_{k-1}(\beta^{m-1}(t),t)\\
&= f(\beta^m(t))+\sum_{k=1}^{m-1}(-1)^k D_\beta^k f(\beta^{m-k}(t))h_k(\beta^{m-1}(t),t)\\
&\quad +(-1)^m(\beta(t)-t)D_\beta^m f(t)h_m(\beta^{m-1}(t),t)\\
&\quad +(\beta(t)-t)\sum_{k=0}^{m-1}(-1)^{k+1} D_\beta^{k+1} f(\beta^{m-k-1}(t))h_k(\beta^{m-1}(t),t)\\
&= \sum_{k=0}^{m-1}(-1)^k D_\beta^k f(\beta^{m-k}(t))h_k(\beta^{m-1}(t),t)\\
&\quad -(\beta(t)-t)\sum_{k=0}^{m-1}(-1)^k D_\beta^{k+1} f(\beta^{m-k-1}(t))h_k(\beta^{m-1}(t),t)\\
&= \sum_{k=0}^{m-1}(-1)^k\left(D_\beta^k f(\beta^{m-k}(t))-(\beta(t)-t)D_\beta^{k+1} f(\beta^{m-k-1}(t))\right)
\end{aligned}$$

$$\times h_k(\beta^{m-1}(t),t)$$

$$= \sum_{k=0}^{m-1}(-1)^k D_\beta^k f\left(\beta^{m-1-k}(t)\right) h_k(\beta^{m-1}(t),t)$$

$$= f(t), \quad t \in I.$$

This completes the proof.

Theorem 2.6.3 *Let $n \in \mathbb{N}$. If f is n times β- differentiable and p_k, $0 \leq k \leq n-1$, are β-differentiable at some $t \in I$ with*

$$D_\beta p_{k+1}(t) = p_k(t), \quad 0 \leq k \leq n-2, \quad n \in \mathbb{N}\backslash 1,$$

then

$$D_\beta\left(\sum_{k=0}^{n-1}(-1)^k D_\beta^k f\left(\beta^{-k}(\cdot)\right) p_k(\cdot)\right)(t)$$

$$= (-1)^{n-1} D_\beta^n f\left(\beta^{-(n-1)}(t)\right) p_{n-1}(t) + f(\beta(t)) D_\beta p_0(t), \quad t \in I.$$

Proof 2.6.4 *We have*

$$D_\beta\left(\sum_{k=0}^{n-1}(-1)^k D_\beta^k f\left(\beta^{-k}(\cdot)\right) p_k(\cdot)\right)(t)$$

$$= \sum_{k=0}^{n-1}(-1)^k D_\beta\left(D_\beta^k f\left(\beta^{-k}(\cdot)\right) p_k(\cdot)\right)(t)$$

$$= \sum_{k=0}^{n-1}(-1)^k\left(D_\beta^{k+1} f\left(\beta^{-k}(t)\right) p_k(t) + D_\beta^k f\left(\beta^{1-k}(t)\right) D_\beta p_k(t)\right)$$

$$= \sum_{k=0}^{n-1}(-1)^k D_\beta^{k+1} f\left(\beta^{-k}(t)\right) p_k(t) + \sum_{k=0}^{n-1}(-1)^k D_\beta^k f\left(\beta^{1-k}(t)\right) D_\beta p_k(t)$$

$$= \sum_{k=0}^{n-2}(-1)^k D_\beta^{k+1} f\left(\beta^{-k}(t)\right) p_k(t) + (-1)^{n-1} D_\beta^n f\left(\beta^{-(n-1)}(t)\right) p_{n-1}(t)$$

$$+ \sum_{k=1}^{n-1}(-1)^k D_\beta^k f\left(\beta^{1-k}(t)\right) D_\beta p_k(t) + f(\beta(t)) D_\beta p_0(t)$$

$$= \sum_{k=0}^{n-2}(-1)^k D_\beta^{k+1} f\left(\beta^{-k}(t)\right) p_k(t) - \sum_{k=0}^{n-2}(-1)^k D_\beta^{k+1} f\left(\beta^{-k}(t)\right) p_{k+1}(t)$$

$$+ (-1)^{n-1} D_\beta^n f\left(\beta^{-(n-1)}(t)\right) p_{n-1}(t) + f(\beta(t)) D_\beta p_0(t)$$

$$= (-1)^{n-1} D_\beta^n f\left(\beta^{-(n-1)}(t)\right) p_{n-1}(t) + f(\beta(t)) D_\beta p_0(t), \quad t \in I.$$

This completes the proof.

Theorem 2.6.5 (The Taylor Formula) *Let $n \in \mathbb{N}$. Suppose that f is n times differentiable on I and let $a, t \in I$. Then*

$$f(t) = \sum_{k=0}^{n-1} (-1)^k D_\beta^k f\left(\beta^{-k}(a)\right) h_k(a,t)$$
$$+(-1)^{n-1} \int_a^{\beta^{n-1}(t)} D_\beta^n f\left(\beta^{-(n-1)}(\tau)\right) h_{n-1}(\tau,t) d_\beta \tau, \quad t \in I.$$

Proof 2.6.6 *Applying Theorem 2.6.3, we get*

$$D_\beta \left(\sum_{k=0}^{n-1} (-1)^k D_\beta^k f\left(\beta^{-k}(\cdot)\right) h_k(\cdot,t) \right)(\tau)$$
$$= (-1)^{n-1} D_\beta^n f\left(\beta^{-(n-1)}(\tau)\right) h_{n-1}(\tau,t) + f(\beta(\tau)) D_\beta h_0(\tau,t)$$
$$= (-1)^{n-1} D_\beta^n f\left(\beta^{-(n-1)}(\tau)\right) h_{n-1}(\tau,t), \quad \tau, t \in I.$$

We integrate the last relation from a to $\beta^{n-1}(t)$, we get

$$\int_a^{\beta^{n-1}(t)} D_\beta \left(\sum_{k=0}^{n-1} (-1)^k D_\beta^k f\left(\beta^{-k}(\cdot)\right) h_k(\cdot,t) \right)(\tau) d_\beta \tau$$
$$= (-1)^{n-1} \int_a^{\beta^{n-1}(t)} D_\beta^n f\left(\beta^{-(n-1)}(\tau)\right) h_{n-1}(\tau,t) d_\beta \tau, \quad t \in I,$$

or

$$\sum_{k=0}^{n-1} (-1)^k D_\beta^k f\left(\beta^{n-1-k}(t)\right) h_k(\beta^{n-1}(t),t)$$
$$-\sum_{k=0}^{n-1} (-1)^k D_\beta^k f\left(\beta^{-k}(a)\right) h_k(a,t)$$
$$= (-1)^{n-1} \int_a^{\beta^{n-1}(t)} D_\beta^n f\left(\beta^{-(n-1)}(\tau)\right) h_{n-1}(\tau,t) d_\beta \tau, \quad t \in I.$$

Applying Theorem 2.6.1, we get

$$f(t) - \sum_{k=0}^{n-1} (-1)^k D_\beta^k f\left(\beta^{-k}(a)\right) h_k(a,t)$$

$$= (-1)^{n-1} \int_a^{\beta^{n-1}(t)} D_\beta^n f\left(\beta^{-(n-1)}(\tau)\right) h_{n-1}(\tau,t) d_\beta \tau, \quad t \in I.$$

This completes the proof.

Theorem 2.6.7 *The functions g_n and h_n satisfy the relationship*

$$h_n(t,s) = (-1)^n g_n(s,t), \quad s,t \in I.$$

Proof 2.6.8 *Take $n \in \mathbb{N}$ arbitrarily. Let*

$$f(\tau) = g_n(\tau,s), \quad \tau,s \in I.$$

We have

$$\begin{aligned} D_\beta f(\tau) &= g_{n-1}(\beta(\tau),s), \\ D_\beta^2 f(\tau) &= g_{n-2}(\beta^2(\tau),s), \\ &\vdots \\ D_\beta^k f(\tau) &= g_{n-k}(\beta^k(\tau),s), \quad 0 \le k \le n-1, \\ D_\beta^n f(\tau) &= g_0(\beta^n(\tau),s) \\ &= 1, \quad \tau,s \in I. \end{aligned}$$

Hence,

$$\begin{aligned} D_\beta^k f(\beta^{-k}(s)) &= g_{n-k}(\beta^k(\beta^{-k}(s)),s) \\ &= g_{n-k}(s,s) \\ &= 0, \quad 0 \le k \le n-1. \end{aligned}$$

Now, we apply Theorem 2.6.5 for $a = s$ and we get

$$\begin{aligned} f(t) &= g_n(t,s) \\ &= \sum_{k=0}^{n} (-1)^k D_\beta^k f(\beta^{-k}(s)) h_k(s,t) \end{aligned}$$

$$
\begin{aligned}
&+(-1)^n \int_s^{\beta^n(t)} D_\beta^{n+1} f(\beta^{-n}(\tau)) h_{n+1}(\tau,t) d_\beta \tau \\
&= \sum_{k=0}^{n-1} (-1)^k D_\beta^k f(\beta^{-k}(s)) h_k(s,t) + (-1)^n D_\beta f(\beta^{-n}(s)) h_n(s,t) \\
&= (-1)^n h_n(s,t).
\end{aligned}
$$

This completes the proof..

By Theorem 2.6.7, we get the following Taylor formula.

Theorem 2.6.9 (The Taylor Formula) *Let $n \in \mathbb{N}$. Suppose that f is n times differentiable on I and let $a, t \in I$. Then*

$$
\begin{aligned}
f(t) &= \sum_{k=0}^{n-1} D_\beta^k f\left(\beta^{-k}(a)\right) g_k(t,a) \\
&\quad + \int_a^{\beta^{n-1}(t)} D_\beta^n f\left(\beta^{-(n-1)}(\tau)\right) g_{n-1}(t,\tau) d_\beta \tau, \quad t \in I.
\end{aligned}
$$

2.7 Improper Integrals of the First Kind

Throughout this section, suppose that $I = [a, \infty)$, where $a \in \mathbb{R}$ and $s_0 \in I$.

Definition 2.7.1 *Suppose that the real-valued function f is defined on $[a, \infty)$, $f : [a, \infty) \to \mathbb{R}$, and integrable from a to any point $A \in I$, $A \geq a$. Assume that the integral*

$$
F(A) = \int_a^A f(t) d_\beta t
$$

approaches a finite limit as $A \to \infty$. Then, we call that limit the improper integral of the first kind from a to ∞ and write

$$
\int_a^\infty f(t) d_\beta t = \lim_{A \to \infty} \int_a^A f(t) d_\beta t. \tag{2.7}
$$

In this case, we say that the improper integral

$$
\int_a^\infty f(t) d_\beta t \tag{2.8}
$$

exists or that it is convergent. If the limit (2.7) do not exist, then the improper integral (2.8) is said to be divergent.

Example 2.7.2 *Let $I=[1,\infty)$ and*

$$\beta(t)=\frac{1}{2}t+1,\quad t\in I.$$

Here $s_0=2\in I$. Consider the integral

$$J=\int_1^\infty \frac{3t+2}{(1+t^2)(t^2+4t+8)}d_\beta t.$$

We will investigate the integral J for convergence. Let

$$f(t)=\frac{1}{1+t^2},\quad t\in I.$$

For $t\neq 2$, we have

$$\begin{aligned}
D_\beta f(t) &= \frac{f(\beta(t))-f(t)}{\beta(t)-t}\\
&= \frac{\frac{1}{(\beta(t))^2+1}-\frac{1}{1+t^2}}{\beta(t)-t}\\
&= \frac{\frac{t^2+1-(\beta(t))^2-1}{(t^2+1)((\beta(t))^2+1)}}{\beta(t)-t}\\
&= -\frac{(\beta(t))^2-t^2}{(\beta(t)-t)(t^2+1)((\beta(t))^2+1)}\\
&= -\frac{(\beta(t)-t)(\beta(t)+t)}{(\beta(t)-t)(t^2+1)((\beta(t))^2+1)}\\
&= -\frac{\beta(t)+t}{(t^2+1)((\beta(t))^2+1)}\\
&= -\frac{\frac{1}{2}t+1+t}{(t^2+1)\left(\left(\frac{1}{2}t+1\right)^2+1\right)}\\
&= -\frac{\frac{3}{2}t+1}{(t^2+1)\left(\frac{1}{4}t^2+t+1+1\right)}\\
&= -\frac{\frac{1}{2}(3t+2)}{\frac{1}{4}(t^2+1)(t^2+4t+8)}\\
&= -\frac{2(3t+2)}{(t^2+1)(t^2+4t+8)}.
\end{aligned}$$

Next,

$$f'(t) = -\frac{2t}{(1+t^2)^2},\quad t\in I,$$

$$f'(2) = -\frac{4}{25}.$$

Therefore

$$D_\beta f(t) = -\frac{2(3t+2)}{(t^2+1)(t^2+4t+8)}, \quad t \in I,$$

and

$$\frac{3t+2}{(1+t^2)(t^2+4t+8)} = -\frac{1}{2} D_\beta f(t), \quad t \in I.$$

Consequently

$$\begin{aligned}
J &= \int_1^\infty \frac{3t+2}{(1+t^2)(t^2+4t+8)} d_\beta t \\
&= \lim_{A\to\infty} \int_1^A \frac{3t+2}{(1+t^2)(t^2+4t+8)} d_\beta t \\
&= -\frac{1}{2} \lim_{A\to\infty} \int_1^A D_\beta f(t) d_\beta t \\
&= -\frac{1}{2} \lim_{A\to\infty} \left(f(t) \Big|_{t=1}^{t=A} \right) \\
&= -\frac{1}{2} \left(\frac{1}{1+t^2} \Big|_{t=1}^{t=A} \right) \\
&= -\frac{1}{2} \lim_{A\to\infty} \left(\frac{1}{1+A^2} - \frac{1}{2} \right) \\
&= -\frac{1}{2} \left(-\frac{1}{2} \right) \\
&= \frac{1}{4}.
\end{aligned}$$

Thus, J is convergent.

Example 2.7.3 *Let $I = [1,\infty)$ and*

$$\beta(t) = \frac{2}{3} t + 1, \quad t \in I.$$

Consider

$$J = \int_1^\infty \frac{5t+3}{t^2(2t+3)^2} d_\beta t.$$

Here $s_0 = 3$. We will investigate J for convergence. Let

$$f(t) = \frac{1}{t^2}, \quad t \in I.$$

For $t \neq 3$, *we have*

$$
\begin{aligned}
D_\beta f(t) &= \frac{f(\beta(t)) - f(t)}{\beta(t) - t} \\
&= \frac{\frac{1}{(\beta(t))^2} - \frac{1}{t^2}}{\beta(t) - t} \\
&= \frac{t^2 - (\beta(t))^2}{t^2(\beta(t))^2(\beta(t) - t)} \\
&= -\frac{(\beta(t) - t)(\beta(t) + t)}{t^2(\beta(t))^2(\beta(t) - t)} \\
&= -\frac{\beta(t) + t}{t^2(\beta(t))^2} \\
&= -\frac{\frac{2}{3}t + 1 + t}{t^2\left(\frac{2}{3}t + 1\right)^2} \\
&= -\frac{\frac{1}{3}(5t + 3)}{\frac{t^2}{9}(2t + 3)^2} \\
&= -\frac{3(5t + 3)}{t^2(2t + 3)^2}.
\end{aligned}
$$

Next,

$$
\begin{aligned}
f'(t) &= -\frac{2}{t^3}, \quad t \in I, \\
f'(3) &= -\frac{2}{27}.
\end{aligned}
$$

Therefore

$$D_\beta f(t) = -\frac{3(5t + 3)}{t^2(2t + 3)^2}, \quad t \in I,$$

and

$$\frac{5t + 3}{t^2(2t + 3)^2} = -\frac{1}{3} D_\beta f(t), \quad t \in I.$$

Consequently

$$
\begin{aligned}
J &= \int_1^\infty \frac{5t + 3}{t^2(2t + 3)^2} d_\beta t \\
&= \lim_{A \to \infty} \int_1^A \frac{5t + 3}{t^2(2t + 3)^2} d_\beta t
\end{aligned}
$$

$$= -\frac{1}{3}\lim_{A\to\infty}\int_1^A D_\beta f(t)d_\beta t$$

$$= -\frac{1}{3}\lim_{A\to\infty}\left(f(t)\Big|_{t=1}^{t=A}\right)$$

$$= -\frac{1}{3}\lim_{A\to\infty}\left(\frac{1}{t^2}\Big|_{t=1}^{t=A}\right)$$

$$= -\frac{1}{3}\lim_{A\to\infty}\left(\frac{1}{A^2}-1\right)$$

$$= \frac{1}{3}.$$

Thus, J is convergent.

Example 2.7.4 *Let $I=[1,\infty)$ and*

$$\beta(t)=\frac{1}{2}t+1,\quad t\in I.$$

Here $s_0=2$. Consider the integral

$$J=\int_1^\infty \frac{\sqrt{2}}{\sqrt{2t}+\sqrt{t+2}}d_\beta t.$$

Let

$$f(t)=\sqrt{t},\quad t\in I.$$

Then, for $t\neq 2$, we have

$$\begin{aligned}
D_\beta f(t) &= \frac{f(\beta(t))-f(t)}{\beta(t)-t}\\
&= \frac{\sqrt{\beta(t)}-\sqrt{t}}{(\sqrt{\beta(t)}-\sqrt{t})(\sqrt{\beta(t)}+\sqrt{t})}\\
&= \frac{1}{\sqrt{\beta(t)}+\sqrt{t}}\\
&= \frac{1}{\sqrt{\frac{t+2}{2}}+\sqrt{t}}\\
&= \frac{\sqrt{2}}{\sqrt{t+2}+\sqrt{2t}}.
\end{aligned}$$

Next,

$$f'(t) = \frac{1}{2\sqrt{t}},\quad t\in I,$$

$$\begin{aligned} f'(2) &= \frac{1}{2\sqrt{2}} \\ &= \frac{\sqrt{2}}{4}. \end{aligned}$$

Therefore

$$D_\beta f(t) = \frac{\sqrt{t}}{\sqrt{t+2}+\sqrt{2t}}, \quad t \in I.$$

Hence,

$$\begin{aligned} J &= \int_1^\infty \frac{\sqrt{2}}{\sqrt{t+2}+\sqrt{2t}} d_\beta t \\ &= \lim_{A\to\infty} \int_1^A \frac{\sqrt{2}}{\sqrt{t+2}+\sqrt{2t}} d_\beta t \\ &= \lim_{A\to\infty} \int_1^A D_\beta f(t) d_\beta t \\ &= \lim_{A\to\infty} f(t)\Big|_{t=1}^{t=A} \\ &= \lim_{A\to\infty} \left(\sqrt{t}\Big|_{t=1}^{t=A} \right) \\ &= \lim_{A\to\infty} (\sqrt{A}-1) \\ &= \infty. \end{aligned}$$

Thus, J is divergent.

Exercise 2.7.5 *Let $I = [1,\infty)$ and*

$$\beta(t) = \frac{2}{3}t+1, \quad t \in I.$$

Investigate for convergence and divergence the following integrals.

1.
$$\int_1^\infty (t^4+2t^2+t+3) d_\beta t.$$

2.
$$\int_1^\infty \frac{3}{(t+1)(2t+3)} d_\beta t.$$

3.

$$\int_1^\infty \frac{3}{(2t+5)(4t+21)} d_\beta t.$$

Theorem 2.7.6 (The Cauchy Criterion) *For the existence of the integral (2.8) necessary and sufficient that for any given $\varepsilon > 0$, there exists $A_0 > a$ such that*

$$\left| \int_{A_1}^{A_2} f(t) d_\beta t \right| < \varepsilon \tag{2.9}$$

for any $A_1, A_2 \in I$ satisfying $A_1 > A_0$, $A_2 > A_0$.

Proof 2.7.7 *The convergence of the integral (2.8) is equivalent to the existence of the limit*

$$\lim_{B \to A} F(B).$$

Using the Cauchy criterion for the existence of the limit of a function, it follows that the existence of the integral (2.8) is equivalent to the condition (2.9). This completes the proof.

Example 2.7.8 *Let $I = \mathbb{R}$ and*

$$\beta(t) = \frac{2}{3}t + 1, \quad t \in I.$$

Consider the integral

$$\int_1^\infty \frac{3}{2(t+5)(t+9)} d_\beta t.$$

Here $s_0 = 3 \in [1, \infty)$. Let

$$f(t) = \frac{1}{t+5}, \quad t \in [1, \infty).$$

Then, for $\neq 3$, we have

$$\begin{aligned} D_\beta f(t) &= \frac{f(\beta(t)) - f(t)}{\beta(t) - t} \\ &= \frac{\frac{1}{\beta(t)+5} - \frac{1}{t+5}}{\beta(t) - t} \\ &= \frac{t+5-\beta(t)-5}{(\beta(t)-t)(t+5)(\beta(t)+5)} \\ &= -\frac{\beta(t)-t}{(\beta(t)-t)(t+5)(\beta(t)+5)} \\ &= -\frac{1}{(t+5)\left(\frac{2}{3}t+6\right)} \end{aligned}$$

$$= -\frac{3}{(t+5)(2t+18)}$$

$$= -\frac{3}{2(t+5)(t+9)}.$$

Next,

$$f'(t) = -\frac{1}{(t+5)^2}, \quad t \in I,$$

$$f'(3) = -\frac{1}{64}.$$

Therefore

$$D_\beta f(t) = -\frac{3}{2(t+5)(t+9)}, \quad t \in I.$$

Let $\varepsilon > 0$ be arbitrarily chosen. Take

$$A > \max\left\{\frac{2-5\varepsilon}{\varepsilon}, 1\right\}.$$

Then $A > 0$ and

$$A > \frac{2-5\varepsilon}{\varepsilon}$$

or

$$A\varepsilon > 2 - 5\varepsilon,$$

or

$$(A+5)\varepsilon > 2,$$

or

$$\frac{2}{A+5} < \varepsilon.$$

Let $A_1, A_2 > A$ be arbitrarily chosen. Then

$$\left|\int_{A_1}^{A_2} \frac{3}{2(t+5)(t+9)} d_\beta t\right| = \left|-\int_{A_1}^{A_2} D_\beta F(t) d_\beta t\right|$$

$$= \left|-F(t)\Big|_{t=A_1}^{t=A_2}\right|$$

$$= |F(A_1) - F(A_2)|$$

$$= \left|\frac{1}{A_1+5} - \frac{1}{A_2+5}\right|$$

$$\begin{aligned} &< \frac{1}{A_1+5}+\frac{1}{A_2+5} \\ &< \frac{1}{A+5}+\frac{1}{A+5} \\ &= \frac{2}{A+5} \\ &< \varepsilon. \end{aligned}$$

Now, applying the Cauchy criterion, it follows that the considered integral is convergent.

Exercise 2.7.9 *Let $I = \mathbb{R}$ and*

$$\beta(t) = \frac{9}{10}t+1, \quad t \in I.$$

Using the Cauchy criterion, prove that the following integrals are convergent.

1.
$$\int_1^{\infty} \frac{1}{(t+2)(9t+30)} d_\beta t.$$

2.
$$\int_1^{\infty} \frac{1}{(2t+3)(9t+25)} d_\beta t.$$

3.
$$\int_1^{\infty} \frac{1}{(t+7)(9t+80)} d_\beta t.$$

Definition 2.7.10 *An integral* (2.8) *is said to be absolutely convergent provided the integral*

$$\int_a^{\infty} |f(t)| d_\beta t \tag{2.10}$$

is convergent.

Theorem 2.7.11 *If the integral* (2.8) *is absolutely convergent, then it is convergent.*

Proof 2.7.12 *Let the integral* (2.8) *is absolutely convergent. Then the integral* (5.12) *is convergent. Suppose that $\varepsilon > 0$ is arbitrarily chosen. Hence, applying the Cauchy criterion, it follows that there is $A > a$ such that for any $A_1, A_2 > A$, we have*

$$\left| \int_{A_1}^{A_2} |f(t)| d_\beta t \right| < \varepsilon.$$

From here, for any $A_1, A_2 > A$,

$$\left| \int_{A_1}^{A_2} f(t) d_\beta t \right| \leq \left| \int_{A_1}^{A_2} |f(t)| d_\beta t \right|$$

$$< \varepsilon.$$

This completes the proof.

Theorem 2.7.13 *An integral (2.8) with $f(t) \geq 0$ for any $t \geq a$ is convergent if and only if there exists a constant $M > 0$ such that*

$$\int_a^A f(t)d_\beta t \leq M, \quad A \geq a.$$

Proof 2.7.14 *1. Let*

$$F(A) = \int_a^A f(t)d_\beta t.$$

Then

$$F(A) \leq M, \quad A \geq a.$$

Hence,

$$\begin{aligned} 0 &\leq \int_a^\infty f(t)d_\beta t \\ &= \lim_{A\to\infty} F(A) \\ &\leq M. \end{aligned}$$

Therefore, the integral (2.8) is convergent.

2. Let the integral (2.8) is convergent. Assume that $F(A)$, $A \geq a$, is unbounded. Then

$$\begin{aligned} \int_a^\infty f(t)d_\beta t &= \lim_{A\to\infty} F(A) \\ &= \infty. \end{aligned}$$

This is a contradiction. This completes the proof.

Theorem 2.7.15 *Let*

$$0 \leq f(t) \leq g(t), \quad t \in [a,\infty). \tag{2.11}$$

Then the convergence of the improper integral

$$\int_a^\infty g(t)d_\beta t \tag{2.12}$$

implies the convergence of the improper integral (2.8), while the divergence of the improper integral (2.8) implies the divergence of the improper integral (2.11).

Proof 2.7.16 *By* (2.11), *it follows that*

$$\begin{aligned} 0 &\leq \int_a^A f(t)d_\beta t \\ &\leq \int_a^A g(t)d_\beta t, \quad A \in [a,\infty). \end{aligned}$$

This completes the proof.

Example 2.7.17 *Let* $I = [1,\infty)$ *and*

$$\beta(t) = \frac{1}{2}t + 1, \quad t \in I.$$

Consider the integral

$$K = \int_1^\infty \frac{3t+2}{(1+t)(1+t^2)^4(t^2+4t+8)^3} d_\beta t.$$

Here

$$f(t) = \frac{3t+2}{(1+t)(1+t^2)^4(t^2+4t+8)^3}, \quad t \in I.$$

Let

$$g(t) = \frac{3t+2}{(1+t^2)(t^2+4t+8)}, \quad t \in I.$$

We have

$$\begin{aligned} 0 &\leq f(t) \\ &\leq g(t), \quad t \in I. \end{aligned}$$

By Example 2.7.2, we have that the integral

$$\int_1^\infty g(t)d_\beta t$$

is convergent. Hence, applying Theorem 2.7.15, we conclude that K *is convergent.*

Example 2.7.18 *Let* $I = [1,\infty)$ *and*

$$\beta(t) = \frac{2}{3}t + 1, \quad t \in I.$$

Consider the integral

$$K = \int_1^\infty \frac{5t}{(t+1)^2(2t+3)^4(t+7)^{10}} d_\beta t.$$

Here

$$f(t) = \frac{5t}{(t+1)^2(2t+3)^4(t+7)^{10}}, \quad t \in I.$$

Let

$$g(t) = \frac{5t+3}{t^2(2t+3)^2}, \quad t \in I.$$

We have

$$\begin{aligned} 0 &\leq f(t) \\ &\leq g(t), \quad t \in I. \end{aligned}$$

By Example 2.7.3, it follows that the integral

$$\int_1^\infty g(t) d_\beta t$$

is convergent. Now, applying Theorem 2.7.15, we conclude that K is convergent.

Example 2.7.19 *Let* $I = [1, \infty)$ *and*

$$\beta(t) = \frac{1}{2}t + 1, \quad t \in I.$$

Consider the integral

$$K = \int_1^\infty \frac{2}{\sqrt{t+2}} d_\beta t.$$

Here

$$f(t) = \frac{2}{\sqrt{t+2}}, \quad t \in I.$$

Let

$$g(t) = \frac{\sqrt{2}}{\sqrt{2t} + \sqrt{t+2}}, \quad t \in I.$$

We have

$$\begin{aligned} 0 &\leq g(t) \\ &\leq f(t), \quad t \in I. \end{aligned}$$

By Example 2.7.4, we obtain that the integral

$$\int_1^\infty g(t) d_\beta t$$

is divergent. Now, applying Theorem 2.7.15, we conclude that the considered integral is divergent.

Example 2.7.20 *Let $I = \mathbb{R}$ and*

$$\beta(t) = \frac{2}{3}t + 1, \quad t \in I.$$

Consider the integral

$$K = \int_1^\infty \frac{1}{(t+5)^{10}(t+9)^8(t^2+1)} d_\beta t.$$

Here

$$f(t) = \frac{1}{(t+5)^{10}(t+9)^8(t^2+1)}, \quad t \in I.$$

Let

$$g(t) = \frac{3}{2(t+5)(t+9)}, \quad t \in I.$$

We have

$$\begin{aligned} 0 &\leq f(t) \\ &\leq g(t), \quad t \in I. \end{aligned}$$

Note that by Example 2.7.8, we have that the integral

$$\int_1^\infty g(t) d_\beta t$$

is convergent. Now, applying Theorem 2.7.15, we conclude that K is convergent.

Exercise 2.7.21 *Let $I = [1, \infty)$ and*

$$\beta(T) = \frac{2}{3}t + 1, \quad t \in I.$$

Investigate for convergence and divergence the following integrals.

1. $$\int_1^\infty (2t^8 + t^4 + 4t^2 + 2t + 6) d_\beta t.$$

2. $$\int_1^\infty \frac{1}{(t+1)^2(50t+3)^2} d_\beta t.$$

3. $$\int_1^\infty \frac{1}{(2t+5)^2(4t+23)^3(t^2+1)} d_\beta t.$$

Hint 2.7.22 *Use Exercise 2.7.5 and Theorem 2.7.15.*

Answer 2.7.23 *1. divergent.*

2. *convergent.*

3. *convergent.*

Theorem 2.7.24 *Let*

$$|f(t)| \leq g(t), \quad t \in I, \quad t \geq a.$$

Then the convergence of the integral (2.11) *implies the convergence of the integral* (2.8).

Proof 2.7.25 *Since the integral* (2.11) *is convergent, we get that the integral* (5.12) *is convergent. Therefore, the integral* (2.8) *is absolutely convergent. Now, applying Theorem 2.7.11, we conclude that the integral* (2.8) *is convergent. This completes the proof.*

Theorem 2.7.26 (Comparison Criterion) *Let the integrals*

$$\int_a^\infty f(t)d_\beta t \quad and \quad \int_a^\infty g(t)d_\beta t$$

be improper integrals of the first kind with positive integrands. Suppose that the limit

$$\lim_{t\to\infty} \frac{f(t)}{g(t)} = L \tag{2.13}$$

exists(finite) and it is not zero. Then the integrals

$$\int_a^\infty f(t)d_\beta t \quad and \quad \int_a^\infty g(t)d_\beta t$$

are simultaneously convergent and divergent.

Proof 2.7.27 *Let* $\varepsilon \in (0, L)$ *be arbitrarily chosen. By* (2.13)*, it follows that there exists* $A_0 > a$ *such that*

$$\begin{aligned} L-\varepsilon &\leq \frac{f(t)}{g(t)} \\ &\leq L+\varepsilon, \quad t \geq A_0, \end{aligned}$$

from where

$$\begin{aligned} (L-\varepsilon)g(t) &\leq f(t) \\ &\leq (L+\varepsilon)g(t), \quad t \geq A_0. \end{aligned}$$

Hence,

$$\begin{aligned} (L-\varepsilon)\int_a^\infty g(t)d_\beta t &\leq \int_a^\infty f(t)d_\beta t \\ &\leq (L+\varepsilon)\int_a^\infty g(t)d_\beta t. \end{aligned} \tag{2.14}$$

We have the following cases.

1. *Let*

$$\int_a^\infty g(t)d_\beta t$$

is convergent. Then

$$(L+\varepsilon)\int_a^\infty g(t)d_\beta t$$

is convergent. Now, using (2.15) *and Theorem 2.7.15, we conclude that*

$$\int_a^\infty f(t)d_\beta t$$

is convergent.

2. *Let*

$$\int_a^\infty f(t)d_\beta t$$

is convergent. Then

$$\frac{1}{L-\varepsilon}\int_a^\infty f(t)d_\beta t$$

is convergent. Hence, using (2.15) *and Theorem 2.7.15, we conclude that*

$$\int_a^b g(t)d_\beta t$$

is convergent.

3. *Let*

$$\int_a^\infty g(t)d_\beta t$$

is divergent. Then

$$(L-\varepsilon)\int_a^\infty g(t)d_\beta t$$

is divergent. Now, using (2.15) *and Theorem 2.7.15, we conclude that*

$$\int_a^\infty f(t)d_\beta t$$

is divergent.

4. *Let*

$$\int_a^\infty f(t)d_\beta t$$

is divergent. Then

$$\frac{1}{L+\varepsilon}\int_a^\infty f(t)d_\beta t$$

is divergent. Hence, using (2.15) *and Theorem 2.7.15, we conclude that*

$$\int_a^b g(t)d_\beta t$$

is divergent. This completes the proof.

Example 2.7.28 *Let $I = [1,\infty)$ and*

$$\beta(t) = \frac{1}{2}t + 1, \quad t \in I.$$

Consider the integral

$$K = \int_1^\infty \frac{t+1}{(2t^2+4)(3t^2+t+8)} d_\beta t.$$

Let

$$\begin{aligned} f(t) &= \frac{t+1}{(2t^2+4)(3t^2+t+8)}, \\ g(t) &= \frac{3t+2}{(1+t^2)(t^2+4t+8)}, \quad t \in I. \end{aligned}$$

Then

$$\frac{f(t)}{g(t)} = \frac{(t+1)(1+t^2)(t^2+4t+8)}{(3t+2)(2t^2+4)(3t^2+t+8)}, \quad t \in I.$$

and

$$\begin{aligned} \lim_{t\to\infty} \frac{f(t)}{g(t)} &= \lim_{t\to\infty} \frac{(t+1)(1+t^2)(t^2+4t+8)}{(3t+2)(2t^2+4)(3t^2+t+8)} \\ &= \frac{1}{18} \\ &\neq 0. \end{aligned}$$

Thus, K and the integral

$$J = \int_1^\infty \frac{3t+2}{(1+t^2)(t^2+4t+8)} d_\beta t$$

are simultaneously convergent and divergent. By Example 2.7.2, we have that J is convergent. Therefore, K is convergent.

Example 2.7.29 *Let $I = [1,\infty)$ and*

$$\beta(t) = \frac{2}{3}t + 1, \quad t \in I.$$

Consider the integral

$$K = \int_1^\infty \frac{t+2}{(t+1)^2(t+4)^2} d_\beta t.$$

Let

$$f(t) = \frac{5t+3}{t^2(2t+3)^2},$$

$$g(t) = \frac{t+2}{(t+1)^2(t+4)^2}, \quad t \in I.$$

Then

$$\frac{f(t)}{g(t)} = \frac{(5t+3)(t+1)^2(t+4)^2}{t^2(t+2)(2t+3)^2}, \quad t \in I.$$

Hence,

$$\begin{aligned} \lim_{t\to\infty} \frac{f(t)}{g(t)} &= \lim_{t\to\infty} \frac{f(t)}{g(t)} = \frac{(5t+3)(t+1)^2(t+4)^2}{t^2(t+2)(2t+3)^2} \\ &= \frac{5}{4} \\ &\neq 0. \end{aligned}$$

Now, applying Theorem 2.7.26, we conclude that K and

$$J = \int_1^\infty \frac{5t+3}{t^2(2t+3)^2} d_\beta t$$

are simultaneously convergent and divergent. By Example 2.7.3, we have that J is convergent. Therefore, K is convergent.

Example 2.7.30 *Let $I = [1, \infty)$ and*

$$\beta(t) = \frac{1}{2}t + 1, \quad t \in I.$$

Consider the integral

$$K = \int_1^\infty \frac{1}{\sqrt{t+2}} d_\beta t.$$

Let

$$\begin{aligned} f(t) &= \frac{1}{\sqrt{t+2}}, \\ g(t) &= \frac{\sqrt{2}}{\sqrt{2t} + \sqrt{t+2}}, \quad t \in I. \end{aligned}$$

Then

$$\frac{f(t)}{g(t)} = \frac{\sqrt{2t} + \sqrt{t+2}}{\sqrt{2t+4}}, \quad t \in I.$$

Hence,

$$\lim_{t\to\infty} \frac{f(t)}{g(t)} = \lim_{t\to\infty} \frac{\sqrt{2t} + \sqrt{t+2}}{\sqrt{2t+4}}$$

$$= \frac{\sqrt{2}+1}{\sqrt{2}}$$

$$\neq 0.$$

Now, applying Theorem 2.7.26, we conclude that K and

$$J = \int_1^{\infty} \frac{\sqrt{2}}{\sqrt{2t}+\sqrt{t+2}} d_{\beta}t$$

are simultaneously convergent and divergent. By Example 2.7.4, we have that J is divergent. Thus, K is divergent.

Exercise 2.7.31 *Let* $I = [1, \infty)$ *and*

$$\beta(t) = \frac{2}{3}t + 1, \quad t \in I.$$

Investigate for convergence and divergence the following integrals.

1.
$$\int_1^{\infty} (2t^4 + 4t^2 + 2t + 1) d_{\beta}t.$$

2.
$$\int_1^{\infty} \frac{1}{(t+1)^2} d_{\beta}t.$$

3.
$$\int_1^{\infty} \frac{1}{(t+5)(t+23)} d_{\beta}t.$$

Theorem 2.7.32 *Let f be* β*-integrable on* $[a, A]$ *for any* $A \geq a$, $A, a \in I$. *Suppose that the integral* (2.8) *is absolutely convergent. Suppose that g is monotone on* $[a, \infty)$ *and*

$$\lim_{t \to \infty} g(t) = 0. \tag{2.15}$$

Then the improper integral of the first kind

$$\int_a^{\infty} f(t) g(t) d_{\beta}t \tag{2.16}$$

is convergent.

Proof 2.7.33 *Let* $\varepsilon > 0$ *be arbitrarily chosen. Since the integral* (2.8) *is absolutely convergent, by the Cauchy criterion, Theorem 2.7.6, it follows that there is* $A_0 \geq 0$ *such that*

$$\left| \int_{A_1}^{A_2} |f(t)| d_{\beta}t \right| < \varepsilon$$

for any $A_1, A_2 \geq A_0$. *By* (2.15), *it follows that there is* $A' \geq a$ *such that*

$$|g(t)| \leq 1, \quad t \geq A'.$$

Let

$$A'' = \max\{A', A_0\}.$$

Then, for any $A_1, A_2 \geq A''$, *we get*

$$\begin{aligned} \left| \int_{A_1}^{A_2} f(t)g(t)d_\beta t \right| &\leq \left| \int_{A_1}^{A_2} |f(t)||g(t)|d_\beta t \right| \\ &\leq \left| \int_{A_1}^{A_2} |f(t)|d_\beta t \right| \\ &< \varepsilon. \end{aligned}$$

Now, applying the Cauchy criterion, we conclude that the integral (2.16) *is convergent. This completes the proof.*

Example 2.7.34 *Let* $I = [1, \infty)$ *and*

$$\beta(t) = \frac{1}{2}t + 1, \quad t \in I.$$

Consider the integral

$$K = \int_1^\infty \frac{3t+2}{(1+t^2)(t^2+4t+8)(t^4+1)} d_\beta t.$$

Let

$$\begin{aligned} f(t) &= \frac{3t+2}{(1+t^2)(t^2+4t+8)}, \\ g(t) &= \frac{1}{1+t^4}, \quad t \in [1, \infty). \end{aligned}$$

We have

$$\begin{aligned} g'(t) &= -\frac{4t^3}{(1+t^4)^2} \\ &< 0, \quad t \in [1, \infty). \end{aligned}$$

Thus, g is monotone decreasing on $[1, \infty)$. *Next,*

$$\begin{aligned} \lim_{t \to \infty} g(t) &= \lim_{t \to \infty} \frac{1}{1+t^4} \\ &= 0. \end{aligned}$$

By Example 2.7.2, we have that

$$\int_1^\infty f(t)d_\beta t$$

is convergent. Hence and Theorem 2.7.32, we conclude that the integral K is convergent.

Example 2.7.35 *Let $I = [1,\infty)$ and*

$$\beta(t) = \frac{2}{3}t + 1, \quad t \in I.$$

Consider the integral

$$K = \int_1^\infty \frac{5t+3}{t^2(2t+3)^2(t^4+1)} d_\beta t.$$

Let

$$f(t) = \frac{5t+3}{t^2(2t+3)^2}, \quad t \in I,$$

and g be as in Example 2.7.34. By Example 2.7.34, it follows that the function g satisfies all conditions of Theorem 2.7.32. By Example 2.7.3, it follows that

$$\int_1^\infty f(t)d_\beta t$$

is absolutely convergent. Now, applying Theorem 2.7.32, we conclude that K is convergent.

Exercise 2.7.36 *Let $I = \mathbb{R}$ and*

$$\beta(t) = \frac{9}{10}t + 1, \quad t \in I.$$

Investigate the following integrals for convergence and divergence.

1.
$$\int_1^\infty \frac{1}{(t+2)(9t+30)t^3} d_\beta t.$$

2.
$$\int_1^\infty \frac{1}{(2t+3)(9t+25)(1+t^8)} d_\beta t.$$

3.
$$\int_1^\infty \frac{1}{(t+7)(9t+80)(1+t^{10})} d_\beta t.$$

2.8 Improper Integrals of the Second Kind

Suppose that $a, b \in I$ and $a < b$.

Definition 2.8.1 *Let the real-valued function f be defined on $[a,b)$ and f is β-integrable on any interval $[a,c]$ with $a < c < b$. Let also, f be unbounded on $[a,b)$. The formal expression*

$$\int_a^b f(t)d_\beta t \tag{2.17}$$

is called the improper integral of the second kind. We say that the integral (2.17) *is improper at $t = b$. We also say that f has a singularity at $t = b$. If the left-sided limit*

$$\lim_{c \to b-} \int_a^c f(t)d_\beta t \tag{2.18}$$

exists as a finite number, then the improper integral (2.17) *is said to be convergent. In such a case, we call this limit the value of the improper integral* (2.17) *and we write*

$$\int_a^b f(t)d_\beta t = \lim_{c \to b-} \int_a^c f(t)d_\beta t.$$

If the limit (2.18) *does not exist, then the improper integral* (2.17) *is said to be divergent.*

Remark 2.8.2 *As above, one can define the integral* (2.17) *as improper integral of the second kind at $t = a$.*

Remark 2.8.3 *All theorems in the previous section have exact analogues for improper integrals of the second kind.*

Example 2.8.4 *Let $I = [0,\infty)$ and*

$$\beta(t) = t^3, \quad t \in I.$$

Consider the integral

$$J = \int_0^{\frac{1}{2}} \frac{1}{t^{\frac{1}{2}}(1+t)} d_\beta t.$$

Here

$$f(t) = \frac{1}{t^{\frac{1}{2}}(t+1)}, \quad t \in I,$$

and $s_0 = 1$. We have that f is unbounded on $(0, \frac{1}{2}]$. Note that J is an improper integral of the second kind. It is improper at 0. *Let*

$$g(t) = \sqrt{t}, \quad t \in I.$$

Then, for $t \in \left[0, \frac{1}{2}\right]$, *we have*

$$\begin{aligned} D_\beta g(t) &= \frac{g(\beta(t)) - g(t)}{\beta(t) - t} \\ &= \frac{\sqrt{\beta(t)} - \sqrt{t}}{\beta(t) - t} \\ &= \frac{\sqrt{\beta(t)} - \sqrt{t}}{(\sqrt{\beta(t)} - \sqrt{t})(\sqrt{\beta(t)} + \sqrt{t})} \\ &= \frac{1}{\sqrt{\beta(t)} + \sqrt{t}} \\ &= \frac{1}{\sqrt{t^3} + \sqrt{t}} \\ &= \frac{1}{\sqrt{t}(t+1)} \\ &= f(t). \end{aligned}$$

Hence,

$$\begin{aligned} J &= \lim_{s \to 0-} \int_s^{\frac{1}{2}} f(t) d_\beta t \\ &= \lim_{s \to 0-} \left(g(t) \Big|_{t=s}^{t=\frac{1}{2}} \right) \\ &= \lim_{s \to 0-} \left(\sqrt{t} \Big|_{t=s}^{t=\frac{1}{2}} \right) \\ &= \lim_{s \to 0-} \left(\frac{1}{\sqrt{2}} - \sqrt{s} \right) \\ &= \frac{1}{\sqrt{2}}. \end{aligned}$$

Thus, J is convergent.

Exercise 2.8.5 *Let* $I = \left[-\frac{1}{2}, \infty\right)$ *and*

$$\beta(t) = t^3, \quad t \in I.$$

Prove that the integral

$$\int_{-\frac{1}{2}}^{0} \frac{1}{t^2 + t\sqrt[3]{t} + t\sqrt[3]{t^2}} d_\beta t$$

is convergent.

2.9 Advanced Practical Problems

Problem 2.9.1 *Let* $I = \mathbb{R}$ *and*

$$\begin{aligned} \beta(t) &= \frac{2}{3}t + 1, \\ f(t) &= \frac{10}{3}t - 1, \\ F(t) &= 2t^2 - 3t - 1, \quad t \in I. \end{aligned}$$

Prove that F *is a* β*-antiderivative of* f.

Problem 2.9.2 *Let* $I = \mathbb{R}$ *and*

$$\beta(t) = \frac{2}{3}t - 1, \quad t \in I.$$

Compute

1. $$\int_{-3}^{-1} (t-1) d_\beta t.$$

2. $$\int_{-3}^{1} (t-1) d_\beta t.$$

3. $$\int_{-1}^{1} (t-1) d_\beta t.$$

Problem 2.9.3 *Let* $I = \mathbb{R}$ *and*

$$\begin{aligned} \beta(t) &= \frac{1}{8}t - 1, \\ f(t) &= \frac{t+1}{t^2+1}, \quad t \in I. \end{aligned}$$

Prove that

$$\int_{1}^{4} D_\beta f(t) d_\beta t = -\frac{12}{17}.$$

Problem 2.9.4 *Let* $I = \mathbb{R}$ *and*

$$\beta(t) = \frac{1}{4}t + 1,$$

$$f(t) = t^2 - t + 2,$$

$$g(t) = \frac{2+t}{1+t^2}, \quad t \in I.$$

Prove that

1.

$$\int_{-1}^{3} f(t) D_\beta g(t) d_\beta t = 2 - \int_{-1}^{3} D_\beta f(t) g(\beta(t)) d_\beta t.$$

2.

$$\int_{-1}^{3} f(\beta(t)) D_\beta g(t) d_\beta t = 2 - \int_{-1}^{3} D_\beta f(t) g(t) d_\beta t.$$

Problem 2.9.5 *Let* $I = [1, \infty)$ *and*

$$\beta(t) = \frac{3}{4}t + 1, \quad t \in I.$$

Investigate for convergence and divergence the following integrals.

1.

$$\int_1^\infty (t^5 - 3t^2 + 1) d_\beta t.$$

2.

$$\int_1^\infty \frac{4}{(t+1)(3t+8)} d_\beta t.$$

3.

$$\int_1^\infty \frac{4}{(t+3)(3t+16)} d_\beta t.$$

Problem 2.9.6 *Let* $I = \mathbb{R}$ *and*

$$\beta(t) = \frac{1}{2}t + 3, \quad t \in I.$$

Using the Cauchy criterion, prove that the following integrals are convergent.

1.

$$\int_1^\infty \frac{1}{(t+1)(t+8)} d_\beta t.$$

2.

$$\int_1^\infty \frac{1}{(2t+1)(t+7)} d_\beta t.$$

3.

$$\int_1^\infty \frac{1}{(t+3)(t+12)} d_\beta t.$$

Problem 2.9.7 *Let $I = \mathbb{R}$ and*

$$\beta(t) = \frac{9}{10}t + 1, \quad t \in I.$$

Investigate for convergence and divergence the following integrals.

1.
$$\int_1^\infty \frac{1}{10(t+2)^4(9t+30)^{50}} d_\beta t.$$

2.
$$\int_1^\infty \frac{1}{(2t+3)^{10}(9t+25)^4} d_\beta t.$$

3.
$$\int_1^\infty \frac{1}{(t^2+49)(9t^2+83)} d_\beta t.$$

Problem 2.9.8 *Let $I = \mathbb{R}$ and*

$$\beta(t) = \frac{1}{2}t + 3, \quad t \in I.$$

Investigate for convergence and divergence the following integrals.

1.
$$\int_1^\infty \frac{1}{(t+1)^2(t+8)^4} d_\beta t.$$

2.
$$\int_1^\infty \frac{1}{(2t+11)^5(t+7)^8} d_\beta t.$$

3.
$$\int_1^\infty \frac{1}{(2t+9)^2(t+12)^4} d_\beta t.$$

Problem 2.9.9 *Let $I = \mathbb{R}$ and*

$$\beta(t) = \frac{9}{10}t + 1, \quad t \in I.$$

Investigate for convergence and divergence the following integrals.

1.
$$\int_1^\infty \frac{1}{10(t+2)^2} d_\beta t.$$

2.
$$\int_1^\infty \frac{1}{(3t+4)(t+25)} d_\beta t.$$

3.
$$\int_1^\infty \frac{1}{(t+49)(t+83)} d_\beta t.$$

Problem 2.9.10 *Let $I = \mathbb{R}$ and*

$$\beta(t) = \frac{1}{2}t + 3, \quad t \in I.$$

Investigate for convergence and divergence the following integrals.

1.
$$\int_1^\infty \frac{1}{(t+3)(2t+11)} d_\beta t.$$

2.
$$\int_1^\infty \frac{1}{(2t+11)(t+9)} d_\beta t.$$

3.
$$\int_1^\infty \frac{1}{(t+9)(t+11)} d_\beta t.$$

Problem 2.9.11 *Let $I = \mathbb{R}$ and*

$$\beta(t) = \frac{1}{2}t + 3, \quad t \in I.$$

Investigate for convergence and divergence the following integrals.

1.
$$\int_1^\infty \frac{1}{(t+1)(t+8)(1+\frac{2}{t}+t^4)} d_\beta t.$$

2.
$$\int_1^\infty \frac{1}{(2t+1)(t+7)(1+t^6)} d_\beta t.$$

3.
$$\int_1^\infty \frac{1}{(t+3)(t+12)(1+t^7)} d_\beta t.$$

Problem 2.9.12 *Let $I = \left[-\frac{1}{2}, \frac{1}{2}\right]$ and*

$$\beta(t) = t^5, \quad t \in I.$$

Prove that the integral

$$\int_{-\frac{1}{2}}^0 \frac{1}{t^{\frac{10}{3}} + t^2 + t^{\frac{2}{3}}} d_\beta t$$

is convergent.

2.10 Notes and References

In this chapter we define β-integral and prove some of its properties. In the chapter are deduced the basic rules for β-integration. It is deducted the Taylor formula. In the chapter they are investigated improper integrals of the first and second kinds. Some of the results in this chapter can be found in [1].

3

β-Elementary Functions

Suppose that $I \subseteq \mathbb{R}$ and β is first kind or second kind general quantum operator.

3.1 β-Regressive Functions

Definition 3.1.1 *A function $f: I \to \mathbb{R}$ is said to be β-regressive if*

$$1+(\beta(t)-t)f(t) \neq 0, \quad t \in I.$$

The set of all β-regressive functions on I is denoted by $\mathscr{R}_\beta$.

Definition 3.1.2 *In $\mathscr{R}_\beta$ define β-circle plus $\oplus_\beta$ as follows*

$$(f \oplus_\beta g)(t) = f(t)+g(t)+(\beta(t)-t)f(t)g(t), \quad t \in I.$$

Example 3.1.3 *Let $I = \mathbb{R}$ and*

$$\begin{aligned} \beta(t) &= \frac{2}{3}t+1, \\ f(t) &= 1+t, \\ g(t) &= t^2, \quad t \in I. \end{aligned}$$

We will find

$$(f \oplus_\beta g)(t), \quad t \in I.$$

We have

$$\begin{aligned} (f \oplus_\beta g)(t) &= f(t)+g(t)+(\beta(t)-t)f(t)g(t) \\ &= 1+t+t^2+\left(\frac{2}{3}t+1-t\right)(1+t)t^2 \end{aligned}$$

DOI: 10.1201/9781003472131-3

$$= 1+t+t^2+\left(1-\frac{1}{3}t\right)(t^3+t^2)$$

$$= 1+t+t^2-\frac{1}{3}t^3+t^3-\frac{1}{3}t^4+t^2$$

$$= 1+t+2t^2+\frac{2}{3}t^3-\frac{1}{3}t^4, \quad t \in I.$$

Example 3.1.4 *Let $I = (-\sqrt{2}, \sqrt{2})$ and*

$$\beta(t) = \frac{1}{2}t^3,$$

$$f(t) = t,$$

$$g(t) = t^2, \quad t \in I.$$

We will find

$$(f \oplus_\beta g)(t), \quad t \in I.$$

We have

$$\begin{aligned}(f \oplus_\beta g)(t) &= f(t)+g(t)+(\beta(t)-t)f(t)g(t)\\ &= t+t^2+\left(\frac{1}{2}t^3-t\right)\cdot t\cdot t^2\\ &= t+t^2+\left(\frac{1}{2}t^3-t\right)t^3\\ &= t+t^2-t^4+\frac{1}{2}t^6, \quad t \in I.\end{aligned}$$

Exercise 3.1.5 *Let $I = (-\sqrt[3]{3}, \sqrt[3]{3})$ and*

$$\beta(t) = \frac{1}{3}t^4,$$

$$f(t) = 2-t,$$

$$g(t) = 1+t, \quad t \in I.$$

Find

$$(f \oplus_\beta g)(t), \quad t \in I.$$

Theorem 3.1.6 *The set $(\mathscr{R}_\beta, \oplus_\beta)$ is an Abelian group.*

Proof 3.1.7 *Let $f, g, h \in \mathscr{R}_\beta$ be arbitrarily chosen. Then*

$$\begin{aligned} 1+(\beta(t)-t)f(t) &\neq 0, \\ 1+(\beta(t)-t)g(t) &\neq 0, \quad t \in I. \end{aligned}$$

Then

$$1+(\beta(t)-t)(f \oplus_\beta g)(t) = 1+(\beta(t)-t)(f(t)+g(t)+(\beta(t)-t)f(t)g(t))$$

$$\begin{aligned} &= 1+(\beta(t)-t)f(t)+(\beta(t)-t)g(t)+(\beta(t)-t)^2 f(t)g(t) \\ &= (1+(\beta(t)-t)f(t))+(\beta(t)-t)g(t)(1+(\beta(t)-t)f(t)) \\ &= (1+(\beta(t)-t)f(t))(1+(\beta(t)-t)g(t)) \\ &\neq 0, \quad t \in I. \end{aligned}$$

Thus, $f \oplus_\beta g \in \mathscr{R}_\beta$. Next,

$$(f \oplus_\beta g) \oplus_\beta h(t) = (f(t)+g(t)+(\beta(t)-t)f(t)g(t))+h(t)$$

$$\begin{aligned} &+(\beta(t)-t)(f(t)+g(t)+(\beta(t)-t)f(t)g(t))h(t) \\ &= f(t)+g(t)+(\beta(t)-t)f(t)g(t)+h(t)+(\beta(t)-t)f(t)h(t) \\ &+(\beta(t)-t)g(t)h(t)+(\beta(t)-t)^2 f(t)g(t)h(t) \\ &= f(t)+(g(t)+h(t)+(\beta(t)-t)g(t)h(t)) \\ &+(\beta(t)-t)(g(t)+h(t)+(\beta(t)-t)g(t)h(t))f(t) \\ &= f(t)+(g \oplus_\beta h)(t)+(\beta(t)-t)(g \oplus_\beta h)(t)f(t) \\ &= f(t) \oplus_\beta (g \oplus_\beta h)(t), \quad t \in I. \end{aligned}$$

Therefore in $(\mathscr{R}_\beta, \oplus_\beta)$ the associativity law holds. Note that $0 \in \mathscr{R}_\beta$ and

$$(0 \oplus_\beta f)(t) = 0+f(t)+(\beta(t)-t)\cdot 0 \cdot f(t)$$

$$= \quad f(t), \quad t \in I.$$

Also,

$$\begin{aligned}(f \oplus_\beta g)(t) &= f(t)+g(t)+(\beta(t)-t)f(t)g(t)\\ &= g(t)+f(t)+(\beta(t)-t)g(t)f(t)\\ &= (g \oplus_\beta f)(t), \quad t \in I,\end{aligned}$$

i.e, the commutativity law holds. Let

$$l(t) = -\frac{f(t)}{1+(\beta(t)-t)f(t)}, \quad t \in I.$$

We will show that $l \in \mathscr{R}_\beta$. *Really, we have*

$$\begin{aligned}1+(\beta(t)-t)l(t) &= 1-\frac{(\beta(t)-t)f(t)}{1+(\beta(t)-t)f(t)}\\ &= \frac{1}{1+(\beta(t)-t)f(t)}\\ &\neq 0, \quad t \in I.\end{aligned}$$

Then

$$\begin{aligned}(f \oplus_\beta l)(t) &= f(t)+l(t)+(\beta(t)-t)f(t)l(t)\\ &= f(t)-\frac{f(t)}{1+(\beta(t)-t)f(t)}-\frac{(\beta(t)-t)(f(t))^2}{1+(\beta(t)-t)f(t)}\\ &= \frac{f(t)+(\beta(t)-t)(f(t))^2-f(t)-(\beta(t)-t)(f(t))^2}{1+(\beta(t)-t)f(t)}\\ &= 0, \quad t \in I.\end{aligned}$$

So, any element of $\mathscr{R}_\beta$ *has inverse with respect to the* β*-circle addition* $\oplus_\beta$. *Therefore,* $(\mathscr{R}_\beta, \oplus_\beta)$ *is an Abelian group. This completes the proof.*

Definition 3.1.8 *The group* $(\mathscr{R}_\beta, \oplus_\beta)$ *is said to be* β*-regressive group.*

Definition 3.1.9 *For* $f \in \mathscr{R}_\beta$, *define* β*-circle minus as follows.*

$$(\ominus_\beta f)(t) = -\frac{f(t)}{1+(\beta(t)-t)f(t)}, \quad t \in I.$$

By the proof of Theorem 3.1.6, it follows that if $f \in \mathscr{R}_\beta$, then $\ominus f \in \mathscr{R}_\beta$.

Example 3.1.10 *Let I, β, f be as in Example 3.1.3. Then*

$$\begin{aligned}
(\ominus_\beta f)(t) &= -\frac{f(t)}{1+(\beta(t)-t)f(t)} \\
&= -\frac{1+t}{1+\left(\frac{2}{3}t+1-t\right)(1+t)} \\
&= -\frac{1+t}{1+\left(1-\frac{1}{3}t\right)(1+t)} \\
&= -\frac{1+t}{1+1+t-\frac{1}{3}t-\frac{1}{3}t^2} \\
&= -\frac{1+t}{2+\frac{2}{3}t-\frac{1}{3}t^2} \\
&= -\frac{3(1+t)}{6+2t-t^2} \\
&= \frac{3(1+t)}{t^2-2t-6}, \quad t \in I.
\end{aligned}$$

Next,

$$\begin{aligned}
(\ominus_\beta g)(t) &= -\frac{g(t)}{1+(\beta(t)-t)g(t)} \\
&= -\frac{t^2}{1+\left(\frac{2}{3}t+1-t\right)t^2} \\
&= -\frac{t^2}{1+\left(1-\frac{1}{3}t\right)t^2} \\
&= -\frac{3t^2}{3+(3-t)t^2} \\
&= -\frac{3t^2}{3+3t^2-t^3} \\
&= \frac{3t^2}{t^3-3t^2-3}, \quad t \in I.
\end{aligned}$$

Example 3.1.11 *Let I, β and f be as in Example 3.1.4. Then*

$$\begin{aligned}
(\ominus_\beta f)(t) &= -\frac{f(t)}{1+(\beta(t)-t)f(t)} \\
&= -\frac{t}{1+\left(\frac{1}{2}t^3-t\right)t}
\end{aligned}$$

$$
\begin{aligned}
&= -\frac{2t}{2+(t^3-2t)t} \\
&= -\frac{2t}{t^4-2t^2+2}, \quad t \in I,
\end{aligned}
$$

and

$$
\begin{aligned}
(\ominus_\beta g)(t) &= -\frac{g(t)}{1+(\beta(t)-t)g(t)} \\
&= -\frac{t^2}{1+\left(\frac{1}{2}t^2-t\right)t^2} \\
&= -\frac{2t^2}{2+(t^3-2t)t^2} \\
&= -\frac{2t^2}{t^5-2t^3+2}, \quad t \in I.
\end{aligned}
$$

Exercise 3.1.12 *Let $I = \mathbb{R}$ and*

$$
\begin{aligned}
\beta(t) &= \frac{1}{2}t, \\
f(t) &= t-t^2, \quad t \in I.
\end{aligned}
$$

Find

$$(\ominus_\beta f)(t), \quad t \in I.$$

Definition 3.1.13 *For $f, g \in \mathscr{R}_\beta$, define*

$$f \ominus_\beta g = f \oplus_\beta (\ominus_\beta g).$$

By the definition, we get

$$
\begin{aligned}
(f \ominus_\beta g)(t) &= f(t) + (\ominus_\beta g)(t) + (\beta(t)-t)f(t)(\ominus_\beta g)(t) \\
&= f(t) - \frac{g(t)}{1+(\beta(t)-t)g(t)} - \frac{(\beta(t)-t)f(t)g(t)}{1+(\beta(t)-t)g(t)} \\
&= \frac{f(t)+(\beta(t)-t)f(t)g(t)-g(t)-(\beta(t)-t)f(t)g(t)}{1+(\beta(t)-t)g(t)} \\
&= \frac{f(t)-g(t)}{1+(\beta(t)-t)g(t)}, \quad t \in I.
\end{aligned}
$$

Example 3.1.14 *Let I, β, f and g be as in Example 3.1.3. Then*

$$(f \ominus_\beta g)(t) = \frac{f(t)-g(t)}{1+(\beta(t)-t)g(t)}$$

$$\begin{aligned}
&= \frac{1+t-t^2}{1+\left(\frac{2}{3}t+1-t\right)t^2}\\
&= \frac{1+t-t^2}{1+\left(1-\frac{1}{3}t\right)t^2}\\
&= \frac{3(1+t-t^2)}{3+(3-t)t^2}\\
&= \frac{3(1+t-t^2)}{3+3t^2-t^3}\\
&= \frac{3(t^2-t-1)}{t^3-3t^2-3}, \quad t\in I,
\end{aligned}$$

and

$$\begin{aligned}
(g\ominus_\beta f)(t) &= \frac{g(t)-f(t)}{1+(\beta(t)-t)f(t)}\\
&= \frac{t^2-t-1}{1+\left(\frac{2}{3}t+1-t\right)(1+t)}\\
&= \frac{t^2-t-1}{1+\left(1-\frac{1}{3}t\right)(t+1)}\\
&= \frac{3(t^2-t-1)}{3+(3-t)(t+1)}\\
&= \frac{3(t^2-t-1)}{3+3+3t-t^2-t}\\
&= \frac{3(t^2-t-1)}{-t^2+2t+6}\\
&= -\frac{3(t^2-t-1)}{t^2-2t-6}, \quad t\in I.
\end{aligned}$$

Example 3.1.15 *Let I, β, f and g be as in Example 3.1.4. Then*

$$\begin{aligned}
(f\ominus_\beta g)(t) &= \frac{f(t)-g(t)}{1+(\beta(t)-t)g(t)}\\
&= \frac{t-t^2}{1+\left(\frac{1}{2}t^3-t\right)t^2}\\
&= \frac{2t(1-t)}{2+(t^3-2t)t^2}
\end{aligned}$$

$$= \frac{2t(1-t)}{t^5-2t^3+2}, \quad t \in I,$$

and

$$\begin{aligned}(g \ominus_\beta f)(t) &= \frac{g(t)-f(t)}{1+(\beta(t)-t)f(t)}\\ &= \frac{t^2-t}{1+\left(\frac{1}{2}t^3-t\right)t}\\ &= \frac{2t(t-1)}{2+(t^3-2t)t}\\ &= \frac{2t(t-1)}{t^4-2t^2+2}, \quad t \in I.\end{aligned}$$

Exercise 3.1.16 *Let $I = \mathbb{R}$ and*

$$\begin{aligned}\beta(t) &= \frac{1}{4}t,\\ f(t) &= 2-t^2,\\ g(t) &= 2+t^2, \quad t \in I.\end{aligned}$$

Find

1. $(f \oplus_\beta g)(t)$, $t \in I$.
2. $(\ominus_\beta f)(t)$, $t \in I$.
3. $(\ominus_\beta g)(t)$, $t \in I$.
4. $(f \ominus_\beta g)(t)$, $t \in I$.
5. $(g \ominus_\beta f)(t)$, $t \in I$.

Let $f, g \in \mathscr{R}_\beta$. Then, we have the following.

1. $\ominus_\beta(\ominus_\beta f) = f$.

Proof 3.1.17 *We have*

$$\begin{aligned}(\ominus_\beta(\ominus_\beta f))(t) &= -\frac{(\ominus_\beta f)(t)}{1+(\beta(t)-t)(\ominus_\beta f)(t)}\\ &= \frac{\frac{f(t)}{1+(\beta(t)-t)f(t)}}{1-\frac{(\beta(t)-t)f(t)}{1+(\beta(t)-t)f(t)}}\end{aligned}$$

$$= \frac{\frac{f(t)}{1+(\beta(t)-t)f(t)}}{\frac{1}{1+(\beta(t)-t)f(t)}}$$

$$= f(t), \quad t \in I.$$

This completes the proof.

2. $f \ominus_\beta f = 0$.

Proof 3.1.18 *We have*

$$(f \ominus_\beta f)(t) = (f \oplus_\beta (\ominus_\beta f))(t)$$

$$= f(t) + \left(-\frac{f(t)}{1+(\beta(t)-t)f(t)}\right) - (\beta(t)-t)\frac{(f(t))^2}{1+(\beta(t)-t)f(t)}$$

$$= \frac{f(t)+(\beta(t)-t)(f(t))^2 - f(t) - (\beta(t)-t)(f(t))^2}{1+(\beta(t)-t)f(t)}$$

$$= 0, \quad t \in I.$$

3. $f \ominus_\beta g \in \mathscr{R}_\beta$.

Proof 3.1.19 *Since* $f, g \in \mathscr{R}_\beta$, *we have that*

$$1+(\beta(t)-t)f(t) \neq 0,$$

$$1+(\beta(t)-t)g(t) \neq 0, \quad t \in I.$$

Then

$$1+(\beta(t)-t)(f \ominus_\beta g)(t) = 1+(\beta(t)-t)\frac{f(t)-g(t)}{1+(\beta(t)-t)g(t)}$$

$$= \frac{1+(\beta(t)-t)g(t)+(\beta(t)-t)f(t)-(\beta(t)-t)g(t)}{1+(\beta(t)-t)g(t)}$$

$$= \frac{1+(\beta(t)-t)f(t)}{1+(\beta(t)-t)g(t)}$$

$$\neq 0, \quad t \in I.$$

Thus, $f \ominus_\beta g \in \mathscr{R}_\beta$. *This completes the proof.*

4. $\ominus_\beta (f \ominus_\beta g) = g \ominus_\beta f$.

Proof 3.1.20*We have*

$$(g \ominus_\beta f)(t) = \frac{g(t) - f(t)}{1 + (\beta(t) - t)f(t)}, \quad t \in I.$$

Hence, using the computations in the previous point, we get

$$\begin{aligned}
(\ominus_\beta (f \ominus_\beta g))(t) &= -\frac{(f \ominus_\beta g)(t)}{1 + \beta(t)(f \ominus_\beta g)(t)} \\
&= -\frac{\frac{f(t)-g(t)}{1+(\beta(t)-t)g(t)}}{1 + \frac{(\beta(t)-t)(f(t)-g(t))}{1+(\beta(t)-t)g(t)}} \\
&= -\frac{f(t) - g(t)}{1 + (\beta(t) - t)g(t) + (\beta(t) - t)(f(t) - g(t))} \\
&= \frac{g(t) - f(t)}{1 + (\beta(t) - t)f(t)} \\
&= (g \ominus_\beta f)(t), \quad t \in I.
\end{aligned}$$

This completes the proof.

5. $(\ominus_\beta (f \oplus_\beta g)) = (\ominus_\beta f) \oplus_\beta (\ominus_\beta g)$.

Proof 3.1.21*We have*

$$(\ominus_\beta (f \oplus_\beta g))(t) = -\frac{(f \oplus_\beta g)(t)}{1 + (\beta(t) - t)(f \oplus_\beta g)(t)}$$

$$\begin{aligned}
&= -\frac{f(t) + g(t) + (\beta(t) - t)f(t)g(t)}{1 + (\beta(t) - t)(f(t) + g(t) + (\beta(t) - t)f(t)g(t))} \\
&= -\frac{f(t) + g(t) + (\beta(t) - t)f(t)g(t)}{1 + (\beta(t) - t)f(t) + (\beta(t) - t)g(t)(1 + (\beta(t) - t)f(t))} \\
&= -\frac{f(t) + g(t) + (\beta(t) - t)f(t)g(t)}{(1 + (\beta(t) - t)f(t))(1 + (\beta(t) - t)g(t))}, \quad t \in I.
\end{aligned}$$

Next,

$$((\ominus_\beta f) \oplus_\beta (\ominus_\beta g))(t) = (\ominus_\beta f)(t) + (\ominus_\beta g)(t)$$

$$+(\beta(t) - t)(\ominus_\beta f)(t)(\ominus_\beta g)(t)$$

$$= -\frac{f(t)}{1 + (\beta(t) - t)f(t)} - \frac{g(t)}{1 + (\beta(t) - t)g(t)}$$

$$+\frac{(\beta(t)-t)f(t)g(t)}{(1+(\beta(t)-t)f(t))(1+(\beta(t)-t)g(t))}$$

$$= -\frac{1}{(1+(\beta(t)-t)f(t))(1+(\beta(t)-t)g(t))}\Big(f(t)$$

$$\times(1+(\beta(t)-t)g(t))$$

$$+g(t)(1+(\beta(t)-t)f(t))-(\beta(t)-t)f(t)g(t)\Big)$$

$$= -\frac{f(t)+g(t)+(\beta(t)-t)f(t)g(t)}{(1+(\beta(t)-t)f(t))(1+(\beta(t)-t)g(t))}$$

$$= (\ominus_\beta(f\oplus_\beta g))(t), \quad t\in I.$$

This completes the proof.

6.

$$f(t)\oplus_\beta \frac{g(t)}{1+(\beta(t)-t)f(t)}, \quad t\in I.$$

Proof 3.1.22*We have*

$$f(t)\oplus_\beta \frac{g(t)}{1+(\beta(t)-t)f(t)} = f(t)+\frac{g(t)}{1+(\beta(t)-t)f(t)}$$

$$+(\beta(t)-t)\frac{f(t)g(t)}{1+(\beta(t)-t)f(t)}$$

$$= f(t)+\frac{1+(\beta(t)-t)f(t)}{1+(\beta(t)-t)f(t)}g(t)$$

$$= f(t)+g(t), \quad t\in I.$$

This completes the proof.

Definition 3.1.23 *For $f\in\mathscr{R}_\beta$, define β-circle square as follows*

$$f^{②_\beta}=(-f)(\ominus_\beta f).$$

For $f\in\mathscr{R}_\beta$, we have

$$f^{②_\beta}(t) = (-f(t))(\ominus_\beta f)(t)$$

$$= -f(t)\frac{-f(t)}{1+(\beta(t)-t)f(t)}$$

$$= \frac{(f(t))^2}{1+(\beta(t)-t)f(t)}, \quad t \in I.$$

Hence,

$$\frac{(f(t))^2}{f^{②_\beta}(t)} = 1+(\beta(t)-t)f(t), \quad t \in I.$$

Example 3.1.24 *Let I, β, f and g be as in Example 3.1.3. Then, we have the following.*

$$\begin{aligned} f^{②_\beta}(t) &= (-f(t))(\ominus_\beta f)(t) \\ &= -(1+t)\left(\frac{3(1+t)}{t^2-2t-6}\right) \\ &= -\frac{3(1+t)^2}{t^2-2t-6}, \quad t \in I, \end{aligned}$$

and

$$\begin{aligned} g^{②_\beta}(t) &= (-g(t))(\ominus_\beta g)(t) \\ &= -t^2\left(\frac{3t^2}{t^3-3t^2-3}\right) \\ &= -\frac{3t^4}{t^3-3t^2-3}, \quad t \in I. \end{aligned}$$

Example 3.1.25 *Let I, β, f and g be as in Example 3.1.4. Then, we have the following.*

$$\begin{aligned} f^{②_\beta}(t) &= (-f(t))(\ominus_\beta f)(t) \\ &= -t\left(-\frac{2t}{t^4-2t^2+2}\right) \\ &= \frac{2t^2}{t^4-2t^2+2}, \quad t \in I, \end{aligned}$$

and

$$g^{②_\beta}(t) = (-g(t))(\ominus_\beta g)(t)$$

$$= -t^2\left(-\frac{2t^2}{t^5-2t^3+2}\right)$$

$$= \frac{2t^4}{t^5-2t^3+2}, \quad t \in I.$$

Exercise 3.1.26 *Let $I = \mathbb{R}$ and*

$$\begin{aligned} \beta(t) &= \frac{7}{8}t+1, \\ f(t) &= t^3, \quad t \in I. \end{aligned}$$

Find

$$f^{②_\beta}(t), \quad t \in I.$$

Below, suppose that $f \in \mathscr{R}_\beta$. Then, we have the following.

1. $(\ominus_\beta f)^{②_\beta} = f^{②_\beta}$.

   Proof 3.1.27 *We have*

$$(\ominus_\beta f)(t) = -\frac{f(t)}{1+(\beta(t)-t)f(t)}, \quad t \in I,$$

and

$$f^{②_\beta}(t) = \frac{(f(t))^2}{1+(\beta(t)-t)f(t)}, \quad t \in I.$$

Hence,

$$\begin{aligned} (\ominus_\beta f)^{②_\beta}(t) &= \frac{\left(-\frac{f(t)}{1+(\beta(t)-t)f(t)}\right)^2}{1-\frac{(\beta(t)-t)f(t)}{1+(\beta(t)-t)f(t)}} \\ &= \frac{\frac{(f(t))^2}{(1+(\beta(t)-t)f(t))^2}}{\frac{1}{1+(\beta(t)-t)f(t)}} \\ &= \frac{(f(t))^2}{1+(\beta(t)-t)f(t)}, \quad t \in I. \end{aligned}$$

This completes the proof.

2. $$f(t)+(\ominus_\beta f)(t) = (\beta(t)-t)f^{②_\beta}(t), \quad t \in I.$$

Proof 3.1.28 *We have*

$$\begin{aligned} f(t)+(\ominus_\beta f)(t) &= f(t)-\frac{f(t)}{1+(\beta(t)-t)f(t)} \\ &= f(t)\left(1-\frac{1}{1+(\beta(t)-t)f(t)}\right) \\ &= f(t)\left(\frac{1+(\beta(t)-t)f(t)-1}{1+(\beta(t)-t)f(t)}\right) \\ &= (\beta(t)-t)\frac{(f(t))^2}{1+(\beta(t)-t)f(t)} \\ &= (\beta(t)-t)f^{②_\beta}(t), \quad t\in I. \end{aligned}$$

3.

$$f\oplus_\beta f^{②_\beta}=f+f^2.$$

Proof 3.1.29 *We have*

$$\begin{aligned} \left(f\oplus_\beta f^{②_\beta}\right)(t) &= f(t)+f^{②_\beta}(t)+(\beta(t)-t)f(t)f^{②_\beta}(t) \\ &= f(t)+\frac{(f(t))^2}{1+(\beta(t)-t)f(t)}+\frac{(\beta(t)-t)(f(t))^3}{1+(\beta(t)-t)f(t)} \\ &= f(t)+\frac{(1+(\beta(t)-t)f(t))(f(t))^2}{1+(\beta(t)-t)f(t)} \\ &= f(t)+(f(t))^2, \quad t\in I. \end{aligned}$$

This completes the proof.

3.2 β-Exponential Functions

Suppose that $f: I\to\mathbb{C}$ is a continuous function at s_0. Then the series

$$\sum_{k=0}^{\infty} f\left(\beta^k(t)\right)(\beta^k(t)-\beta^{k+1}(t))$$

is uniformly convergent on any compact interval $J\subseteq I$ containing s_0. Thus, both products

$$\prod_{k=0}^{\infty}\left(1-f(\beta^k(t))(\beta^k(t)-\beta^{k+1}(t))\right)$$

and

$$\prod_{k=0}^{\infty}\left(1+f\left(\beta^k(t)\right)(\beta^k(t)-\beta^{k+1}(t))\right)$$

are convergent on any compact interval $J \subseteq I$ containing s_0.

Definition 3.2.1 *Define the β-exponential functions $e_{f,\beta}$ and $E_{f,\beta}$ as follows*

$$e_{f,\beta}(t) = \frac{1}{\prod\limits_{k=0}^{\infty}(1-f(\beta^k(t))(\beta^k(t)-\beta^{k+1}(t)))}$$

and

$$E_{f,\beta}(t) = \prod_{k=0}^{\infty}\left(1+f(\beta^k(t))(\beta^k(t)-\beta^{k+1}(t))\right),$$

respectively.

Example 3.2.2 *Let $I = [1,2]$ and*

$$\begin{aligned} \beta(t) &= \frac{1}{2}t+\frac{1}{2}, \\ f(t) &= \frac{2}{t}, \quad t \in I. \end{aligned}$$

Here $s_0 = 1$. Then

$$\begin{aligned} [k]_{\frac{1}{2}} &= \frac{1-\left(\frac{1}{2}\right)^k}{1-\frac{1}{2}} \\ &= 2\left(1-\left(\frac{1}{2}\right)^k\right), \quad k \in \mathbb{N}, \end{aligned}$$

and

$$\begin{aligned} \beta^k(t) &= \left(\frac{1}{2}\right)^k t+2\left(1-\left(\frac{1}{2}\right)^k\right)\frac{1}{2} \\ &= \left(\frac{1}{2}\right)^k t+1-\left(\frac{1}{2}\right)^k \\ &= \left(\frac{1}{2}\right)^k (t-1)+1, \quad k \in \mathbb{N}, \quad t \in I, \end{aligned}$$

and

$$\beta^{k+1}(t) = \left(\frac{1}{2}\right)^{k+1}(t-1)+1, \quad k \in \mathbb{N}_0, \quad t \in I.$$

Hence,

$$\begin{aligned}\beta^k(t)-\beta^{k+1}(t) &= \left(\frac{1}{2}\right)^k (t-1)+1-\left(\frac{1}{2}\right)^{k+1}(t-1)-1\\ &= \left(\frac{1}{2}\right)^{k+1}(t-1), \quad k\in\mathbb{N}, \quad t\in I,\end{aligned}$$

and

$$\begin{aligned}f\left(\beta^k(t)\right) &= \frac{2}{\beta^k(t)}\\ &= \frac{2}{\left(\frac{1}{2}\right)^k t+1-\left(\frac{1}{2}\right)^k}\\ &= \frac{2}{\left(\frac{1}{2}\right)^k (t-1)+1}\\ &= \frac{2^{k+1}}{t-1+2^k}, \quad k\in\mathbb{N}, \quad t\in I.\end{aligned}$$

Therefore

$$\begin{aligned}\prod_{k=0}^{\infty}\left(1-f(\beta^k(t))(\beta^k(t)-\beta^{k+1}(t))\right) &= \prod_{k=0}^{\infty}\left(1-\frac{2^{k+1}}{t-1+2^k}\cdot\frac{t-1}{2^{k+1}}\right)\\ &= \prod_{k=0}^{\infty}\left(1-\frac{t-1}{t-1+2^k}\right), \quad t\in I,\end{aligned}$$

and

$$\begin{aligned}\prod_{k=0}^{\infty}\left(1+f(\beta^k(t))(\beta^k(t)-\beta^{k+1}(t))\right) &= \prod_{k=0}^{\infty}\left(1+\frac{2^{k+1}}{t-1+2^k}\cdot\frac{t-1}{2^{k+1}}\right)\\ &= \prod_{k=0}^{\infty}\left(1+\frac{t-1}{t-1+2^k}\right), \quad t\in I,\end{aligned}$$

Consequently

$$e_{\frac{2}{t},\frac{1}{2}t+\frac{1}{2}}(t)=\frac{1}{\prod\limits_{k=0}^{\infty}\left(1-\frac{t-1}{t-1+2^k}\right)}, \quad t\in I,$$

and

$$E_{\frac{2}{t},\frac{1}{2}t+\frac{1}{2}}(t)=\prod_{k=0}^{\infty}\left(1+\frac{t-1}{t-1+2^k}\right), \quad t\in I.$$

Example 3.2.3 *Let* $I=[0,2]$ *and*

$$\beta(t) = \frac{1}{2}t,$$

$$f(t) = (1+i)t, \quad t \in I.$$

Then

$$\beta^k(t) = \left(\frac{1}{2}\right)^k t, \quad t \in I, \quad k \in \mathbb{N},$$

and

$$\begin{aligned}\beta^k(t) - \beta^{k+1}(t) &= \left(\frac{1}{2}\right)^k t - \left(\frac{1}{2}\right)^{k+1} t \\ &= \left(\frac{1}{2}\right)^{k+1} t, \quad t \in I, \quad k \in \mathbb{N},\end{aligned}$$

and

$$\begin{aligned}f\left(\beta^k(t)\right)(\beta^k(t) - \beta^{k+1}(t)) &= (1+i)\left(\frac{1}{2}\right)^k t \left(\frac{1}{2}\right)^{k+1} t \\ &= \frac{(1+i)t^2}{2^{2k+1}}, \quad t \in I, \quad k \in \mathbb{N}.\end{aligned}$$

Hence,

$$\prod_{k=0}^{\infty}\left(1 - f(\beta^k(t))(\beta^k(t) - \beta^{k+1}(t))\right) = \prod_{k=0}^{\infty}\left(1 - \frac{(1+i)t^2}{2^{2k+1}}\right),$$

$t \in I$, *and*

$$\prod_{k=0}^{\infty}\left(1 + f(\beta^k(t))(\beta^k(t) - \beta^{k+1}(t))\right) = \prod_{k=0}^{\infty}\left(1 + \frac{(1+i)t^2}{2^{2k+1}}\right),$$

$t \in I$. *Consequently*

$$e_{(1+i)t, \frac{1}{2}t}(t) = \frac{1}{\prod\limits_{k=0}^{\infty}\left(1 - \frac{(1+i)t^2}{2^{2k+1}}\right)}, \quad t \in I,$$

and

$$E_{(1+i)t, \frac{1}{2}t}(t) = \prod_{k=0}^{\infty}\left(1 + \frac{(1+i)t^2}{2^{2k+1}}\right), \quad t \in I.$$

Exercise 3.2.4 *Let* $I = [0,3]$ *and*

$$\begin{aligned}\beta(t) &= \frac{1}{3}t, \\ \|f(t) &= 7t, \quad t \in I.\end{aligned}$$

Find

1. $e_{7t,\frac{1}{3}t}$, $t \in I$.
2. $E_{7t,\frac{1}{3}t}(t)$, $t \in I$.

Below, we will deduct some of the properties of the β-exponential functions.

1. $e_{f,\beta}(s_0) = E_{f,\beta}(s_0) = 1$.

Proof 3.2.5 *We have*

$$\beta^k(s_0) = s_0, \quad k \in \mathbb{N}.$$

Then

$$\begin{aligned}\prod_{k=0}^{\infty}\left(1 - f(\beta^k(s_0))(\beta^k(s_0) - \beta^{k+1}(s_0))\right) &= \prod_{k=0}^{\infty}(1 - f(s_0)(s_0 - s_0)) \\ &= \prod_{k=0}^{\infty} 1 \\ &= 1\end{aligned}$$

and

$$\begin{aligned}\prod_{k=0}^{\infty}\left(1 + f(\beta^k(s_0))(\beta^k(s_0) - \beta^{k+1}(s_0))\right) &= \prod_{k=0}^{\infty}(1 + f(s_0)(s_0 - s_0)) \\ &= \prod_{k=0}^{\infty} 1 \\ &= 1.\end{aligned}$$

Hence,

$$\begin{aligned}e_{f,\beta}(s_0) &= \frac{1}{\prod\limits_{k=0}^{\infty}(1 - f(\beta^k(s_0))(\beta^k(s_0) - \beta^{k+1}(s_0)))} \\ &= \frac{1}{1} \\ &= 1\end{aligned}$$

and

$$\begin{aligned}E_{f,\beta}(s_0) &= \prod_{k=0}^{\infty}\left(1 + f(\beta^k(s_0))(\beta^k(s_0) - \beta^{k+1}(s_0))\right) \\ &= 1.\end{aligned}$$

This completes the proof.

2. $D_\beta e_{f,\beta}(t) = f(t)e_{f,\beta}(t), t \in I.$

Proof 3.2.6*Let $t \neq s_0$. Then*

$$
\begin{aligned}
D_\beta e_{f,\beta}(t) &= \frac{1}{\beta(t)-t}(e_{f,\beta}(\beta(t)) - e_{f,\beta}(t)) \\
&= \frac{1}{\beta(t)-t}\left(\frac{1}{\prod\limits_{k=0}^{\infty}(1-f(\beta^{k+1}(t))(\beta^{k+1}(t)-\beta^{k+2}(t)))}\right. \\
&\quad \left. - \frac{1}{\prod\limits_{k=0}^{\infty}(1-f(\beta^{k}(t))(\beta^{k}(t)-\beta^{k+1}(t)))}\right) \\
&= \frac{1}{\beta(t)-t}\left(\frac{1}{\prod\limits_{k=1}^{\infty}(1-f(\beta^{k}(t))(\beta^{k}(t)-\beta^{k+1}(t)))}\right. \\
&\quad \left. - \frac{1}{\prod\limits_{k=0}^{\infty}(1-f(\beta^{k}(t))(\beta^{k}(t)-\beta^{k+1}(t)))}\right) \\
&= \frac{1}{\beta(t)-t}\left(\frac{1}{\prod\limits_{k=1}^{\infty}(1-f(\beta^{k}(t))(\beta^{k}(t)-\beta^{k+1}(t)))}\right. \\
&\quad \left. - \frac{1}{(1-f(t)(t-\beta(t)))\prod\limits_{k=1}^{\infty}(1-f(\beta^{k}(t))(\beta^{k}(t)-\beta^{k+1}(t)))}\right) \\
&= \frac{1}{(\beta(t)-t)\prod\limits_{k=1}^{\infty}(1-f(\beta^{k}(t))(\beta^{k}(t)-\beta^{k+1}(t)))} \\
&\quad \times\left(1-\frac{1}{(1-f(t)(t-\beta(t)))}\right) \\
&= \frac{1}{(\beta(t)-t)\prod\limits_{k=1}^{\infty}(1-f(\beta^{k}(t))(\beta^{k}(t)-\beta^{k+1}(t)))} \\
&\quad \times\frac{1-f(t)(t-\beta(t))-1}{(1-f(t)(t-\beta(t)))} \\
&= f(t)\frac{1}{\prod\limits_{k=0}^{\infty}(1-f(\beta^{k}(t))(\beta^{k}(t)-\beta^{k+1}(t)))} \\
&= f(t)e_{f,\beta}(t), \quad t \in I.
\end{aligned}
$$

Let $t = s_0$. *Then*

$$\lim_{h\to 0+}\frac{e_{f,\beta}(s_0+h)-e_{f,\beta}(s_0)}{h} = \lim_{h\to 0+}\frac{e_{f,\beta}(s_0+h)-1}{h}$$

$$= \lim_{h\to 0+}\frac{1}{h}\left(\frac{1}{\prod\limits_{k=0}^{\infty}(1-f(\beta^k(s_0+h))(\beta^k(s_0+h)-\beta^{k+1}(s_0+h)))}-1\right)$$

$$= \lim_{h\to 0+}\left(\frac{1-\prod\limits_{k=0}^{\infty}(1-f(\beta^k(s_0+h))(\beta^k(s_0+h)-\beta^{k+1}(s_0+h)))}{\prod\limits_{k=0}^{\infty}(1-f(\beta^k(s_0+h))(\beta^k(s_0+h)-\beta^{k+1}(s_0+h)))}\right)$$

$$= \lim_{h\to 0+}\left(\frac{1}{h}\cdot\frac{1-(1-f(s_0)h)}{1-f(s_0)h}\right)$$

$$= \lim_{h\to 0+}\frac{f(s_0)h}{h(1-f(s_0)h)}$$

$$= \lim_{h\to 0+}\frac{f(s_0)}{1-f(s_0)h}$$

$$= f(s_0).$$

As above,

$$\lim_{h\to 0-}\frac{e_{f,\beta}(s_0+h)-e_{f,\beta}(s_0)}{h} = f(s_0).$$

This completes the proof.

3. $D_\beta E_{f,\beta}(t) = f(t)E_{f,\beta}(\beta(t))$.

Proof 3.2.7*Let* $t \neq s_0$. *Then*

$$D_\beta E_{f,\beta}(t) = \frac{E_{f,\beta}(\beta(t))-E_{f,\beta}(t)}{\beta(t)-t}$$

$$= \frac{1}{\beta(t)-t}\left(\prod_{k=0}^{\infty}\left(1+f(\beta^{k+1}(t))(\beta^{k+1}(t)-\beta^{k+2}(t))\right)\right.$$

$$\left.-\prod_{k=0}^{\infty}\left(1+f(\beta^{k}(t))(\beta^{k}(t)-\beta^{k+1}(t))\right)\right)$$

$$= \frac{1}{\beta(t)-t}\left(\prod_{k=1}^{\infty}\left(1+f(\beta^{k}(t))(\beta^{k}(t)-\beta^{k+1}(t))\right)\right.$$

$$\left.-(1+f(t)(t-\beta(t)))\prod_{k=1}^{\infty}\left(1+f(\beta^{k}(t))(\beta^{k}(t)-\beta^{k+1}(t))\right)\right)$$

$$\begin{aligned}
&= \frac{1}{\beta(t)-t}\prod_{k=1}^{\infty}\left(1-f(\beta^k(t))(\beta^k(t)-\beta^{k+1}(t))\right)\\
&\quad \times(1-(1+f(t)(t-\beta(t))))\\
&= \frac{f(t)(\beta(t)-t)}{\beta(t)-t}\prod_{k=1}^{\infty}\left(1+f(\beta^k(t))(\beta^k(t)-\beta^{k+1}(t))\right)\\
&= f(t)E_{f,\beta}(\beta(t)).
\end{aligned}$$

Let $t = s_0$. *Then*

$$\lim_{h\to 0+}\frac{E_{f,\beta}(s_0+h)-E_{f,\beta}(s_0)}{h}$$

$$\begin{aligned}
&= \lim_{h\to 0+}\frac{1}{h}\left(\prod_{k=0}^{\infty}\left(1-f(\beta^k(s_0+h))(\beta^k(s_0+h)-\beta^{k+1}(s_0+h))\right)-1\right)\\
&= \lim_{h\to 0+}\frac{1}{h}(1+f(s_0)(s_0+h-s_0)-1)\\
&= \lim_{h\to 0+}\frac{f(s_0)h}{h}\\
&= f(s_0).
\end{aligned}$$

As above,

$$\lim_{h\to 0-}\frac{E_{f,\beta}(s_0+h)-E_{f,\beta}(s_0)}{h} = f(s_0).$$

This completes the proof.

4. $e_{f,\beta}(\beta(t)) = (1+(\beta(t)-t)f(t))e_{f,\beta}(t),\ t\in I.$

   Proof 3.2.8*We have*

$$\begin{aligned}
e_{f,\beta}(\beta(t)) &= e_{f,\beta}(t)+(\beta(t)-t)D_\beta e_{f,\beta}(t)\\
&= e_{f,\beta}(t)+(\beta(t)-t)f(t)e_{f,\beta}(t)\\
&= (1+(\beta(t)-t)f(t))e_{f,\beta}(t), \quad t\in I.
\end{aligned}$$

   This completes the proof.

5. $E_{f,\beta}(t) = (1-(\beta(t)-t)f(t))E_{f,\beta}(\beta(t)),\ t\in I.$

Proof 3.2.9 *We have*

$$\begin{aligned} E_{f,\beta}(\beta(t)) &= E_{f,\beta}(t) + (\beta(t)-t)D_\beta E_{f,\beta}(t) \\ &= E_{f,\beta}(t) + (\beta(t)-t)f(t)E_{f,\beta}(\beta(t)), \end{aligned}$$

whereupon

$$\begin{aligned} E_{f,\beta}(t) &= E_{f,\beta}(\beta(t)) - (\beta(t)-t)f(t)E_{f,\beta}(\beta(t)) \\ &= (1-(\beta(t)-t)f(t))E_{f,\beta}(\beta(t)), \quad t \in I. \end{aligned}$$

This completes the proof.

6. $\dfrac{1}{e_{f,\beta}(t)} = e_{\ominus_\beta f,\beta}(t), t \in I.$

Proof 3.2.10 *We have*

$$\begin{aligned} e_{f,\beta}(t) &= \frac{1}{\prod\limits_{k=0}^{\infty}(1-f(\beta^k(t))(\beta^k(t)-\beta^{k+1}(t)))} \\ &= \prod_{k=0}^{\infty}\frac{1}{(1-f(\beta^k(t))(\beta^k(t)-\beta^{k+1}(t)))}, \quad t \in I. \end{aligned}$$

Hence,

$$\frac{1}{e_{f,\beta}(t)} = \prod_{k=0}^{\infty}\left(1-f(\beta^k(t))(\beta^k(t)-\beta^{k+1}(t))\right),$$

$t \in I$, and

$$\begin{aligned} e_{\ominus_\beta f,\beta}(t) &= \prod_{k=0}^{\infty}\frac{1}{1-\ominus_\beta f(\beta^k(t))(\beta^k(t)-\beta^{k+1}(t))} \\ &= \prod_{k=0}^{\infty}\frac{1}{1+\frac{f(\beta^k(t))}{1+(\beta^{k+1}(t)-\beta^k(t))f(\beta^k(t))}(\beta^k(t)-\beta^{k+1}(t))} \\ &= \prod_{k=0}^{\infty}\left(1-f(\beta^k(t))(\beta^k(t)-\beta^{k+1}(t))\right) \\ &= \frac{1}{e_{f,\beta}(t)}, \quad t \in I. \end{aligned}$$

This completes the proof.

7. $e_{f,\beta}(t)e_{g,\beta}(t) = e_{f\oplus_\beta g}(t), t \in I.$

Proof 3.2.11 *We have*

$$
\begin{aligned}
e_{f,\beta}(t)e_{g,\beta}(t) &= \left(\prod_{k=0}^{\infty}\frac{1}{(1-f(\beta^k(t))(\beta^k(t)-\beta^{k+1}(t)))}\right)\\
&\quad\times\left(\prod_{k=0}^{\infty}\frac{1}{(1-g(\beta^k(t))(\beta^k(t)-\beta^{k+1}(t)))}\right)\\
&= \prod_{k=0}^{\infty}\frac{1}{(1-f(\beta^k(t))(\beta^k(t)-\beta^{k+1}(t)))}\\
&\quad\times\prod_{k=0}^{\infty}1\left(1-g(\beta^k(t))(\beta^k(t)-\beta^{k+1}(t))\right)\\
&= \prod_{k=0}^{\infty}\frac{1}{(1-(f\oplus_\beta g)(\beta^k(t))(\beta^k(t)-\beta^{k+1}(t)))}\\
&= e_{f\oplus_\beta g,\beta}(t), \quad t\in I.
\end{aligned}
$$

This completes the proof.

8. $\dfrac{e_{f,\beta}(t)}{e_{g,\beta}(t)} = e_{f\ominus_\beta g,\beta}(t),\, t\in I.$

Proof 3.2.12 *We have*

$$
\begin{aligned}
\frac{e_{f,\beta}(t)}{e_{g,\beta}(t)} &= e_{f,\beta}(t)e_{\ominus_\beta g}(t)\\
&= e_{f\oplus_\beta(\ominus_\beta g),\beta}(t)\\
&= e_{f\ominus_\beta g,\beta}(t), \quad t\in I.
\end{aligned}
$$

This completes the proof.

3.3 β-Trigonometric Functions

In this section, we will define β-trigonometric functions and we will deduct some of their properties. Let $f : I \to \mathbb{R}$.

Definition 3.3.1 *Define β-trigonometric functions as follows.*

$$
\sin_{f,\beta}(t) = \frac{e_{if,\beta}(t)-e_{-if,\beta}(t)}{2i},
$$

$$\cos_{f,\beta}(t) = \frac{e_{if,\beta}(t)+e_{-if,\beta}(t)}{2},$$

$$Sin_{f,\beta}(t) = \frac{E_{if,\beta}(t)-E_{-if,\beta}(t)}{2i},$$

$$Cos_{f,\beta}(t) = \frac{E_{if,\beta}(t)+E_{-if,\beta}(t)}{2}, \quad t \in I.$$

Now, we will deduct some basic properties of the β-trigonometric functions.

1. $D_\beta \sin_{f,\beta}(t) = f(t)\cos_{f,\beta}(t), t \in I.$

   Proof 3.3.2 *We have*

$$\begin{aligned} D_\beta \sin_{f,\beta}(t) &= \frac{D_\beta e_{if,\beta}(t) - D_\beta e_{-if,\beta}(t)}{2i} \\ &= \frac{if(t)e_{if,\beta}(t) + if(t)e_{-if,\beta}(t)}{2i} \\ &= f(t)\frac{e_{if,\beta}(t)+e_{-if,\beta}(t)}{2} \\ &= f(t)\cos_{f,\beta}(t), \quad t \in I. \end{aligned}$$

   This completes the proof.

2. $D_\beta \cos_{f,\beta}(t) = -f(t)\sin_{f,\beta}(t), t \in I.$

   Proof 3.3.3 *We have*

$$\begin{aligned} D_\beta \cos_{f,\beta}(t) &= \frac{D_\beta e_{if,\beta}(t) + D_\beta e_{-if,\beta}(t)}{2} \\ &= \frac{if(t)e_{if,\beta}(t) - if(t)e_{-if,\beta}(t)}{2} \\ &= -f(t)\frac{e_{if,\beta}(t)-e_{-if,\beta}(t)}{2i} \\ &= -f(t)\sin_{f,\beta}(t), \quad t \in I. \end{aligned}$$

   This completes the proof.

3. $D_\beta Sin_{f,\beta}(t) = f(t)Cos_{f,\beta}(\beta(t)), t \in I.$

Proof 3.3.4*We have*

$$\begin{aligned}
D_\beta Sin_{f,\beta}(t) &= \frac{D_\beta E_{if,\beta}(t) - D_\beta E_{-if,\beta}(t)}{2i} \\
&= \frac{if(t)E_{if,\beta}(\beta(t)) + if(t)E_{-if,\beta}(\beta(t))}{2i} \\
&= f(t)\frac{E_{if,\beta}(\beta(t)) + E_{-if,\beta}(\beta(t))}{2} \\
&= f(t)Cos_{f,\beta}(\beta(t)), \quad t \in I.
\end{aligned}$$

This completes the proof.

4. $D_\beta Cos_{f,\beta}(t) = -f(t)Sin_{f,\beta}(\beta(t)), t \in I.$

Proof 3.3.5*We have*

$$\begin{aligned}
D_\beta Cos_{f,\beta}(t) &= \frac{D_\beta E_{if,\beta}(t) + D_\beta E_{-if,\beta}(t)}{2} \\
&= \frac{if(t)E_{if,\beta}(\beta(t)) - if(t)E_{-if,\beta}(\beta(t))}{2} \\
&= -f(t)\frac{E_{if,\beta}(\beta(t)) - E_{-if,\beta}(\beta(t))}{2i} \\
&= -f(t)Sin_{f,\beta}(\beta(t)), \quad t \in I.
\end{aligned}$$

This completes the proof.

5. $(\cos_{f,\beta}(t))^2 + (\sin_{f,\beta}(t))^2 = e_{if,\beta}(t)e_{-if,\beta}(t), t \in I.$

Proof 3.3.6*We have*

$$\begin{aligned}
&(\cos_{f,\beta}(t))^2 + (\sin_{f,\beta}(t))^2 \\
&= \left(\frac{e_{if,\beta}(t) + e_{-if,\beta}(t)}{2}\right)^2 + \left(\frac{e_{if,\beta}(t) - e_{-if,\beta}(t)}{2i}\right)^2 \\
&= \frac{(e_{if,\beta}(t))^2 + 2e_{if,\beta}(t)e_{-if,\beta}(t) + (e_{-if,\beta}(t))^2}{4} \\
&\quad - \frac{(e_{if,\beta}(t))^2 - 2e_{if,\beta}(t)e_{-if,\beta}(t) + (e_{-if,\beta}(t))^2}{4} \\
&= e_{if,\beta}(t)e_{-if,\beta}(t), \quad t \in I.
\end{aligned}$$

This completes the proof.

6. $(Cos_{f,\beta}(t))^2 + (Sin_{f,\beta}(t))^2 = E_{if,\beta}(t)E_{-if,\beta}(t),\, t \in I.$

Proof 3.3.7 *We have*

$$\begin{aligned}
&(Cos_{f,\beta}(t))^2 + (Sin_{f,\beta}(t))^2 \\
&= \left(\frac{E_{if,\beta}(t) + E_{-if,\beta}(t)}{2}\right)^2 + \left(\frac{E_{if,\beta}(t) - E_{-if,\beta}(t)}{2i}\right)^2 \\
&= \frac{(E_{if,\beta}(t))^2 + 2E_{if,\beta}(t)E_{-if,\beta}(t) + (E_{-if,\beta}(t))^2}{4} \\
&\quad - \frac{(E_{if,\beta}(t))^2 - 2E_{if,\beta}(t)E_{-if,\beta}(t) + (E_{-if,\beta}(t))^2}{4} \\
&= E_{if,\beta}(t)E_{-if,\beta}(t), \quad t \in I.
\end{aligned}$$

This completes the proof.

7. $\cos_{f,\beta}(t) + i\sin_{f,\beta}(t) = e_{if,\beta}(t),\, t \in I.$

Proof 3.3.8 *We have*

$$\begin{aligned}
\cos_{f,\beta}(t) + i\sin_{f,\beta}(t) &= \frac{e_{if,\beta}(t) + e_{-if,\beta}(t)}{2} + i\frac{e_{if,\beta}(t) - e_{-if,\beta}(t)}{2i} \\
&= \frac{e_{if,\beta}(t) + e_{-if,\beta}(t) + e_{if,\beta}(t) - e_{-if,\beta}(t)}{2} \\
&= e_{if,\beta}(t), \quad t \in I.
\end{aligned}$$

This completes the proof.

8. $Cos_{f,\beta}(t) + iSin_{f,\beta}(t) = E_{if,\beta}(t),\, t \in I.$

Proof 3.3.9 *We have*

$$\begin{aligned}
Cos_{f,\beta}(t) + iSin_{f,\beta}(t) &= \frac{E_{if,\beta}(t) + E_{-if,\beta}(t)}{2} + i\frac{E_{if,\beta}(t) - E_{-if,\beta}(t)}{2i} \\
&= \frac{E_{if,\beta}(t) + E_{-if,\beta}(t) + E_{if,\beta}(t) - E_{-if,\beta}(t)}{2} \\
&= E_{if,\beta}(t), \quad t \in I.
\end{aligned}$$

This completes the proof.

Exercise 3.3.10 *Prove that*

$$\sin_{f,\beta}(t)Sin_{f,\beta}(t)+\cos_{f,\beta}(t)Cos_{f,\beta}(t)=1, \quad t\in I.$$

3.4 β-Hyperbolic Functions

In this section, we will define β-hyperbolic functions and we will deduct some of their properties. Let $f: I \to \mathbb{R}$.

Definition 3.4.1 *Define β-hyperbolic functions as follows.*

$$\begin{aligned}
\sinh_{f,\beta}(t) &= \frac{e_{f,\beta}(t)-e_{-f,\beta}(t)}{2},\\
\cosh_{f,\beta}(t) &= \frac{e_{f,\beta}(t)+e_{-f,\beta}(t)}{2},\\
Sinh_{f,\beta}(t) &= \frac{E_{f,\beta}(t)-E_{-f,\beta}(t)}{2},\\
Cosh_{f,\beta}(t) &= \frac{E_{f,\beta}(t)+E_{-f,\beta}(t)}{2}, \quad t\in I.
\end{aligned}$$

Now, we will deduct some basic properties of the β-hyperbolic functions.

1. $D_\beta \sinh_{f,\beta}(t) = f(t)\cosh_{f,\beta}(t)$, $t\in I$.

   Proof 3.4.2*We have*

$$\begin{aligned}
D_\beta \sinh_{f,\beta}(t) &= \frac{D_\beta e_{f,\beta}(t)-D_\beta e_{-f,\beta}(t)}{2}\\
&= \frac{f(t)e_{f,\beta}(t)+f(t)e_{-f,\beta}(t)}{2}\\
&= f(t)\frac{e_{f,\beta}(t)+e_{-f,\beta}(t)}{2}\\
&= f(t)\cosh_{f,\beta}(t), \quad t\in I.
\end{aligned}$$

   This completes the proof.

2. $D_\beta \cosh_{f,\beta}(t) = f(t)\sinh_{f,\beta}(t)$, $t\in I$.

   Proof 3.4.3*We have*

$$D_\beta \cosh_{f,\beta}(t) = \frac{D_\beta e_{f,\beta}(t)+D_\beta e_{-f,\beta}(t)}{2}$$

$$= \frac{f(t)e_{f,\beta}(t) - f(t)e_{-f,\beta}(t)}{2}$$

$$= f(t)\frac{e_{f,\beta}(t) - e_{-f,\beta}(t)}{2}$$

$$= f(t)\sinh_{f,\beta}(t), \quad t \in I.$$

This completes the proof.

3. $D_\beta Sinh_{f,\beta}(t) = f(t)Cosh_{f,\beta}(\beta(t)), t \in I.$

   Proof 3.4.4*We have*

$$\begin{aligned} D_\beta Sinh_{f,\beta}(t) &= \frac{D_\beta E_{f,\beta}(t) - D_\beta E_{-f,\beta}(t)}{2} \\ &= \frac{f(t)E_{f,\beta}(\beta(t)) + f(t)E_{-f,\beta}(\beta(t))}{2} \\ &= f(t)\frac{E_{f,\beta}(\beta(t)) + E_{-f,\beta}(\beta(t))}{2} \\ &= f(t)Cosh_{f,\beta}(\beta(t)), \quad t \in I. \end{aligned}$$

   This completes the proof.

4. $D_\beta Cosh_{f,\beta}(t) = f(t)Sinh_{f,\beta}(\beta(t)), t \in I.$

   Proof 3.4.5*We have*

$$\begin{aligned} D_\beta Cosh_{f,\beta}(t) &= \frac{D_\beta E_{f,\beta}(t) + D_\beta E_{-f,\beta}(t)}{2} \\ &= \frac{f(t)E_{f,\beta}(\beta(t)) - f(t)E_{-f,\beta}(\beta(t))}{2} \\ &= f(t)\frac{E_{f,\beta}(\beta(t)) - E_{-f,\beta}(\beta(t))}{2} \\ &= f(t)Sinh_{f,\beta}(\beta(t)), \quad t \in I. \end{aligned}$$

   This completes the proof.

5. $(\cosh_{f,\beta}(t))^2 - (\sinh_{f,\beta}(t))^2 = e_{f,\beta}(t)e_{-f,\beta}(t), t \in I.$

Proof 3.4.6 *We have*

$$\begin{aligned}
&(\cosh_{f,\beta}(t))^2-(\sinh_{f,\beta}(t))^2\\
&= \left(\frac{e_{f,\beta}(t)+e_{-f,\beta}(t)}{2}\right)^2-\left(\frac{e_{f,\beta}(t)-e_{-f,\beta}(t)}{2}\right)^2\\
&= \frac{(e_{f,\beta}(t))^2+2e_{f,\beta}(t)e_{-f,\beta}(t)+(e_{-f,\beta}(t))^2}{4}\\
&\quad -\frac{(e_{f,\beta}(t))^2-2e_{f,\beta}(t)e_{-f,\beta}(t)+(e_{-f,\beta}(t))^2}{4}\\
&= e_{f,\beta}(t)e_{-f,\beta}(t), \quad t\in I.
\end{aligned}$$

This completes the proof.

6. $(Cosh_{f,\beta}(t))^2-(Sinh_{f,\beta}(t))^2=E_{f,\beta}(t)E_{-f,\beta}(t), t\in I.$

Proof 3.4.7 *We have*

$$\begin{aligned}
&(Cosh_{f,\beta}(t))^2-(Sinh_{f,\beta}(t))^2\\
&= \left(\frac{E_{f,\beta}(t)+E_{-f,\beta}(t)}{2}\right)^2-\left(\frac{E_{f,\beta}(t)-E_{-f,\beta}(t)}{2}\right)^2\\
&= \frac{(E_{f,\beta}(t))^2+2E_{f,\beta}(t)E_{-f,\beta}(t)+(E_{-f,\beta}(t))^2}{4}\\
&\quad -\frac{(E_{f,\beta}(t))^2-2E_{f,\beta}(t)E_{-f,\beta}(t)+(E_{-f,\beta}(t))^2}{4}\\
&= E_{f,\beta}(t)E_{-f,\beta}(t), \quad t\in I.
\end{aligned}$$

This completes the proof.

7. $\cosh_{f,\beta}(t)+\sinh_{f,\beta}(t)=e_{f,\beta}(t), t\in I.$

Proof 3.4.8 *We have*

$$\begin{aligned}
\cosh_{f,\beta}(t)+\sinh_{f,\beta}(t) &= \frac{e_{f,\beta}(t)+e_{-f,\beta}(t)}{2}+\frac{e_{f,\beta}(t)-e_{-f,\beta}(t)}{2}\\
&= \frac{e_{f,\beta}(t)+e_{-f,\beta}(t)+e_{f,\beta}(t)-e_{-f,\beta}(t)}{2}\\
&= e_{f,\beta}(t), \quad t\in I.
\end{aligned}$$

This completes the proof.

8. $Cosh_{f,\beta}(t) + Sinh_{f,\beta}(t) = E_{f,\beta}(t), t \in I.$

Proof 3.4.9 *We have*

$$\begin{aligned} Cosh_{f,\beta}(t) + Sinh_{f,\beta}(t) &= \frac{E_{f,\beta}(t) + E_{-f,\beta}(t)}{2} + \frac{E_{f,\beta}(t) - E_{-f,\beta}(t)}{2} \\ &= \frac{E_{f,\beta}(t) + E_{-f,\beta}(t) + E_{f,\beta}(t) - E_{-f,\beta}(t)}{2} \\ &= E_{f,\beta}(t), \quad t \in I. \end{aligned}$$

This completes the proof.

Exercise 3.4.10 *Prove that*

$$\cosh_{f,\beta}(t) - \sinh_{f,\beta}(t) = e_{-f,\beta}(t), \quad t \in I.$$

3.5 Advanced Practical Problems

Problem 3.5.1 *Let* $I = \mathbb{R}$ *and*

$$\begin{aligned} \beta(t) &= \frac{3}{4}t + 1, \\ f(t) &= 1 - t, \\ g(t) &= 1 + t, \quad t \in I. \end{aligned}$$

Find

1. $$(f \oplus_\beta g)(t), \quad t \in I.$$

2. $$(\ominus_\beta f)(t), \quad t \in I.$$

3. $$(\ominus_\beta g)(t), \quad t \in I.$$

4. $$(f \ominus_\beta g)(t), \quad t \in I.$$

5.
$$(g \ominus_\beta f)(t), \quad t \in I.$$

6.
$$f^{②_\beta}(t), \quad t \in I.$$

7.
$$g^{②_\beta}(t), \quad t \in I.$$

Problem 3.5.2 *Let $I = [0,4]$ and*

$$\begin{aligned} \beta(t) &= \frac{1}{4}t, \\ \|f(t) &= 9t, \quad t \in I. \end{aligned}$$

Find

1. $e_{9t,\frac{1}{4}t}(t)$, $t \in I$.
2. $E_{9t,\frac{1}{4}t}(t)$, $t \in I$.

Problem 3.5.3 *Prove that*

$$\sin_{f,\beta}(t)Cos_{f,\beta}(t) - \cos_{f,\beta}(t)Sin_{f,\beta}(t) = 0, \quad t \in I.$$

Problem 3.5.4 *Prove that*

$$Cosh_{f,\beta}(t) - Sinh_{f,\beta}(t) = E_{-f,\beta}(t), \quad t \in I.$$

3.6 Notes and References

This chapter is devoted on some β-elementary functions. They are defined β-exponential, β-trigonometric and β-hyperbolic functions. Some of their basic properties are deduced. Some of the results in this chapter can be found in [2].

4

General Quantum Polynomial Interpolation

In this chapter we consider the problem of general quantum polynomial interpolation for finding a polynomial that agrees with some information that we have for a given real-valued function f of a single real-variable x. We construct the Lagrange, β-Lagrange, Hermite and β-Hermite interpolation polynomial for a given real-valued function f.

Let $I \subseteq \mathbb{R}$ and $\beta : I \to I$ be first kind or second kind general quantum operator.

4.1 General Quantum Lagrange Interpolation

Let $\mathscr{P}_n$, $n \in \mathbb{N}_0$, denote the set of all polynomials of degree $\leq n$ defined over the set $\mathbb{R}$ of real numbers. Let $n \in \mathbb{N}$ and $x_i \in \mathbb{T}$, $i \in \{0,1,\dots,n\}$, are distinct and y_i, $i \in \{0,1,\dots,n\}$, be real numbers. We will find $p_n \in \mathscr{P}_n$ such that $p_n(x_i) = y_i$, $i \in \{0,1,\dots,n\}$.

Theorem 4.1.1 *Suppose that $n \in \mathbb{N}$. Then there exist polynomials $L_k \in \mathscr{P}_n$, $k \in \{0,1,\dots,n\}$, such that*

$$L_k(x_i) = \begin{cases} 1 & if \quad i = k \\ 0 & if \quad i \neq k, \end{cases}$$

$i,k \in \{0,1,\dots,n\}$. Moreover,

$$p_n(x) = \sum_{k=0}^{n} L_k(x) y_k$$

satisfies the condition $p_n(x_i) = y_i$, $i \in \{0,1,\dots,n\}$, $p_n \in \mathscr{P}_n$.

Proof 4.1.2 *Define*

$$L_k(x) = C_k \sum_{i=0, i\neq k}^{n} (x - x_i),$$

DOI: 10.1201/9781003472131-4

where $C_k \in \mathbb{R}$, $k \in \{0,1,\ldots,n\}$, *will be determined below. We have* $L_k(x_i) = 0$, $i \in \{0,1,\ldots,n\}$, $i \neq k$, *and*

$$\begin{aligned} L_k(x_k) &= C_k \prod_{i=0, i\neq k}^{n} (x_k - x_i) \\ &= 1, \quad k \in \{0,1,\ldots,n\}. \end{aligned}$$

Thus,

$$C_k = \frac{1}{\prod_{i=0,i\neq k}^{n}(x_k - x_i)}, \quad k \in \{0,1,\ldots,n\},$$

and

$$L_k(x) = \prod_{i=0,i\neq k}^{n} \frac{x - x_i}{x_k - x_i}, \quad k \in \{0,1,\ldots,n\}. \tag{4.1}$$

We have that $L_k \in \mathscr{P}_n$, $k \in \{0,1,\ldots,n\}$, *and* $p_n \in \mathscr{P}_n$. *This completes the proof.*

Theorem 4.1.3 *Assume that* $n \in \mathbb{N}_0$. *Let* $x_i \in \mathbb{T}$, $i \in \{0,1,\ldots,n\}$, *be distinct and* $y_i \in \mathbb{R}$, $i \in \{0,1,\ldots,n\}$. *Then there exists a unique polynomial* $p_n \in \mathscr{P}_n$ *such that*

$$p_n(x_i) = y_i, \quad i \in \{0,1,\ldots,n\}.$$

Proof 4.1.4 *The existence of the polynomial* p_n *follows by Theorem 4.1.1. Suppose that there exist two polynomials* $p_n, q_n \in \mathscr{P}_n$ *such that*

$$p_n(x_i) = q_n(x_i) = y_i, \quad i \in \{0,1,\ldots,n\}.$$

Then the polynomial $h_n = p_n - q_n$ *has* $n+1$ *distinct roots. Therefore,* $h_n \equiv 0$ *or* $p_n \equiv q_n$. *This completes the proof.*

Definition 4.1.5 *Assume that* $n \in \mathbb{N}_0$. *Let* $x_i \in \mathbb{T}$, $i \in \{0,1,\ldots,n\}$, *be distinct and* $y_i \in \mathbb{R}$, $i \in \{0,1,\ldots,n\}$. *The polynomial*

$$p_n(x) = \sum_{k=0}^{n} L_k(x) y_k,$$

where L_k, $k \in \{0,1,\ldots,n\}$, *are defined with (4.1), will be called the Lagrange interpolation polynomial of degree n with interpolation points* x_i, $i \in \{0,1,\ldots,n\}$.

Definition 4.1.6 *Assume that* $n \in \mathbb{N}_0$. *Let* $x_i \in [a,b] \subset \mathbb{T}$, $i \in \{0,1,\ldots,n\}$, *be distinct and* $f : [a,b] \to \mathbb{R}$ *be a given function. The polynomial*

$$p_n(x) = \sum_{k=0}^{n} L_k(x) f(x_k),$$

where L_k, $k \in \{0,1,\ldots,n\}$, *are defined with (4.1), will be called the general quantum Lagrange interpolation polynomial of degree n with interpolation points* x_i, $i \in \{0,1,\ldots,n\}$, *for the function* f.

Example 4.1.7 *Let $I = [-3, \infty)$ and $\beta(x) = 2x+3$, $x \in I$. Then $s_0 = -3$. We will construct the general quantum Lagrange interpolation polynomial for the set*

$$\{(-3,0),(-2,2),(0,1),(1,0)\}.$$

Here

$$\begin{aligned} x_0 &= -3, \quad x_1 = -2, \quad x_2 = 0, \quad x_3 = 1, \\ y_0 &= 0, \quad y_1 = 2, \quad y_2 = 1, \quad y_3 = 1, \quad n = 3. \end{aligned}$$

Then

$$\begin{aligned} L_0(x) &= \prod_{i=1}^{3} \frac{x - x_i}{x_0 - x_i} \\ &= \frac{(x-x_1)(x-x_2)(x-x_3)}{(x_0-x_1)(x_0-x_2)(x_0-x_3)} \\ &= \frac{(x+2)x(x-1)}{-1 \cdot (-3) \cdot (-4)} \\ &= -\frac{(x-1)x(x+2)}{12}, \\ L_1(x) &= \prod_{i=0, i \neq 1}^{3} \frac{x - x_i}{x_1 - x_i} \\ &= \frac{(x-x_0)(x-x_2)(x-x_3)}{(x_1-x_0)(x_1-x_2)(x_1-x_3)} \\ &= \frac{(x+3)x(x-1)}{1 \cdot (-2) \cdot (-3)} \\ &= \frac{(x-1)x(x+3)}{6}, \\ L_2(x) &= \prod_{i=0, i \neq 2}^{3} \frac{x - x_i}{x_2 - x_i} \\ &= \frac{(x-x_0)(x-x_1)(x-x_3)}{(x_2-x_0)(x_2-x_1)(x_2-x_3)} \\ &= \frac{(x+3)(x+2)(x-1)}{3 \cdot 2 \cdot (-1)} \\ &= -\frac{(x-1)(x+2)(x+3)}{6}, \end{aligned}$$

$$\begin{aligned} L_3(x) &= \prod_{i=0}^{2} \frac{x-x_i}{x_3-x_i} \\ &= \frac{(x-x_0)(x-x_1)(x-x_2)}{(x_3-x_0)(x_3-x_1)(x_3-x_2)} \\ &= \frac{(x+3)(x+2)x}{4\cdot 3\cdot 1} \\ &= \frac{x(x+2)(x+3)}{12}, \quad x \in \mathbb{T}. \end{aligned}$$

Hence,

$$\begin{aligned} p_3(x) &= y_0L_0(x)+y_1L_1(x)+y_2L_2(x)+y_3L_3(x) \\ &= 2\left(\frac{x(x-1)(x+3)}{6}\right) - \frac{(x-1)(x+2)(x+3)}{6} \\ &\quad + \frac{x(x+2)(x+3)}{12} \\ &= \frac{(x-1)(x+3)(2x-x-2)}{6} + \frac{x(x+2)(x+3)}{12} \\ &= \frac{(x-2)(x-1)(x+3)}{6} + \frac{x(x+2)(x+3)}{12} \\ &= \frac{(x+3)(2(x-1)(x-2)+x(x+2))}{12} \\ &= \frac{(x+3)(2x^2-6x+4+x^2+2x)}{12} \\ &= \frac{(x+3)(3x^2-4x+4)}{12}, \quad x \in \mathbb{T}. \end{aligned}$$

Exercise 4.1.8 *Let* $I=[1,\infty)$ *and* $\beta(x)=\frac{3}{4}x+\frac{1}{2}$, $x \in I$. *Construct the general quantum Lagrange interpolation polynomial for the set*

$$\{(1,-1),(4,0),(8,1),(16,2)\}.$$

Definition 4.1.9 *Suppose that* $n \in \mathbb{N}_0$ *and* $x_j \in I$, $j \in \{0,1,\ldots,n\}$, *are distinct points. For* $x \in I$, *define the polynomials*

$$\pi_{n+1}(x) = \prod_{j=0}^{n}(x-x_j), \quad \Pi_{n+1}^k(x) = D_\beta^k \pi_{n+1}(x), \quad k \in \mathbb{N}_0.$$

Example 4.1.10 *Let* $I = [1, \infty)$, $\beta(x) = 2x - 1$, $x \in I$, $x_0 = 1$, $x_0 = 2$. *Here*

$$n = 1, \quad s_0 = 1.$$

Then

$$\begin{aligned}\pi_2(x) &= (x - x_0)(x - x_1)\\ &= (x-1)(x-2)\\ &= x^2 - 3x + 2, \quad x \in I,\end{aligned}$$

and

$$\begin{aligned}\Pi_2^1(x) &= D_\beta \pi_2(x)\\ &= \beta(x) + x - 3\\ &= 2x - 1 + x - 3\\ &= 3x - 4,\\ \Pi_2^2(x) &= D_\beta^2 \pi_2(x)\\ &= 3, \quad x \in \mathbb{T}.\end{aligned}$$

Example 4.1.11 *Let* $I = [1, \infty)$, $\beta(x) = 3x - 2$, $x \in I$, *and*

$$x_0 = 1, \quad x_1 = 2, \quad x_2 = 3.$$

Then

$$\begin{aligned}\pi_3(x) &= (x - x_0)(x - x_1)(x - x_2)\\ &= (x-1)(x-2)(x-3)\\ &= (x^2 - 3x + 2)(x - 3)\\ &= x^3 - 3x^2 - 3x^2 + 9x + 2x - 6\end{aligned}$$

$$= \quad x^3 - 6x^2 + 11x - 6, \quad x \in I.$$

We have

$$s_0 = 1,$$

and

$$\begin{aligned}
\Pi_3^1(x) &= D_\beta \pi_3(x) \\
&= (\beta(x))^2 + x\beta(x) + x^2 - 6(\beta(x) + x) + 11 \\
&= (3x-2)^2 + x(3x-2) + x^2 - 6(3x-2+x) + 11 \\
&= 9x^2 - 12x + 4 + 3x^2 - 2x + x^2 - 24x + 12 + 11 \\
&= 13x^2 - 38x + 27, \\
\Pi_3^2(x) &= D_\beta^2 \pi_3(x) \\
&= 13(\beta(x) + x) - 38 \\
&= 13(3x - 2 + x) - 38 \\
&= 39x - 26 - 38 \\
&= 39x - 64, \\
\Pi_3^3(x) &= D_\beta^3 \pi_3(x) \\
&= 39, \quad x \in I.
\end{aligned}$$

Exercise 4.1.12 *Let* $I = \left[\frac{1}{64}, \infty\right)$, $\beta(x) = \frac{1}{2}x + \frac{1}{32}$, $x \in I$, *and*

$$x_0 = \frac{1}{64}, \quad x_1 = \frac{1}{16}, \quad x_2 = \frac{1}{4}, \quad x_3 = 1.$$

Find $\pi_4(x)$, $\Pi_4^2(x)$, $\Pi_4^4(x)$, $x \in \mathbb{T}$.

Theorem 4.1.13 *Suppose that $n \in \mathbb{N}_0$, $a,b \in I$, $a < b$, $x_j \in [a,b]$, $j \in \{0,1,\dots,n\}$, are distinct and $f : [a,b] \to \mathbb{R}$, $D_\beta^k f(x)$ exist for any $x \in [a,b]$ and for any $k \in \{1,\dots,n+1\}$. Then for any $x \in [a,b]$ there exists $\xi = \xi(x) \in (a,b)$ such that*

$$f(x) - p_n(x) = \frac{D_\beta^{n+1} f(\xi)}{\Pi_{n+1}^{n+1}(\xi)} \pi_{n+1}(x)$$

or

$$F_{min,n+1}(\xi) \le \frac{f(x) - p_n(x)}{\pi_{n+1}(x)} \le F_{max,n+1}(\xi),$$

where

$$\begin{aligned} F_{max,n+1}(\xi) &= \max\left\{ \frac{D_\beta^{n+1} f(\xi)}{\Pi_{n+1}^{n+1}(\xi)}, \frac{D_\beta^{n+1} f(\beta^{-1}(\xi))}{\Pi_{n+1}^{n+1}(\beta^{-1}(\xi))} \right\}, \\ F_{min,n+1}(\xi) &= \min\left\{ \frac{D_\beta^{n+1} f(\xi)}{\Pi_{n+1}^{n+1}(\xi)}, \frac{D_\beta^{n+1} f(\beta^{-1}(\xi))}{\Pi_{n+1}^{n+1}(\beta^{-1}(\xi))} \right\}. \end{aligned}$$

Proof 4.1.14 *Let p_n be the general quantum Lagrange interpolation polynomial for the function f with interpolation points x_j, $j \in \{0,1,\dots,n\}$. Define the function*

$$\phi(t) = f(t) - p_n(t) - \frac{f(x) - p_n(x)}{\pi_{n+1}(x)} \pi_{n+1}(t), \quad t \in [a,b].$$

Then

$$\begin{aligned} \phi(x_j) &= f(x_j) - p_n(x_j) - \frac{f(x) - p_n(x)}{\pi_{n+1}(x)} \pi_{n+1}(x_j) \\ &= f(x_j) - f(x_j) \\ &= 0, \quad j \in \{0,1,\dots,n\}, \end{aligned}$$

and $\phi(x) = 0$. Thus, $\phi : [a,b] \to \mathbb{R}$ has at least $n+2$ general zeros (GZs). Hence and the β-Rolle theorem, it follows that $D_\beta^{n+1}\phi$ has at least one GZ on (a,b). Therefore, there exists an $\xi = \xi(x) \in (a,b)$ such that

$$D_\beta^{n+1}\phi(\xi) = 0 \quad or \quad D_\beta^{n+1}\phi(\beta^{-1}(\xi)) D_\beta^{n+1}\phi(\xi) < 0.$$

Note that

$$D_\beta^{n+1}\phi(t) = D_\beta^{n+1} f(t) - \frac{f(x) - p_n(x)}{\pi_{n+1}(x)} D_\beta^{n+1}\pi_{n+1}(t), \quad t \in [a,b].$$

1. *Let $D_\beta^{n+1}\phi = 0$. Then*

$$D_\beta^{n+1} f(\xi) = \frac{f(x) - p_n(x)}{\pi_{n+1}(x)} D_\beta^{n+1}\pi_{n+1}(\xi)$$

or

$$\begin{aligned} f(x) - p_n(x) &= \frac{D_\beta^{n+1} f(\xi)}{D_\beta^{n+1} \pi_{n+1}(\xi)} \pi_{n+1}(x) \\ &= \frac{D_\beta^{n+1} f(\xi)}{\Pi_{n+1}^{n+1}(\xi)} \pi_{n+1}(x). \end{aligned}$$

2. *Let*

$$D_\beta^{n+1} \phi(\beta^{-1}(\xi)) D_\beta^{n+1} \phi(\xi) < 0.$$

Then

$$\begin{aligned} D_\beta^{n+1} \phi(\beta^{-1}(\xi)) &= D_\beta^{n+1} f(\beta^{-1}(\xi)) \\ &\quad - \frac{f(x) - p_n(x)}{\pi_{n+1}(x)} D_\beta^{n+1} \pi_{n+1}(\beta^{-1}(\xi)) \\ &= D_\beta^{n+1} f(\beta^{-1}(\xi)) - \frac{f(x) - p_n(x)}{\pi_{n+1}(x)} \Pi_{n+1}^{n+1}(\beta^{-1}(\xi)), \end{aligned}$$

and

$$D_\beta^{n+1} \phi(\xi) = D_\beta^{n+1} f(\xi) - \frac{f(x) - p_n(x)}{\pi_{n+1}(x)} \Pi_{n+1}^{n+1}(\xi).$$

Hence,

$$\begin{aligned} 0 &> D_\beta^{n+1} \phi(\beta^{-1}(\xi)) D_\beta^{n+1} \phi(\xi) \\ &= \left(D_\beta^{n+1} f(\beta^{-1}(\xi)) - \frac{f(x) - p_n(x)}{\pi_{n+1}(x)} \Pi_{n+1}^{n+1}(\beta^{-1}(\xi)) \right) \\ &\quad \times \left(D_\beta^{n+1} f(\xi) - \frac{f(x) - p_n(x)}{\pi_{n+1}(x)} \Pi_{n+1}^{n+1}(\xi) \right) \\ &= \left(\frac{f(x) - p_n(x)}{\pi_{n+1}(x)} \right)^2 \Pi_{n+1}^{n+1}(\beta^{-1}(\xi)) \Pi_{n+1}^{n+1}(\xi) \\ &\quad - \frac{f(x) - p_n(x)}{\pi_{n+1}(x)} \left(\Pi_{n+1}^{n+1}(\beta^{-1}(\xi)) D_\beta^{n+1} f(\xi) \right. \\ &\quad \left. + \Pi_{n+1}^{n+1}(\xi) D_\beta^{n+1} f(\beta^{-1}(\xi)) \right) \\ &\quad + D_\beta^{n+1} f(\beta^{-1}(\xi)) D_\beta^{n+1} f(\xi). \end{aligned}$$

Hence,

$$F_{min,n+1}(\xi) \le \frac{f(x) - p_n(x)}{\pi_{n+1}(x)} \le F_{max,n+1}(\xi).$$

This completes the proof.

Remark 4.1.15 *Suppose that all conditions of Theorem 4.1.13 hold. If*

$$\lim_{n\to\infty} \max_{x\in[a,b]} \left(\frac{D_\beta^{n+1} f(\xi)}{\Pi_{n+1}^{n+1}(\xi)} \pi_{n+1}(x) \right) = 0$$

and

$$\lim_{n\to\infty} \max_{x\in[a,b]} \left(\frac{D_\beta^{n+1} f(\beta^{-1}(\xi))}{\Pi_{n+1}^{n+1}(\beta^{-1}(\xi))} \pi_{n+1}(x) \right) = 0,$$

then

$$\lim_{n\to\infty} \max_{x\in[a,b]} |f(x) - p_n(x)| = 0.$$

4.2 β-Lagrange Interpolation

In this section, we will show that a given function can be approximated with so called β-Lagrange polynomials. We will show that there are classes of first kind general quantum operators for which the Lagrange interpolation polynomials and the β-Lagrange interpolation polynomials are different and there are classes of first kind general quantum operators for which the Lagrange interpolation polynomials and the β-Lagrange interpolation polynomials coincide. With $\mathscr{P}_n^\beta$, $n \in \mathbb{N}_0$, we will denote the set of all polynomials in the form

$$a_n(\beta(x))^n + a_{n-1}(\beta(x))^{n-1} + \cdots + a_1(x)\beta(x) + a_0,$$

where $a_j \in \mathbb{R}$, $j \in \{0,1,\ldots,n\}$. Let $a, b \in I$, $a < b$.

Example 4.2.1 *Let $I = [-1,\infty)$, $\beta(x) = 3x - 1$, $x \in I$, and $a = -1$, $b = 5$. Take the points*

$$x_0 = -1, \quad x_1 = 1, \quad x_2 = 3.$$

Then

$$\beta(x_0) = -4, \quad \beta(x_1) = 2, \quad \beta(x_2) = 8.$$

Example 4.2.2 *Let $I = [1,\infty)$, $\beta(x) = \frac{1}{5}x + 1$, $x \in I$, $a = 1$, $b = 16$. Take the points*

$$x_0 = 1, \quad x_1 = 2, \quad x_2 = 4.$$

Then

$$\beta(x_0) = \frac{6}{4}, \quad \beta(x_1) = \frac{7}{5}, \quad \beta(x_2) = \frac{9}{5}.$$

As in the previous section, one can prove the following result.

Theorem 4.2.3 *Suppose that $n \in \mathbb{N}$ and $x_j \in I$, $j \in \{0,1,\ldots,n\}$, are distinct. Then there exist unique polynomials $L_{\beta k} \in \mathscr{P}_n^\beta$, $k \in \{0,1,\ldots,n\}$, such that*

$$L_{\beta k}(x_i) = \begin{cases} 1 & \text{if} \quad i = k \\ 0 & \text{if} \quad i \neq k, \end{cases}$$

$i,k \in \{0,1,\ldots,n\}$. Moreover,

$$\begin{aligned} p_{\beta n}(x) &= \sum_{k=0}^{n} L_{\beta k}(x) y_k \\ &= \sum_{k=0}^{n} \left(\prod_{j=0, j\neq k}^{n} \frac{\beta(x) - \beta(x_j)}{\beta(x_k) - \beta(x_j)} \right) y_k. \end{aligned}$$

satisfies the condition $p_{\beta n}(x_i) = y_i$, $i \in \{0,1,\ldots,n\}$, $p_{\beta n} \in \mathscr{P}_n^\beta$.

Definition 4.2.4 *Assume that $n \in \mathbb{N}_0$. Let $x_i \in I$, $i \in \{0,1,\ldots,n\}$, be distinct and $y_i \in \mathbb{R}$, $i \in \{0,1,\ldots,n\}$. The polynomial*

$$p_{\beta n}(x) = \sum_{k=0}^{n} L_{\beta k}(x) y_k,$$

where $L_{\beta k}$, $k \in \{0,1,\ldots,n\}$, are defined in Theorem 4.2.3, will be called the β-Lagrange interpolation polynomial of degree n with interpolation points x_i, $i \in \{0,1,\ldots,n\}$.

Definition 4.2.5 *Assume that $n \in \mathbb{N}_0$. Let $x_i \in [a,b] \subset I$, $i \in \{0,1,\ldots,n\}$, be distinct and $f : [a,b] \to \mathbb{R}$ be a given function. The polynomial*

$$p_{\beta n}(x) = \sum_{k=0}^{n} L_{\beta k}(x) f(x_k),$$

where $L_{\beta k}$, $k \in \{0,1,\ldots,n\}$, are defined in Theorem 4.2.3, will be called the β-Lagrange interpolation polynomial of degree n with interpolation points x_i, $i \in \{0,1,\ldots,n\}$, for the function f.

Example 4.2.6 *Let $I = \{-2,-1,0,3,7\}$ and $\beta : I \to \mathbb{R}$ be a first kind general quantum operator so that*

$$\beta(-2) = -1, \quad \beta(-1) = 0, \quad \beta(0) = 3, \quad \beta(3) = 4, \quad \beta(7) = 7.$$

$$a = -2, \quad b = 7, \quad x_0 = -2, \quad x_1 = 0,$$

$f : I \to \mathbb{R}$ be defined by

$$f(t) = t + 3, \quad t \in I.$$

We will find the β-Lagrange interpolation polynomial for the function f with interpolation points x_0, x_1. We have

$$\begin{aligned}
\beta(x_0) &= \beta(-2) = -1, \\
\beta(x_1) &= \beta(0) = 3, \\
L_{\beta 0}(x) &= \frac{\beta(x) - \beta(x_1)}{\beta(x_0) - \beta(x_1)} \\
&= \frac{\beta(x) + 1}{-1 - 3} \\
&= -\frac{1}{4}(\beta(x) + 1), \\
L_{\beta 1}(x) &= \frac{\beta(x) - \beta(x_0)}{\beta(x_1) - \beta(x_0)} \\
&= \frac{\beta(x) + 1}{3 + 1} \\
&= \frac{1}{4}(\beta(x) + 1), \\
f(x_0) &= f(-2) \\
&= -2 + 3 \\
&= 1, \\
f(x_1) &= f(0) \\
&= 0 + 3 \\
&= 3.
\end{aligned}$$

Thus,

$$\begin{aligned}
p_{\beta 1}(x) &= f(x_0)L_{\beta 0}(x) + f(x_1)L_{\beta 1}(x) \\
&= -\frac{1}{4}(\beta(x) + 1) + \frac{3}{4}(\beta(x) + 1)
\end{aligned}$$

$$= \frac{1}{4}(3\beta(x)+3-\beta(x)-1)$$

$$= \frac{1}{2}(\beta(x)+1), \quad x \in I.$$

Next,

$$\begin{aligned} p_{\beta 1}(0) &= \frac{1}{2}(\beta(0)+1) \\ &= \frac{1}{2}(3+1) \\ &= 2, \\ p_{\beta 1}(-1) &= \frac{1}{2}(\beta(-1)+1) \\ &= \frac{1}{2}(0+1) \\ &= \frac{1}{2}. \end{aligned}$$

Now, we will find the Lagrange interpolation polynomial for the function f with interpolation points x_0, x_1. We have

$$\begin{aligned} L_0(x) &= \frac{x-x_1}{x_0-x_1} \\ &= -\frac{1}{2}x, \\ L_1(x) &= \frac{x-x_0}{x_1-x_0} \\ &= \frac{x+2}{2}, \quad x \in I. \end{aligned}$$

Therefore

$$\begin{aligned} p_1(x) &= f(x_0)L_0(x)+f_1(x)L_1(x) \\ &= -\frac{1}{2}x+3\frac{x+2}{2} \\ &= \frac{3x+6-x}{2} \end{aligned}$$

$$= \quad x+3, \quad x \in I.$$

Then

$$p_1(0) = 3, \quad p_1(-1) = 2.$$

We see that

$$p_1(0) \neq p_{\beta 1}(0), \quad p_1(-1) \neq p_{\beta 1}(-1).$$

Remark 4.2.7 *By the above example we see that there are cases for which the Lagrange interpolation polynomial and the β-Lagrange interpolation polynomial for a function f are different.*

Exercise 4.2.8 *Let $I = [-1, 7]$, $\beta(x) = \frac{1}{10}x + 1$, $x \in I$,*

$$a = -1, \quad b = 7, \quad x_1 = -\frac{1}{8}, \quad x_2 = 3.$$

Find the β-Lagrange and Lagrange interpolation polynomials for the function $f : I \to \mathbb{R}$ defined by

$$f(t) = \frac{t+1}{t^2 - t + 1} + 3t, \quad t \in [-1, 7],$$

with interpolation and interpolation points x_0, x_1, x_2.

Example 4.2.9 *Let $I = \mathbb{R}$, $\beta(x) = \frac{1}{2}x - 2$, $x \in I$,*

$$a = x_0 = -1, \quad x_1 = 1, \quad x_2 = 3,$$

$f : I \to \mathbb{R}$ be defined by

$$f(t) = t^2 + t + 1, \quad t \in I.$$

We will find β-Lagrange interpolation polynomial with interpolation points x_0, x_1, x_2. We have

$$\begin{aligned} \beta(x_0) &= \beta(-1) \\ &= -\frac{1}{2} - 2 \\ &= -\frac{5}{2}, \\ \beta(x_1) &= \beta(1) \\ &= \frac{1}{2} - 2 \\ &= -\frac{3}{2}, \end{aligned}$$

$$\begin{aligned}\beta(x_2) &= \beta(3)\\ &= \frac{3}{2}-2\\ &= -\frac{1}{2}.\end{aligned}$$

Then

$$\begin{aligned}L_{\beta 0}(x) &= \frac{(\beta(x)-\beta(x_1))(\beta(x)-\beta(x_2))}{(\beta(x_0)-\beta(x_1))(\beta(x_0)-\beta(x_2))}\\ &= \frac{\left(\frac{1}{2}x-2+\frac{3}{2}\right)\left(\frac{1}{2}x-2+\frac{1}{2}\right)}{\left(-\frac{5}{2}+\frac{3}{2}\right)\left(-\frac{5}{2}+\frac{1}{2}\right)}\\ &= \frac{\left(\frac{1}{2}x-\frac{1}{2}\right)\left(\frac{1}{2}x-\frac{3}{2}\right)}{-1\cdot(-2)}\\ &= \frac{(x-1)(x-3)}{8},\\ L_{\beta 1}(x) &= \frac{(\beta(x)-\beta(x_0))(\beta(x)-\beta(x_2))}{(\beta(x_1)-\beta(x_0))(\beta(x_1)-\beta(x_2))}\\ &= \frac{\left(\frac{1}{2}x-2+\frac{5}{2}\right)\left(\frac{1}{2}x-2+\frac{1}{2}\right)}{\left(-\frac{3}{2}+\frac{5}{2}\right)\left(-\frac{3}{2}+\frac{1}{2}\right)}\\ &= \frac{\left(\frac{1}{2}x+\frac{1}{2}\right)\left(\frac{1}{2}x-\frac{1}{2}\right)}{1\cdot(-1)}\\ &= -\frac{(x+1)(x-3)}{4},\\ L_{\beta 2}(x) &= \frac{(\beta(x)-\beta(x_0))(\beta(x)-\beta(x_1))}{(\beta(x_2)-\beta(x_0))(\beta(x_2)-\beta(x_0))}\\ &= \frac{\left(\frac{1}{2}x-2+\frac{5}{2}\right)\left(\frac{1}{2}x-2+\frac{3}{2}\right)}{\left(-\frac{1}{2}+\frac{5}{2}\right)\left(-\frac{1}{2}+\frac{3}{2}\right)}\\ &= \frac{\left(\frac{1}{2}x+\frac{1}{2}\right)\left(\frac{1}{2}x-\frac{1}{2}\right)}{2\cdot 1}\\ &= \frac{(x-1)(x+1)}{8}, \quad x\in I,\end{aligned}$$

$$\begin{aligned}
f(x_0) &= f(-1) \\
&= (-1)^2 - 1 + 1 \\
&= 1, \\
f(x_1) &= f(1) \\
&= 1^2 + 1 + 1 \\
&= 3, \\
f(x_2) &= f(3) \\
&= 3^2 + 3 + 1 \\
&= 13.
\end{aligned}$$

Hence,

$$\begin{aligned}
p_{\beta 2}(x) &= f(x_0)L_{\beta 0}(x) + f(x_1)L_{\beta 1}(x) + f(x_2)L_{\beta 2}(x) \\
&= 1 \cdot \frac{(x-1)(x-3)}{8} - 3 \cdot \frac{(x+1)(x-3)}{4} + 13 \cdot \frac{(x+1)(x-1)}{8} \\
&= \frac{(x-1)(x-3) - 6(x+1)(x-3) + 13(x+1)(x-1)}{8} \\
&= \frac{x^2 - 4x + 3 - 6(x^2 - 2x - 3) + 13(x^2 - 1)}{8} \\
&= \frac{x^2 - 4x + 3 - 6x^2 + 12x + 18 + 13x^2 - 13}{8} \\
&= \frac{8x^2 + 8x + 8}{8} \\
&= x^2 + x + 1, \quad x \in I.
\end{aligned}$$

is the β-Lagrange interpolation polynomial for the function f. Now, we will find the Lagrange interpolation polynomial for the function f with interpolation points x_0,

x_1, x_2. We have

$$\begin{aligned} L_0(x) &= \frac{(x-x_1)(x-x_2)}{(x_0-x_1)(x_0-x_2)} \\ &= \frac{(x-1)(x-3)}{(-1-1)(-1-3)} \\ &= \frac{(x-1)(x-3)}{8}, \\ L_1(x) &= \frac{(x-x_0)(x-x_2)}{(x_1-x_0)(x_1-x_2)} \\ &= \frac{(x+1)(x-3)}{(1+1)(1-3)} \\ &= -\frac{(x+1)(x-3)}{4}, \\ L_2(x) &= \frac{(x-x_0)(x-x_1)}{(x_2-x_0)(x_2-x_1)} \\ &= \frac{(x+1)(x-1)}{(3+1)(3-1)} \\ &= \frac{(x+1)(x-1)}{8}, \quad x \in I. \end{aligned}$$

Hence,

$$\begin{aligned} p_2(x) &= f(x_0)L_0(x) + f(x_1)L_1(x) + f(x_2)L_2(x) \\ &= \frac{(x-1)(x-3)}{8} - 3\frac{(x+1)(x-3)}{4} + 13\frac{(x+1)(x-1)}{8} \\ &= \frac{x^2-4x+3-6(x^2-2x-3)+13(x^2-1)}{8} \\ &= \frac{x^2-4x+3-6x^2+12x+18+13x^2-13}{8} \\ &= \frac{8x^2+8x+8}{8} \\ &= x^2+x+1, \quad x \in I. \end{aligned}$$

Consequently

$$p_2(x) = p_{\beta 2}(x), \quad x \in I.$$

Remark 4.2.10 *By the above example we see that there are cases for which the Lagrange interpolation polynomial and the β-Lagrange interpolation polynomial for a function f coincide.*

Exercise 4.2.11 *Let $I = \mathbb{R}$, $\beta(x) = \frac{1}{15}x$, $x \in I$,*

$$a = 1 = x_0, \quad x_1 = 2, \quad x_2 = 8, \quad x_3 = 32 = b,$$

$f : I \to \mathbb{R}$ is defined by

$$f(t) = t^4 - 3t^2 + 2t + 1, \quad t \in I.$$

Find the β-Lagrange interpolation polynomial for the function f with interpolation points x_0, x_1, x_2.

Now, we will describe some classes of first kind general quantum operators for which the β-Lagrange and Lagrange interpolation polynomials coincide.

Theorem 4.2.12 *Let $I = \mathbb{R}$, $\beta(t) = ct$ for any $t \in I$ and for some constant $c \in (0,1)$, or $\beta(t) = at + d$ for any $t \in I$ and for some constants $a \in (0,1)$ and $d > 0$. Let also, $n \in \mathbb{N}$ and*

$$a_1 = x_0 < x_1 < \ldots < x_n = b_1, \quad x_j \in I, \quad j \in \{0,1,\ldots,n\},$$

be interpolation points. Then

$$L_k(x) = L_{\beta k}(x), \quad x \in [a_1, b_1], \quad k \in \{0,1,\ldots,n\}.$$

Proof 4.2.13 *1. Let $\beta(t) = ct$, $c \in (0,1)$, for any $t \in I$. Then*

$$\begin{aligned} L_{\beta k}(x) &= \prod_{j=0, j\neq k}^{n} \frac{\beta(x) - \beta(x_j)}{\beta(x_k) - \beta(x_j)} \\ &= \prod_{j=0, j\neq k}^{n} \frac{cx - cx_j}{cx_k - cx_j} \\ &= \prod_{j=0, j\neq k}^{n} \frac{x - x_j}{x_k - x_j} \\ &= L_k(x), \quad k \in \{0,1,\ldots,n\}, \quad x \in [a_1, b_1]. \end{aligned}$$

2. Let $\beta(t) = at + d$ for any $t \in I$, $a \in (0,1)$, $d \neq 0$. Then

$$\begin{aligned} L_{\beta k}(x) &= \prod_{j=0, j\neq k}^{n} \frac{\beta(x) - \beta(x_j)}{\beta(x_k) - \beta(x_j)} \\ &= \prod_{j=0, j\neq k}^{n} \frac{ax + d - (ax_j + d)}{ax_k + d - (ax_j + d)} \end{aligned}$$

$$\begin{aligned} &= \prod_{j=0, j\neq k}^{n} \frac{x-x_j}{x_k - x_j} \\ &= L_k(x), \quad k \in \{0,1,\ldots,n\}, \quad x \in [a_1, b_1]. \end{aligned}$$

This completes the proof.

Definition 4.2.14 *Suppose that $n \in \mathbb{N}_0$ and $x_j \in I$, $j \in \{0,1,\ldots,n\}$, are distinct points. For $x \in I$, define the polynomials*

$$\pi_{\beta n+1}(x) = \prod_{j=0}^{n} (\beta(x) - \beta(x_j)), \quad \Pi^k_{\beta n+1}(x) = D^k_\beta \pi_{\beta n+1}(x), \quad k \in \mathbb{N}_0.$$

Example 4.2.15 *Let $I = [0,\infty)$ and $\beta(x) = \frac{1}{2}x + 1$, $x \in I$. Take*

$$a = 0, \quad b = 1, \quad x_0 = 0, \quad x_1 = \frac{1}{3}.$$

Then

$$\begin{aligned} \beta(x_0) &= \beta(0) \\ &= 1, \\ \beta(x_1) &= \beta\left(\frac{1}{3}\right) \\ &= \frac{1}{2}\cdot\frac{1}{3} + 1 \\ &= \frac{1}{6} + 1 \\ &= \frac{7}{6}, \\ \beta\left(\frac{1}{2}\right) &= \frac{1}{2}\cdot\frac{1}{2} + 1 \\ &= \frac{1}{4} + 1 \\ &= \frac{5}{4}, \end{aligned}$$

$$\begin{aligned}\beta(1) &= \frac{1}{2}\cdot 1+1\\ &= \frac{1}{2}+1\\ &= \frac{3}{2}.\end{aligned}$$

Then

$$\begin{aligned}\pi_{\beta 2}(x) &= (\beta(x)-\beta(x_0))(\beta(x)-\beta(x_1))\\ &= \left(\frac{1}{2}x+1-1\right)\left(\frac{1}{2}x+1-\frac{7}{6}\right)\\ &= \frac{1}{2}x\left(\frac{1}{2}x-\frac{1}{6}\right)\\ &= \frac{1}{12}x(3x-1), \quad x\in I,\end{aligned}$$

and

$$\begin{aligned}\pi_{\beta 2}\left(\frac{1}{2}\right) &= \frac{1}{12}\cdot\frac{1}{2}\cdot\left(3\cdot\frac{1}{2}-1\right)\\ &= \frac{1}{24}\left(\frac{3}{2}-1\right)\\ &= \frac{1}{24}\cdot\frac{1}{2}\\ &= \frac{1}{48}.\end{aligned}$$

Next,

$$\begin{aligned}\Pi^1_{\beta 2}(x) &= D_\beta \pi_{\beta 2}(x)\\ &= D_\beta\left(\frac{1}{12}(3x^2-x)\right)\\ &= \frac{1}{12}(3(\beta(x)+x)-1)\\ &= \frac{1}{12}\left(3\left(\frac{1}{2}x+1+x\right)-1\right)\end{aligned}$$

$$
\begin{aligned}
&= \frac{1}{12}\left(3\left(\frac{3}{2}x+1\right)-1\right)\\
&= \frac{1}{12}\left(\frac{9}{2}x+3-1\right)\\
&= \frac{1}{12}\left(\frac{9}{2}x+2\right)\\
&= \frac{1}{24}(9x+4), \quad x\in I, \quad x\neq 2.
\end{aligned}
$$

Hence,

$$
\begin{aligned}
\Pi^1_{\beta 2}\left(\frac{1}{2}\right) &= \frac{1}{24}\left(9\cdot\frac{1}{2}+4\right)\\
&= \frac{1}{24}\left(\frac{9+8}{2}\right)\\
&= \frac{17}{48}.
\end{aligned}
$$

Exercise 4.2.16 *Let* $I=[0,\infty)$, $\beta(x)=\frac{3}{8}x+\frac{1}{3}$, $x\in I$, *and*

$$a=0, \quad b=18, x_0=0, \quad x_1=3, \quad x_2=9.$$

Find $\Pi^2_{\beta 3}(3)$.

Theorem 4.2.17 *Suppose that* $n\in\mathbb{N}_0$, $a,b\in I$, $a<b$, $x_j\in[a,b]$, $j\in\{0,1,\ldots,n\}$, *are distinct and* $f:[a,b]\to\mathbb{R}$, $D^k_\beta f(x)$ *exist for any* $x\in[a,b]$ *and for any* $k\in\{1,\ldots,n+1\}$. *Then for any* $x\in[a,b]$ *there exists* $\xi=\xi(x)\in(a,b)$ *such that*

$$f(x)-p_{\beta n}(x)=\frac{D^{n+1}_\beta f(\xi)}{\Pi^{n+1}_{\beta n+1}(\xi)}\pi_{\beta n+1}(x)$$

or

$$F_{\beta min,n+1}(\xi)\leq\frac{f(x)-p_{\beta n}(x)}{\pi_{\beta n+1}(x)}\leq F_{\beta max,n+1}(\xi),$$

where

$$
\begin{aligned}
F_{\beta max,n+1}(\xi) &= \max\left\{\frac{D^{n+1}_\beta f(\xi)}{\Pi^{n+1}_{\beta n+1}(\xi)}, \frac{D^{n+1}_\beta f(\beta^{-1}(\xi))}{\Pi^{n+1}_{\beta n+1}(\beta^{-1}(\xi))}\right\},\\
F_{\beta min,n+1}(\xi) &= \min\left\{\frac{D^{n+1}_\beta f(\xi)}{\Pi^{n+1}_{\beta n+1}(\xi)}, \frac{D^{n+1}_\beta f(\beta^{-1}(\xi))}{\Pi^{n+1}_{\beta n+1}(\beta^{-1}(\xi))}\right\}.
\end{aligned}
$$

Proof 4.2.18 *Let $p_{\beta n}$ be the β-Lagrange interpolation polynomial for the function f with interpolation points x_j, $j \in \{0,1,\ldots,n\}$. Define the function*

$$\phi(t) = f(t) - p_{\beta n}(t) - \frac{f(x) - p_{\beta n}(x)}{\pi_{\beta n+1}(x)} \pi_{\beta n+1}(t), \quad t \in [a,b].$$

From here, the proof repeats the corresponding proof of Theorem 4.1.13 and we omit it. This completes the proof.

Remark 4.2.19 *Suppose that all conditions of Theorem 4.2.17 hold. If*

$$\lim_{n\to\infty} \max_{x\in[a,b]} \left(\frac{D_\beta^{n+1} f(\xi)}{\Pi_{\beta n+1}^{n+1}(\xi)} \pi_{\beta n+1}(x) \right) = 0$$

and

$$\lim_{n\to\infty} \max_{x\in[a,b]} \left(\frac{D_\beta^{n+1} f(\beta^{-1}(\xi))}{\Pi_{\beta n+1}^{n+1}(\beta^{-1}(\xi))} \pi_{\beta n+1}(x) \right) = 0,$$

then

$$\lim_{n\to\infty} \max_{x\in[a,b]} |f(x) - p_{\beta n}(x)| = 0.$$

4.3 Hermite Interpolation

The idea of Lagrange interpolation and β-Lagrange interpolation can be generalized in various ways. Here, in this section we consider one simple extension where a polynomial p is required to take given values and β- derivative values at the interpolation points. For given distinct points x_j, $j \in \{0,1,\ldots,n\}$, and two sets of real numbers y_j, z_j, $j \in \{0,1,\ldots,n\}$, $n \in \mathbb{N}_0$, we need to find a polynomial $p_{2n+1} \in \mathscr{P}_{2n+1}$ satisfying the conditions

$$p_{2n+1}(x_j) = y_j, \quad D_\beta p_{2n+1}(x_j) = z_j, \quad j \in \{0,1,\ldots,n\}.$$

The construction of a such polynomial is similar to that of the Lagrange interpolation polynomial and it is given in the following theorem.

Theorem 4.3.1 (Hermite Interpolation Theorem) *Let $n \in \mathbb{N}_0$ and let $a, b \in I$, $a < b$, and $x_j \in I$, $j \in \{0,1,\ldots,n\}$, be distinct. Let also, $y_j, z_j \in \mathbb{R}$, $j \in \{0,1,\ldots,n\}$. Then there exists a unique polynomial $p_{2n+1} \in \mathscr{P}_{2n+1}$ such that*

$$p_{2n+1}(x_j) = y_j, \quad D_\beta p_{2n+1}(x_j) = z_j, \quad j \in \{0,1,\ldots,n\}. \tag{4.2}$$

Proof 4.3.2 *For $x \in [a,b]$, define the polynomial*

$$M_k(x) = \prod_{j=0, j\neq k}^{n} \frac{x-\beta(x_j)}{x_k-\beta(x_j)}, \quad k \in \{0,1,\ldots,n\}.$$

We have

$$M_k(\beta(x_j)) = 0, \quad M_k(x_k) = 1, \quad j,k \in \{0,1,\ldots,n\}, \quad j \neq k.$$

We will search a polynomial $p_{2n+1} \in \mathscr{P}_{2n+1}$ in the following form

$$p_{2n+1}(x) = \sum_{j=0}^{n} (y_j + (x-x_j)(\alpha_j y_j + \beta_j z_j)) M_j(x) L_j(x), \quad x \in [a,b],$$

where α_j, $\beta_j \in \mathbb{R}$, $j \in \{0,1,\ldots,n\}$, will be determined by the conditions (4.2). *We have*

$$\begin{aligned}
p_{2n+1}(x_k) &= \sum_{j=0}^{n} (y_j + (x_k-x_j)(\alpha_j y_j + \beta_j z_j)) M_j(x_k) L_j(x_k) \\
&= y_k, \\
D_\beta p_{2n+1}(x) &= \sum_{j=0}^{n} (\alpha_j y_j + \beta_j z_j) M_j(\beta(x)) L_j(\beta(x)) \\
&\quad + \sum_{j=0}^{n} \Bigg((y_j + (x-x_j)(\alpha_j y_j + \beta_j z_j)) \\
&\quad \times \left(D_\beta M_j(x) L_j(x) + M_j(\beta(x)) D_\beta L_j(x)\right) \Bigg), \\
D_\beta p_{2n+1}(x_k) &= \sum_{j=0}^{n} (\alpha_j y_j + \beta_j z_j) M_j(\beta(x_k)) L_j(\beta(x_k)) \\
&\quad + \sum_{j=0}^{n} \Bigg((y_j + (x_k-x_j)(\alpha_j y_j + \beta_j z_j)) \\
&\quad \times \left(D_\beta M_j(x_k) L_j(x_k) + M_j(\beta(x_k)) D_\beta L_j(x_k)\right) \Bigg) \\
&= (\alpha_k y_k + \beta_k z_k) M_k(\beta(x_k)) L_k(\beta(x_k)) \\
&\quad + y_k \left(D_\beta M_k(x_k) + M_k(\beta(x_k)) D_\beta L_k(x_k)\right) \\
&= z_k
\end{aligned}$$

or

$$(\alpha_k y_k + \beta_k z_k) M_k(\beta(x_k)) L_k(\beta(x_k))$$
$$= z_k - y_k \left(D_\beta M_k(x_k) + M_k(\beta(x_k)) D_\beta L_k(x_k) \right),$$

or

$$\alpha_k y_k + \beta_k z_k = \frac{z_k - y_k \left(D_\beta M_k(x_k) + M_k(\beta(x_k)) D_\beta L_k(x_k) \right)}{M_k(\beta(x_k)) L_k(\beta(x_k))},$$

and

$$p_{2n+1}(x)$$

$$= \sum_{j=0}^{n} \left(y_j + \frac{z_j - y_j \left(D_\beta M_j(x_j) + M_j(\beta(x_j)) D_\beta L_j(x_j) \right)}{M_j(\beta(x_j)) L_j(\beta(x_j))} (x - x_j) \right)$$
$$\times M_j(x) L_j(x)$$
$$= \sum_{j=0}^{n} \left(\left(1 - \frac{D_\beta M_j(x_j) + M_j(\beta(x_j)) D_\beta L_j(x_j)}{M_j(\beta(x_j)) L_j(\beta(x_j))} (x - x_j) \right) y_j \right.$$
$$\left. + \frac{z_j}{M_j(\beta(x_j)) L_j(\beta(x_j))} (x - x_j) \right) M_j(x) L_j(x),$$

$x \in [a,b]$. *Now, suppose that there are two polynomials such that* $p_{2n+1}, q_{2n+1} \in \mathscr{P}_{2n+1}$ *and*

$$p_{2n+1}(x_k) = q_{2n+1}(x_k) = y_k, \quad D_\beta p_{2n+1}(x_k) = D_\beta q_{2n+1}(x_k) = z_k,$$

$k \in \{0, 1, \ldots, n\}$. *Let* $h_{2n+1} = p_{2n+1} - q_{2n+1}$. *Then* $h_{2n+1} \in \mathscr{P}_{2n+1}$ *and it has at least* $2n+2$ *GZs. Thus,*

$$h_{2n+1} \equiv 0 \quad \text{or} \quad p_{2n+1} \equiv q_{2n+1} \quad \text{on} \quad [a,b].$$

This completes the proof.

Definition 4.3.3 *Let* $n \in \mathbb{N}_0$ *and let* a, $b \in I$, $a < b$, *and* $x_j \in I$, $j \in \{0, 1, \ldots, n\}$, *be* β*-distinct. Let also,* $y_j, z_j \in \mathbb{R}$, $j \in \{0, 1, \ldots, n\}$. *Then the polynomial*

$$p_{2n+1}(x) = \sum_{j=0}^{n} \left(\left(1 - \frac{D_\beta M_j(x_j) + M_j(\beta(x_j)) D_\beta L_j(x_j)}{M_j(\beta(x_j)) L_j(\beta(x_j))} (x - x_j) \right) y_j \right.$$
$$\left. + \frac{z_j}{M_j(\beta(x_j)) L_j(\beta(x_j))} (x - x_j) \right) M_j(x) L_j(x),$$

$x \in [a,b]$, *is called Hermite interpolation polynomial for the set of values given in*

$$\{(x_j, y_j, z_j) : j \in \{0, 1, \ldots, n\}\}.$$

Example 4.3.4 *Let* $I = [0, \infty)$, $\beta(x) = \frac{1}{2}x$, $x \in I$, *and*

$$\begin{aligned} x_0 &= 1, \quad x_1 = 4, \\ y_0 &= 1, \quad y_1 = -1, \\ z_0 &= 1, \quad z_1 = 2. \end{aligned}$$

Here $n = 1$. *Then*

$$\begin{aligned} \beta(x_0) &= \beta(1) \\ &= \frac{1}{2}, \\ \beta(x_1) &= \beta(4) \\ &= \frac{1}{2} \cdot 4 \\ &= 2 \end{aligned}$$

and

$$\begin{aligned} L_0(x) &= \frac{x - x_1}{x_0 - x_1} \\ &= \frac{x-4}{1-4} \\ &= -\frac{1}{3}(x-4), \\ D_\beta L_0(x) &= -\frac{1}{3}, \\ L_0(\beta(x_0)) &= L_0\left(\frac{1}{2}\right) \\ &= -\frac{1}{3}\left(\frac{1}{2} - 4\right) \\ &= -\frac{1}{3}\left(-\frac{7}{2}\right) \end{aligned}$$

$$
\begin{aligned}
&= \frac{7}{6}, \\
L_1(x) &= \frac{x - x_0}{x_1 - x_0} \\
&= \frac{x-1}{4-1} \\
&= \frac{1}{3}(x-1), \\
D_\beta L_1(x) &= \frac{1}{3}, \\
L_1(\beta(x_1)) &= L_1(2) \\
&= \frac{1}{3}(2-1) \\
&= \frac{1}{3}, \\
M_0(x) &= \frac{x - \beta(x_1)}{x_0 - \beta(x_1)} \\
&= \frac{x-2}{\frac{1}{2}-2} \\
&= -\frac{2}{3}(x-2), \\
D_\beta M_0(x) &= -\frac{2}{3}, \\
M_0(\beta(x_0)) &= M_0\left(\frac{1}{2}\right) \\
&= -\frac{2}{3}\left(\frac{1}{2}-2\right) \\
&= -\frac{2}{3}\left(-\frac{3}{2}\right) \\
&= 1, \\
M_1(x) &= \frac{x - \beta(x_0)}{x_1 - \beta(x_0)}
\end{aligned}
$$

$$\begin{aligned}
&= \frac{x-\frac{1}{2}}{2-\frac{1}{2}} \\
&= \frac{1}{3}(2x-1), \\
D_\beta M_1(x) &= \frac{2}{3}, \\
M_1(\beta(x_1)) &= M_1(2) \\
&= \frac{1}{3}(2\cdot 2-1) \\
&= 1, \quad x \in I.
\end{aligned}$$

Hence,

$$\begin{aligned}
p_3(x) &= \left(y_0 + \frac{z_0 - y_0\left(D_\beta M_0(x_0) + M_0(\beta(x_0))D_\beta L_0(x_0)\right)}{M_0(\beta(x_0))L_0(\beta(x_0))}(x-x_0)\right) \\
&\quad \times M_0(x)L_0(x) \\
&\quad + \left(y_1 + \frac{z_1 - y_1\left(D_\beta M_1(x_1) + M_1(\beta(x_1))D_\beta L_1(x_1)\right)}{M_1(\beta(x_1))L_1(\beta(x_1))}(x-x_1)\right) \\
&\quad \times M_1(x)L_1(x) \\
&= \left(1 + \frac{1-(-\frac{2}{3}+1\cdot(-\frac{1}{3}))}{1\cdot\frac{7}{6}}(x-1)\right) \\
&\quad \times \left(-\frac{1}{3}(x-4)\right)\left(-\frac{2}{3}(x-2)\right) \\
&\quad + \left(-1 + \frac{2+1\left(\frac{2}{3}+1\cdot\frac{1}{3}\right)}{1\cdot\frac{1}{3}}(x-4)\right) \\
&\quad \times \left(\frac{1}{3}(2x-1)\right)\left(\frac{1}{3}(x-1)\right) \\
&= \frac{2}{9}\left(1 + \frac{1+1}{\frac{7}{6}}(x-1)\right)(x-2)(x-4)
\end{aligned}$$

$$
\begin{aligned}
& +\frac{1}{9}\left(-1+\frac{2+1}{\frac{1}{3}}(x-4)\right)(2x-1)(x-1) \\
= \; & \frac{2}{9}\left(1+\frac{12}{7}(x-1)\right)(x-2)(x-4) \\
& +\frac{1}{9}(-1+9(x-4))(2x-1)(x-1) \\
= \; & \frac{2}{63}(7+12(x-1))(x-2)(x-4)+\frac{1}{9}(9x-37)(2x-1)(x-1) \\
= \; & \frac{2}{63}(12x-5)(x-2)(x-4)+\frac{1}{9}(9x-37)(2x-1)(x-1) \\
= \; & \frac{1}{63}(2(12x-5)(x-2)(x-4)+7(9x-37)(2x-1)(x-1)) \\
= \; & \frac{1}{63}(2(12x-5)(x^2-6x+8)+7(9x-37)(2x^2-3x+1)) \\
= \; & \frac{1}{63}(2(12x^3-72x^2+96x-5x^2+30x-40) \\
& +7(18x^3-27x^2+9x-74x^2+111x-37)) \\
= \; & \frac{1}{63}(2(12x^3-77x^2+126x-40)+7(18x^3-101x^2+120x-37)) \\
= \; & \frac{1}{63}(24x^3-154x^2+252x-80+126x^3-707x^2+840x-259) \\
= \; & \frac{1}{63}(150x^3-861x^2+1092x-339) \\
= \; & \frac{1}{21}(50x^3-287x^2+364x-113), \quad x \in I.
\end{aligned}
$$

Exercise 4.3.5 *Let $I=\mathbb{R}$, $\beta(x)=\frac{3}{5}x+\frac{1}{10}$, $x \in I$. Find the Hermite interpolation polynomial for the set*

$$
\begin{aligned}
x_0 &= -4, \quad x_1 = 0, \quad x_2 = 4, \\
y_0 &= -1, \quad y_1 = 1, \quad y_2 = -1, \\
z_0 &= 1, \quad z_1 = -1, \quad z_2 = 1.
\end{aligned}
$$

Definition 4.3.6 *Let $n \in \mathbb{N}_0$ and let $a,\ b \in I,\ a < b$, and $x_j \in I,\ j \in \{0, 1, \ldots, n\}$, be β-distinct. Let also, $f : [a,b] \to \mathbb{R}$ be delta differentiable on $[a,b]$. Then the polynomial*

$$\begin{aligned} p_{2n+1}(x) &= \sum_{j=0}^{n} \left(\left(1 - \frac{D_\beta M_j(x_j) + M_j(\beta(x_j)) D_\beta L_j(x_j)}{M_j(\beta(x_j)) L_j(\beta(x_j))} (x - x_j) \right) f(x_j) \right. \\ &\quad \left. + \frac{D_\beta f(x_j)}{M_j(\beta(x_j)) L_j(\beta(x_j))} (x - x_j) \right) M_j(x) L_j(x), \end{aligned}$$

$x \in [a,b]$, is called Hermite interpolation polynomial for the function f.

Exercise 4.3.7 *Let $I = \mathbb{R}$, $\beta(x) = \frac{1}{8}x$, $x \in I$, $f : I \to \mathbb{R}$ be given by*

$$f(x) = 2x^3 - x^2 + \sin_{2,\beta}(x), \quad x \in I.$$

Let also,

$$x_0 = 1, \quad x_1 = 9, \quad x_2 = 81.$$

Find the Hermite interpolation polynomial for the function f.

Definition 4.3.8 *Let $n \in \mathbb{N}_0$ and $a,\ b \in I,\ a < b$, and $x_j \in I,\ j \in \{0, 1, \ldots, n\}$, be distinct. Define the polynomials*

$$\zeta_n(x) = \prod_{j=0}^{n} (x - \beta(x_j)).$$

Theorem 4.3.9 *Suppose that $n \in \mathbb{N}_0$, $a, b \in I$, $a < b$, $x_j \in [a,b]$, $j \in \{0, 1, \ldots, n\}$, are distinct and $f : [a,b] \to \mathbb{R}$, $D_\beta^k f(x)$ exist for any $x \in [a,b]$ and for any $k \in \{1, \ldots, 2n+2\}$. Then for any $x \in [a,b]$ there exists $\xi = \xi(x) \in (a,b)$ such that*

$$f(x) - p_{2n+1}(x) = \frac{D_\beta^{2n+2} f(\xi)}{\pi_{n+1}(x)\zeta_{n+1}(x)} D_\beta^{2n+2} (\pi_{n+1}\zeta_{n+1})(\xi)$$

or

$$G_{min,2n+2}(\xi) \leq \frac{f(x) - p_{2n+1}(x)}{\pi_{n+1}(x)\zeta_{n+1}(x)} \leq G_{max,2n+2}(\xi),$$

where

$$\begin{aligned} G_{max,2n+2}(\xi) &= \max \left\{ \frac{D_\beta^{2n+2} f(\xi)}{D_\beta^{2n+2} (\pi_{n+1}\zeta_{n+1})(\xi)}, \right. \\ &\qquad \left. \frac{D_\beta^{2n+2} f(\beta^{-1}(\xi))}{D_\beta^{2n+2} (\pi_{n+1}\zeta_{n+1})(\beta^{-1}(\xi))} \right\}, \end{aligned}$$

$$G_{min,2n+2}(\xi) = \min\left\{\frac{D_\beta^{2n+2}f(\xi)}{D_\beta^{2n+2}(\pi_{n+1}\zeta_{n+1})(\xi)}, \frac{D_\beta^{2n+2}f(\beta^{-1}(\xi))}{D_\beta^{2n+2}(\pi_{n+1}\zeta_{n+1})(\beta^{-1}(\xi))}\right\}.$$

Proof 4.3.10 *Let p_{2n+1} be the Hermite interpolation polynomial for the function f with interpolation points x_j, $j \in \{0,1,\ldots,n\}$. Define the function*

$$\psi(t) = f(t) - p_{2n+1}(t) - \frac{f(x) - p_{2n+1}(x)}{\pi_{n+1}(x)\zeta_{n+1}(x)}\pi_{n+1}(t)\zeta_{n+1}(t), \quad t \in [a,b].$$

Then

$$\begin{aligned} \psi(x_j) &= f(x_j) - p_{2n+1}(x_j) - \frac{f(x) - p_{2n+1}(x)}{\pi_{n+1}(x)\zeta_{n+1}(x)}\pi_{n+1}(x_j)\zeta_{n+1}(x_j) \\ &= f(x_j) - f(x_j) \\ &= 0, \quad j \in \{0,1,\ldots,n\}, \end{aligned}$$

and $\psi(x) = 0$. Thus, $\psi : [a,b] \to \mathbb{R}$ has at least $n+2$ GZs. Hence and the β- Rolle theorem, it follows that $D_\beta\phi$ has at least $n+1$ GZs on (a,b) that do not coincide with x_j, $j \in \{0,1,\ldots,n\}$. Next,

$$\begin{aligned} D_\beta\psi(t) &= D_\beta f(t) - D_\beta p_{2n+1}(t) - \frac{f(x) - p_{2n+1}(x)}{\pi_{n+1}(x)\zeta_{n+1}(x)}D_\beta(\pi_{n+1}\zeta_{n+1})(t) \\ &= D_\beta f(t) - D_\beta p_{2n+1}(t) \\ &\quad - \frac{f(x) - p_{2n+1}(x)}{\pi_{n+1}(x)\zeta_{n+1}(x)} \\ &\quad \times \left(D_\beta\pi_{n+1}(t)\zeta_{n+1}(\beta(t)) + \pi_{n+1}(t)D_\beta\zeta_{n+1}(t)\right), \end{aligned}$$

$t \in [a,b]$. Hence,

$$\begin{aligned} D_\beta\psi(x_j) &= D_\beta f(x_j) - D_\beta p_{2n+1}(x_j) \\ &\quad - \frac{f(x) - p_{2n+1}(x)}{\pi_{n+1}(x)\zeta_{n+1}(x)} \\ &\quad \times \left(D_\beta\pi_{n+1}(x_j)\zeta_{n+1}(\beta(x_j)) + \pi_{n+1}(x_j)D_\beta\zeta_{n+1}(x_j)\right) \end{aligned}$$

$$= \quad 0, \quad j \in \{0,1,\ldots,n\},$$

i.e., $D_\beta \psi$ *has at least* $n+1$ *GZs at* x_j, $j \in \{0,1,\ldots,n\}$. *Therefore,* $D_\beta \psi$ *has at least* $2n+2$ *GZs in* $[a,b]$. *Then* $D_\beta^{2n+2}\psi$ *has at least one GZ in* (a,b) *and there exists an* $\xi = \xi(x) \in (a,b)$ *such that*

$$D_\beta^{2n+2}\psi(\xi) = 0 \quad or \quad D_\beta^{2n+2}\psi(\beta^{-1}(\xi))D_\beta^{2n+2}\psi(\xi) < 0.$$

Observe that

$$D_\beta^{2n+2}\phi(t) = D_\beta^{2n+2}f(t) - \frac{f(x)-p_{2n+1}(x)}{\pi_{n+1}(x)\zeta_{n+1}(x)}D_\beta^{2n+2}(\pi_{n+1}\zeta_{n+1})(t), \quad t \in [a,b].$$

1. *Let* $D_\beta^{2n+2}\psi(\xi) = 0$. *Then*

$$D_\beta^{2n+2}f(\xi) = \frac{f(x)-p_{2n+1}(x)}{\pi_{n+1}(x)\zeta_{n+1}(x)}D_\beta^{2n+2}(\pi_{n+1}\zeta_{n+1})(\xi)$$

or

$$f(x) - p_{2n+1}(x) \quad = \quad \frac{D_\beta^{2n+2}f(\xi)}{D_\beta^{2n+2}(\pi_{n+1}\zeta_{n+1})(\xi)}\pi_{n+1}(x)\zeta_{n+1}(x).$$

2. *Let*

$$D_\beta^{2n+2}\psi(\beta^{-1}(\xi))D_\beta^{2n+2}\psi(\xi) < 0.$$

Then

$$\begin{aligned} D_\beta^{2n+2}\psi(\beta^{-1}(\xi)) \quad &= \quad D_\beta^{2n+2}f(\beta^{-1}(\xi)) \\ &\quad - \frac{f(x)-p_{2n+1}(x)}{\pi_{n+1}(x)\zeta_{n+1}(x)}D_\beta^{2n+2}(\pi_{n+1}\zeta_{n+1})(\beta^{-1}(\xi)) \end{aligned}$$

and

$$D_\beta^{2n+2}\psi(\xi) = D_\beta^{2n+2}f(\xi) - \frac{f(x)-p_{2n+1}(x)}{\pi_{n+1}(x)\zeta_{n+1}(x)}D_\beta^{2n+2}(\pi_{n+1}\zeta_{n+1})(\xi).$$

Hence,

$$0 \quad > \quad D_\beta^{2n+2}\psi(\beta^{-1}\xi))D_\beta^{2n+2}\psi(\xi)$$

$$= \quad \Bigg(D_\beta^{2n+2}f(\beta^{-1}(\xi))$$

$$- \frac{f(x)-p_{2n+1}(x)}{\pi_{n+1}(x)\zeta_{n+1}(x)}D_\beta^{2n+2}(\pi_{n+1}\zeta_{n+1})(\beta^{-1}(\xi))\Bigg)$$

$$\times\left(D_\beta^{2n+2}f(\xi)-\frac{f(x)-p_{2n+1}(x)}{\pi_{n+1}(x)\zeta_{n+1}(x)}D_\beta^{2n+2}(\pi_{n+1}\zeta_{n+1})(\xi)\right)$$

$$=\left(\frac{f(x)-p_{2n+1}(x)}{\pi_{n+1}(x)\zeta_{n+1}(x)}\right)^2$$

$$\times D_\beta^{2n+2}(\pi_{n+1}\zeta_{n+1})(\xi)D_\beta^{2n+2}(\pi_{n+1}\zeta_{n+1})(\beta^{-1}(\xi))$$

$$-\frac{f(x)-p_{2n+1}(x)}{\pi_{n+1}(x)\zeta_{n+1}(x)}$$

$$\times\left(D_\beta^{2n+2}(\pi_{n+1}\zeta_{n+1})(\beta^{-1}(\xi))D_\beta 2n+2f(\xi)\right.$$

$$\left.+D_\beta^{2n+2}(\pi_{n+1}\zeta_{n+1})(\xi)D_\beta^{2n+2}f(\beta^{-1}(\xi))\right)$$

$$+D_\beta^{2n+2}f(\beta^{-1}(\xi))D_\beta^{2n+2}f(\xi).$$

Hence,

$$G_{min,2n+2}(\xi)\le\frac{f(x)-p_{2n+1}(x)}{\pi_{n+1}(x)\zeta_{n+1}(x)}\le G_{max,2n+2}(\xi).$$

This completes the proof.

Remark 4.3.11 *Suppose that all conditions of Theorem 4.3.9 hold. If*

$$\lim_{n\to\infty}\max_{x\in[a,b]}\left(\frac{D_\beta^{2n+2}f(\xi)}{D_\beta^{2n+2}(\pi_{n+1}\zeta_{n+1})(\xi)}\pi_{n+1}(x)\zeta_{n+1}(x)\right)=0$$

and

$$\lim_{n\to\infty}\max_{x\in[a,b]}\left(\frac{D_\beta^{2n+2}f(\beta^{-1}(\xi))}{D_\beta^{2n+2}(\pi_{n+1}\zeta_{n+1})(\beta^{-1}(\xi))}\pi_{n+1}(x)\zeta_{n+1}(x)\right)=0,$$

then

$$\lim_{n\to\infty}\max_{x\in[a,b]}|f(x)-p_{2n+1}(x)|=0.$$

4.4 β-Hermite Interpolation

In this section, we will construct the β-Hermite interpolation polynomials. We will demonstrate the difference between the Hermite interpolation polynomial and β-Hermite interpolation polynomial.

Theorem 4.4.1 *Let $n \in \mathbb{N}_0$ and let $a, b \in I$, $a < b$, and $x_j \in I$, $j \in \{0,1,\dots,n\}$, be distinct, the operator β is β-differentiable on $[a,b]$ and $D_\beta\beta(x_j) \neq 0$, $j \in \{0,1,\dots,n\}$. Let also, $y_j, z_j \in \mathbb{R}$, $j \in \{0,1,\dots,n\}$. Then there exists a unique polynomial $p_{\beta 2n+1} \in \mathscr{P}^\beta_{2n+1}$ such that*

$$p_{\beta 2n+1}(x_j) = y_j, \quad D_\beta p_{\beta 2n+1}(x_j) = z_j, \quad j \in \{0,1,\dots,n\}.$$

Proof 4.4.2 *Let M_k, $k \in \{0,1,\dots,n\}$, be the polynomials in the proof of Theorem 4.3.1. We will find a polynomial $p_{\beta 2n+1} \in \mathscr{P}^\beta_{2n+1}$ in the following manner*

$$p_{\beta 2n+1}(x) = \sum_{j=0}^{n} (y_j + (\beta(x) - \beta(x_j))(\alpha_j y_j + \beta_j z_j)) M_j(x) L_j(x), \quad x \in [a,b],$$

where α_j, $\beta_j \in \mathbb{R}$, $j \in \{0,1,\dots,n\}$, will be determined below. We have

$$\begin{aligned}
p_{\beta 2n+1}(x_k) &= \sum_{j=0}^{n} (y_j + (\beta(x_k) - \beta(x_j))(\alpha_j y_j + \beta_j z_j)) M_j(x_k) L_j(x_k) \\
&= y_k, \\
D_\beta p_{\beta 2n+1}(x) &= \sum_{j=0}^{n} D_\beta\beta(x)(\alpha_j y_j + \beta_j z_j) M_j(\beta(x)) L_j(\beta(x)) \\
&\quad + \sum_{j=0}^{n} \Big((y_j + (\beta(x) - \beta(x_j))(\alpha_j y_j + \beta_j z_j)) \\
&\quad \times (D_\beta M_j(x) L_j(x) + M_j(\beta(x)) D_\beta L_j(x)) \Big), \\
D_\beta p_{\beta 2n+1}(x_k) &= \sum_{j=0}^{n} D_\beta\beta(x_k)(\alpha_j y_j + \beta_j z_j) M_j(\beta(x_k)) L_j(\beta(x_k)) \\
&\quad + \sum_{j=0}^{n} \Big((y_j + (\beta(x_k) - \beta(x_j))(\alpha_j y_j + \beta_j z_j)) \\
&\quad \times (D_\beta M_j(x_k) L_j(x_k) + M_j(\beta(x_k)) D_\beta L_j(x_k)) \Big) \\
&= D_\beta\beta(x_k)(\alpha_k y_k + \beta_k z_k) M_k(\beta(x_k)) L_k(\beta(x_k)) \\
&\quad + y_k (D_\beta M_k(x_k) + M_k(\beta(x_k)) D_\beta L_k(x_k)) \\
&= z_k
\end{aligned}$$

or

$$\begin{aligned} &D_\beta\beta(x_k)\left(\alpha_k y_k+\beta_k z_k\right)M_k(\beta(x_k))L_k(\beta(x_k))\\ &= z_k - y_k\left(D_\beta M_k(x_k)+M_k(\beta(x_k))D_\beta L_k(x_k)\right), \end{aligned}$$

or

$$\alpha_k y_k+\beta_k z_k = \frac{z_k - y_k\left(D_\beta M_k(x_k)+M_k(\beta(x_k))D_\beta L_k(x_k)\right)}{D_\beta\beta(x_k)M_k(\beta(x_k))L_k(\beta(x_k))},$$

and

$$\begin{aligned} &p_{\beta 2n+1}(x)\\ &= \sum_{j=0}^{n}\left(y_j+\frac{z_j-y_j\left(D_\beta M_j(x_j)+M_j(\beta(x_j))D_\beta L_j(x_j)\right)}{D_\beta\beta(x_j)M_j(\beta(x_j))L_j(\beta(x_j))}(\beta(x)-\beta(x_j))\right)\\ &\quad\times M_j(x)L_j(x)\\ &= \sum_{j=0}^{n}\left(\left(1-\frac{D_\beta M_j(x_j)+M_j(\beta(x_j))D_\beta L_j(x_j)}{D_\beta\beta(x_j)M_j(\beta(x_j))L_j(\beta(x_j))}(\beta(x)-\beta(x_j))\right)y_j\right.\\ &\quad\left.+\frac{z_j}{D_\beta\beta(x_j)M_j(\beta(x_j))L_j(\beta(x_j))}(\beta(x)-\beta(x_j))\right)M_j(x)L_j(x), \end{aligned}$$

$x\in[a,b]$. *Now, suppose that there are two polynomials such that* $p_{\beta 2n+1}, q_{\beta 2n+1} \in \mathscr{P}^{\beta}_{2n+1}$ *and*

$$\begin{aligned} p_{\beta 2n+1}(x_k) &= q_{\beta 2n+1}(x_k)\\ &= y_k,\\ D_\beta p_{\beta 2n+1}(x_k) &= D_\beta q_{\beta 2n+1}(x_k)\\ &= z_k, \quad k\in\{0,1,\ldots,n\}. \end{aligned}$$

Let $h_{\beta 2n+1} = p_{\beta 2n+1} - q_{\beta 2n+1}$. *Then* $h_{\beta 2n+1} \in \mathscr{P}^{\beta}_{2n+1}$ *and it has at least* $2n+2$ *GZs. Thus,*

$$h_{\beta 2n+1}\equiv 0 \quad \text{or} \quad p_{\beta 2n+1}\equiv q_{\beta 2n+1} \quad \text{on} \quad [a,b].$$

This completes the proof.

Definition 4.4.3 *Let* $n\in\mathbb{N}_0$ *and let* $a,\ b\in I,\ a<b$, *and* $x_j\in I,\ j\in\{0,1,\ldots,n\}$, *be distinct, the graininess operator* β *is* β*-differentiable on* $[a,b]$ *and* $D_\beta\beta(x_j)\neq 0$, $j\in\{0,1,\ldots,n\}$. *Let also,* $y_j, z_j\in\mathbb{R}$, $j\in\{0,1,\ldots,n\}$. *The polynomial*

$$p_{\beta 2n+1}(x)$$

$$= \sum_{j=0}^{n} \left(\left(1 - \frac{D_\beta M_j(x_j) + M_j(\beta(x_j)) D_\beta L_j(x_j)}{D_\beta \beta(x_j) M_j(\beta(x_j)) L_j(\beta(x_j))} (\beta(x) - \beta(x_j)) \right) y_j \right.$$

$$\left. + \frac{z_j}{D_\beta \beta(x_j) M_j(\beta(x_j)) L_j(\beta(x_j))} (\beta(x) - \beta(x_j)) \right) M_j(x) L_j(x),$$

$x \in [a,b]$, will be called β-Hermite interpolation polynomial for the set $\{(x_j, y_j, z_j) : j \in \{0,1,\ldots,n\}\}$.

Definition 4.4.4 *Let $n \in \mathbb{N}_0$ and let $a,\ b \in I,\ a < b$, and $x_j \in I,\ j \in \{0,1,\ldots,n\}$, be distinct, the graininess operator β is β-differentiable on $[a,b]$ and $D_\beta \beta(x_j) \neq 0$, $j \in \{0,1,\ldots,n\}$. Let also, $f : [a,b] \to \mathbb{R}$ be β-differentiable. The polynomial*

$$p_{\beta 2n+1}(x)$$

$$= \sum_{j=0}^{n} \left(\left(1 - \frac{D_\beta M_j(x_j) + M_j(\beta(x_j)) D_\beta L_j(x_j)}{D_\beta \beta(x_j) M_j(\beta(x_j)) L_j(\beta(x_j))} (\beta(x) - \beta(x_j)) \right) f(x_j) \right.$$

$$\left. + \frac{D_\beta f(x_j)}{D_\beta \beta(x_j) M_j(\beta(x_j)) L_j(\beta(x_j))} (\beta(x) - \beta(x_j)) \right) M_j(x) L_j(x),$$

$x \in [a,b]$, will be called β-Hermite interpolation polynomial for the function f.

Example 4.4.5 *Let $I = \mathbb{R}$, $\beta(x) = \frac{4}{3}x + \frac{7}{3}$, $x \in I$,*

$$a = -1, \quad b = 9, \quad n = 1, \quad x_0 = -1, \quad x_1 = 5, \quad y_0 = y_1 = 0, \quad z_0 = z_1 = 1.$$

We have

$$\begin{aligned} \beta(x_0) &= \beta(-1) \\ &= \frac{4}{3} \cdot (-1) + \frac{7}{3} \\ &= 1, \\ \beta(x_1) &= \beta(5) \\ &= \frac{4}{3} \cdot 5 + \frac{7}{3} \\ &= 9, \end{aligned}$$

$$
\begin{aligned}
L_0(x) &= \frac{x-x_1}{x_0-x_1} \\
&= -\frac{1}{6}(x-5), \\
L_0(\beta(x_0)) &= L_0(1) \\
&= \frac{2}{3}, \\
D_\beta L_0(x) &= -\frac{1}{6}, \\
L_1(x) &= \frac{x-x_0}{x_1-x_0} \\
&= \frac{1}{6}(x+1), \\
L_1(\beta(x_1)) &= L_1(9) \\
&= \frac{5}{3}, \\
D_\beta L_1(x) &= \frac{1}{6}, \\
M_0(x) &= \frac{x-\beta(x_1)}{x_0-\beta(x_1)} \\
&= -\frac{1}{10}(x-9), \\
M_0(\beta(x_0)) &= M_0(1) \\
&= \frac{4}{5}, \\
D_\beta M_0(x) &= -\frac{1}{10}, \\
M_1(x) &= \frac{x-\beta(x_0)}{x_1-\beta(x_0)} \\
&= \frac{1}{4}(x-1),
\end{aligned}
$$

$$\begin{aligned}
M_1(\beta(x_1)) &= M_1(9)\\
&= 2,\\
D_\beta M_1(x) &= \frac{1}{4},\\
D_\beta\beta(x_0) &= \frac{\beta(\beta(x_0))-\beta(x_0)}{\beta(x_0)-x_0}\\
&= \frac{\beta(1)-1}{1-(-1)}\\
&= \frac{\frac{11}{3}-1}{2}\\
&= \frac{\frac{8}{3}}{2}\\
&= \frac{4}{3},\\
D_\beta\beta(x_1) &= \frac{\beta(\beta(x_1))-\beta(x_1)}{\beta(x_1)-x_1}\\
&= \frac{\beta(9)-9}{9-5}\\
&= \frac{\frac{43}{3}-9}{4}\\
&= \frac{\frac{16}{3}}{4}\\
&= \frac{4}{3}.
\end{aligned}$$

Hence, using that $y_0 = y_1 = 0$, $z_0 = z_1 = 1$, *we get*

$$\begin{aligned}
p_{\beta 3}(x) &= \frac{z_0}{D_\beta\beta(x_0)M_0(\beta(x_0))L_0(\beta(x_0))}(\beta(x)-\beta(x_0))M_0(x)L_0(x)\\
&\quad+\frac{z_1}{D_\beta\beta(x_1)M_1(\beta(x_1))L_1(\beta(x_1))}(\beta(x)-\beta(x_1))M_1(x)L_1(x)\\
&= \frac{1}{\frac{4}{3}\cdot\frac{4}{5}\cdot\frac{2}{3}}(\beta(x)-1)\left(-\frac{1}{10}(x-9)\right)\left(-\frac{1}{6}(x-5)\right)
\end{aligned}$$

$$
\begin{aligned}
& +\frac{1}{\frac{4}{3}\cdot 2\cdot \frac{5}{3}}(\beta(x)-9)\left(\frac{1}{6}(x+1)\right)\left(\frac{1}{4}(x-1)\right) \\
= \; & \frac{3}{128}(\beta(x)-1)(x-9)(x-5) \\
& +\frac{3}{320}(\beta(x)-9)(x+1)(x-1) \\
= \; & \frac{3}{128}\left(\frac{4}{3}x+\frac{7}{3}-1\right)(x-9)(x-5) \\
& +\frac{3}{320}\left(\frac{4}{3}x+\frac{7}{3}-9\right)(x+1)(x-1) \\
= \; & \frac{1}{30}(x+1)(x-9)(x-5)+\frac{1}{320}(4x-20)(x+1)(x-1) \\
= \; & \frac{1}{30}(x+1)(x-9)(x-5)+\frac{1}{80}(x-5)(x+1)(x-1) \\
= \; & \frac{1}{240}(x+1)(x-5)(8(x-9)+3(x-1)) \\
= \; & \frac{1}{240}(x+1)(x-5)(8x-72+3x-3) \\
= \; & \frac{1}{240}(x+1)(x-5)(11x-75), \quad x\in I.
\end{aligned}
$$

Exercise 4.4.6 *Let $I=\mathbb{R}$, $\beta(x)=\frac{3}{8}x+12$, $x\in I$, and*

$$a=-4, \quad b-7, \quad x_0=-4, \quad x_1=0, \quad x_2=\frac{5}{6},$$

the function $f: I\to\mathbb{R}$ be defined by

$$f(t)=\frac{t+1}{t^2+4t+7}+1, \quad t\in I.$$

Find the β-Hermite interpolation polynomial for the function f.

Remark 4.4.7 *Above example shows that the Hermite and β-Hermite interpolation polynomial are different.*

In the next theorem we will give some criteria for coincidence of the Hermite and β-Hermite interpolation polynomials.

Theorem 4.4.8 *Assume I be a time scale such that $\beta(t)=ct$ for any $t\in I$ and for some real constant $c\in(0,1)$ or $\beta(t)=t+d$ for any $t\in I$ and for some real constant* *d. Let $n \in \mathbb{N}_0$ and let $a, b \in I$, $a < b$, and $x_j \in I$, $j \in \{0, 1, \ldots, n\}$, be distinct. Let also, $y_j, z_j \in \mathbb{R}$, $j \in \{0, 1, \ldots, n\}$. Then*

$$p_{\beta 2n+1} \equiv p_{2n+1}.$$

Proof 4.4.9 *1. Let $\beta(t) = ct$ for any $t \in I$ and for some real constant $c \in (0,1)$. Then*

$$\begin{aligned}
& p_{\beta 2n+1}(x) \\
= & \sum_{j=0}^{n} \Bigg(\bigg(1 - \frac{D_\beta M_j(x_j) + M_j(\beta(x_j)) D_\beta L_j(x_j)}{D_\beta \beta(x_j) M_j(\beta(x_j)) L_j(\beta(x_j))} (\beta(x) - \beta(x_j)) \bigg) y_j \\
& + \frac{z_j}{D_\beta \beta(x_j) M_j(\beta(x_j)) L_j(\beta(x_j))} (\beta(x) - \beta(x_j)) \Bigg) M_j(x) L_j(x) \\
= & \sum_{j=0}^{n} \Bigg(\bigg(1 - \frac{D_\beta M_j(x_j) + M_j(\beta(x_j)) D_\beta L_j(x_j)}{c M_j(\beta(x_j)) L_j(\beta(x_j))} (cx - cx_j) \bigg) y_j \\
& + \frac{z_j}{c M_j(\beta(x_j)) L_j(\beta(x_j))} (cx - cx_j) \Bigg) M_j(x) L_j(x) \\
= & \sum_{j=0}^{n} \Bigg(\bigg(1 - \frac{D_\beta M_j(x_j) + M_j(\beta(x_j)) D_\beta L_j(x_j)}{M_j(\beta(x_j)) L_j(\beta(x_j))} (x - x_j) \bigg) y_j \\
& + \frac{z_j}{M_j(\beta(x_j)) L_j(\beta(x_j))} (x - x_j) \Bigg) M_j(x) L_j(x) \\
= & p_{2n+1}(x), \quad x \in [a, b].
\end{aligned}$$

2. Let $\beta(t) = t + d$ for any $t \in I$ and for some real constant d. Then

$$\begin{aligned}
& p_{\beta 2n+1}(x) \\
= & \sum_{j=0}^{n} \Bigg(\bigg(1 - \frac{D_\beta M_j(x_j) + M_j(\beta(x_j)) D_\beta L_j(x_j)}{D_\beta \beta(x_j) M_j(\beta(x_j)) L_j(\beta(x_j))} (\beta(x) - \beta(x_j)) \bigg) y_j \\
& + \frac{z_j}{D_\beta \beta(x_j) M_j(\beta(x_j)) L_j(\beta(x_j))} (\beta(x) - \beta(x_j)) \Bigg) M_j(x) L_j(x) \\
= & \sum_{j=0}^{n} \Bigg(\bigg(1 - \frac{D_\beta M_j(x_j) + M_j(\beta(x_j)) D_\beta L_j(x_j)}{M_j(\beta(x_j)) L_j(\beta(x_j))} (x + d - x_j - d) \bigg) y_j \\
& + \frac{z_j}{M_j(\beta(x_j)) L_j(\beta(x_j))} (x + d - x_j - d) \Bigg) M_j(x) L_j(x)
\end{aligned}$$

$$= \sum_{j=0}^{n}\left(\left(1-\frac{D_\beta M_j(x_j)+M_j(\beta(x_j))D_\beta L_j(x_j)}{M_j(\beta(x_j))L_j(\beta(x_j))}(x-x_j)\right)y_j\right.$$

$$\left.+\frac{z_j}{M_j(\beta(x_j))L_j(\beta(x_j))}(x-x_j)\right)M_j(x)L_j(x)$$

$$= p_{2n+1}(x), \quad x\in[a,b].$$

This completes the proof.

As we have proved Theorem 4.3.9, one can deduct the following result.

Theorem 4.4.10 *Suppose that $n\in\mathbb{N}_0$, $a,b\in I$, $a<b$, $x_j\in[a,b]$, $j\in\{0,1,\dots,n\}$, are distinct and $f:[a,b]\to\mathbb{R}$, $D_\beta^k f(x)$ exist for any $x\in[a,b]$ and for any $k\in\{1,\dots,2n+2\}$. Then for any $x\in[a,b]$ there exists $\xi=\xi(x)\in(a,b)$ such that*

$$f(x)-p_{\beta 2n+1}(x)=\frac{D_\beta^{2n+2}f(\xi)}{\pi_{n+1}(x)\zeta_{n+1}(x)}D_\beta^{2n+2}(\pi_{n+1}\zeta_{n+1})(\xi)$$

or

$$G_{min,2n+2}(\xi)\le\frac{f(x)-p_{\beta 2n+1}(x)}{\pi_{n+1}(x)\zeta_{n+1}(x)}\le G_{max,2n+2}(\xi),$$

where $G_{min,2n+2}(\xi)$ and $G_{max,2n+2}(\xi)$ are defined as in Theorem 4.3.9.

4.5 β-Differentiation

Theorem 4.5.1 *Suppose that $n\ge 0$, $a,b\in I$, $a<b$, $x_j\in[a,b]$, $j\in\{0,1,\dots,n\}$, are distinct and $f:[a,b]\to\mathbb{R}$, $D_\beta^k f(x)$ exist for any $x\in[a,b]$ and for any $k\in\{1,\dots,n+1\}$. Then for any $x\in[a,b]$ there exists $\xi=\xi(x)\in(a,b)$ and distinct points η_j, $j\in\{1,\dots,n\}$, in (a,b), such that*

$$D_\beta f(x)-D_\beta p_n(x)=\frac{D_\beta^{n+1}f(\xi)}{D_\beta^{n+1}\pi_n^*(\xi)}\pi_n^*(x)$$

or

$$H_{min,n+1}(\xi)\le\frac{D_\beta f(x)-D_\beta p_n(x)}{\pi_n^*(x)}\le H_{max,n+1}(\xi),$$

where

$$H_{max,n+1}(\xi) = \max\left\{\frac{D_\beta^{n+1}f(\xi)}{D_\beta^{n+1}\pi_n^*(\xi)},\frac{D_\beta^{n+1}f(\beta^{-1}(\xi))}{D_\beta^{n+1}\pi_n^*(\beta^{-1}(\xi))}\right\},$$

$$H_{min,n+1}(\xi) = \min\left\{ \frac{D_\beta^{n+1} f(\xi)}{D_\beta^{n+1}\pi_n^*(\xi)}, \frac{D_\beta^{n+1} f(\beta^{-1}(\xi))}{D_\beta^{n+1}\pi_n^*(\beta^{-1}(\xi))} \right\},$$

$$\pi_n^*(x) = (x-\eta_1)\dots(x-\eta_n).$$

Proof 4.5.2 *Let p_n be the Lagrange interpolation polynomial for the function f with interpolation points x_j, $j \in \{0,1,\dots,n\}$. Then the function $f - p_n$ has at least $n+1$ GZs in $[a,b]$. Hence and the β- Rolle theorem, it follows that there exist η_j, $j \in \{1,\dots,n\}$, in (a,b) which are GZs of the function $D_\beta f - D_\beta p_n$. Define the function*

$$\chi(t) = D_\beta f(t) - D_\beta p_n(t) - \frac{D_\beta f(x) - D_\beta p_n(x)}{\pi_n^*(x)}\pi_n^*(t), \quad t \in [a,b].$$

Then

$$\begin{aligned} \chi(\eta_j) &= D_\beta f(\eta_j) - D_\beta p_n(\eta_j) - \frac{D_\beta f(x) - D_\beta p_n(x)}{\pi_n^*(x)}\pi_n^*(\eta_j) \\ &= D_\beta f(\eta_j) - D_\beta p_n(\eta_j) \\ &= 0, \quad j \in \{0,1,\dots,n\}, \end{aligned}$$

and $\chi(x) = 0$. Thus, $\chi : [a,b] \to \mathbb{R}$ has at least $n+1$ GZs. Hence and the β-Rolle theorem, it follows that $D_\beta^n \chi$ has at least one GZ on (a,b). Therefore, there exists an $\xi = \xi(x) \in (a,b)$ such that

$$D_\beta^n \chi(\xi) = 0 \quad or \quad D_\beta^n \chi(\beta^{-1}(\xi)) D_\beta^n \chi(\xi) < 0.$$

Note that

$$D_\beta^n \chi(t) = D_\beta^{n+1} f(t) - \frac{D_\beta f(x) - D_\beta p_n(x)}{\pi_n^*(x)} D_\beta^n \pi_n^*(t), \quad t \in [a,b].$$

1. Let $D_\beta^n \chi(\xi) = 0$. Then

$$D_\beta^{n+1} f(\xi) = \frac{D_\beta f(x) - D_\beta p_n(x)}{\pi_n^*(x)} D_\beta^n \pi_n^*(\xi)$$

or

$$\begin{aligned} D_\beta f(x) - D_\beta p_n(x) &= \frac{D_\beta^{n+1} f(\xi)}{D_\beta^n \pi_n^*(\xi)}\pi_n^*(x) \\ &= \frac{D_\beta^{n+1} f(\xi)}{D_\beta^n \pi_n^*(\xi)}\pi_n^*(x). \end{aligned}$$

2. *Let*

$$D_\beta^n \chi(\beta^{-1}(\xi)) D_\beta^n \chi(\xi) < 0.$$

Then

$$\begin{aligned} D_\beta^n \chi(\beta^{-1}(\xi)) &= D_\beta^{n+1} f(\beta^{-1}(\xi)) \\ &\quad - \frac{D_\beta f(x) - D_\beta p_n(x)}{\pi_n^*(x)} D_\beta^n \pi_n^*(\beta^{-1}(\xi)) \end{aligned}$$

and

$$D_\beta^n \chi(\xi) = D_\beta^{n+1} f(\xi) - \frac{D_\beta f(x) - D_\beta p_n(x)}{\pi_n^*(x)} D_\beta^n \pi_n^*(\xi).$$

Hence,

$$\begin{aligned} 0 &> D_\beta^n \chi(\beta^{-1}(\xi)) D_\beta^n \chi(\xi) \\ &= \Bigg(D_\beta^{n+1} f(\beta^{-1}(\xi)) \\ &\quad - \frac{D_\beta f(x) - D_\beta p_n(x)}{\pi_n^*(x)} D_\beta^n \pi_n^*(\beta^{-1}(\xi)) \Bigg) \\ &\quad \times \left(D_\beta^{n+1} f(\xi) - \frac{D_\beta f(x) - D_\beta p_n(x)}{\pi_n^*(x)} D_\beta^n \pi_n^*(\xi) \right) \\ &= \left(\frac{D_\beta f(x) - D_\beta p_n(x)}{\pi_n^*(x)} \right)^2 D_\beta^n \pi_n^*(\beta^{-1}(\xi)) D_\beta^n \pi_n^*(\xi) \\ &\quad - \frac{D_\beta f(x) - D_\beta p_n(x)}{\pi_n^*(x)} \\ &\quad \times \left(D_\beta^n \pi_n^*(\beta^{-1}(\xi)) D_\beta^{n+1} f(\xi) + D_\beta^n \pi_n^*(\xi) D_\beta^{n+1} f(\beta^{-1}(\xi)) \right) \\ &\quad + D_\beta^{n+1} f(\beta^{-1}(\xi)) D_\beta^{n+1} f(\xi). \end{aligned}$$

Hence,

$$H_{min,n+1}(\xi) \le \frac{D_\beta f(x) - D_\beta p_n(x)}{\pi_n^*(x)} \le H_{max,n+1}(\xi).$$

This completes the proof.

Remark 4.5.3 *Suppose that all conditions of Theorem 4.5.1 hold. If*

$$\lim_{n\to\infty} \max_{x\in[a,b]} \left(\frac{D_\beta^{n+1} f(\xi)}{D_\beta^n \pi_n^*(\xi)} \pi_n^*(x) \right) = 0$$

and

$$\lim_{n\to\infty}\max_{x\in[a,b]}\left(\frac{D_\beta^{n+1}f(\beta^{-1}(\xi))}{D_\beta^n\pi_n^*(\beta^{-1}(\xi))}\pi_n^*(x)\right)=0,$$

then

$$\lim_{n\to\infty}\max_{x\in[a,b]}|D_\beta f(x)-D_\beta p_n(x)|=0.$$

4.6 Advanced Practical Problems

Problem 4.6.1 *Let* $I=\mathbb{R}$, $\beta(x)=\frac{1}{8}x+6$, $x\in I$. *Construct the Lagrange interpolation polynomial for the set*

$$\left\{\left(\frac{1}{27},1\right),\left(\frac{1}{9},0\right),\left(\frac{1}{3},-1\right),(1,2)\right\}.$$

Problem 4.6.2 *Let* $I=\mathbb{R}$, $\beta(x)=\frac{1}{4}x+1$, $x\in I$. *Let also,* $f:I\to\mathbb{R}$ *be defined by*

$$f(x)=\sin_{2,\beta}(x)+\cos_{3,\beta}(x)+x^3+x^2,\quad x\in I.$$

Find the Lagrange interpolation polynomial for the function f *with interpolation points* 1, 4, 16, 64, 256.

Problem 4.6.3 *Let* $I=\mathbb{R}$, $\beta(x)=\frac{1}{2}x+2$, $x\in I$. *Let also,*

$$x_0=1,\quad x_1=3,\quad x_2=9$$

Find $\pi_3(x)$, $\Pi_3^2(x)$, $x\in I$.

Problem 4.6.4 *Let* $I=\mathbb{R}$, $\beta(x)=\frac{1}{3}x-2$, $x\in I$. *Let also,*

$$b=1=x_2,\quad x_1=\frac{1}{9},\quad x_0=\frac{1}{81},$$

$f:I\to\mathbb{R}$ *be defined by*

$$f(t)=\frac{t^5-2t^3+t^2+4}{t^7+t^6+t+2},\quad t\in I.$$

Find the β*-Lagrange interpolation polynomial for the function* f *with interpolation points* x_0, x_1, x_2.

Problem 4.6.5 *Let $I = \mathbb{R}$, $\beta(x) = \frac{2}{5}x + 2$, $x \in I$. Let also,*

$$a = 0, \quad b = 12, \quad x_0 = 0, \quad x_1 = \frac{1}{3}, \quad x_2 = 4.$$

Find the β-Lagrange and Lagrange interpolation polynomials for the function $f : I \to \mathbb{R}$ defined by

$$f(t) = \frac{t+1}{t^2 - 2t + 7} + t + 1 + \sin_{1,\beta}(t), \quad t \in I,$$

with interpolation and interpolation points x_0, x_1, x_2.

Problem 4.6.6 *Let $I = \mathbb{R}$, $\beta(x) = \frac{1}{8}x + 6$, $x \in I$. Let also,*

$$a = 0, \quad b = 108, \quad x_0 = 0, \quad x_1 = \frac{1}{2}, \quad x_2 = 11.$$

Find $\Pi^3_{\beta 3}(19)$.

Problem 4.6.7 *Let $I = \mathbb{R}$, $\beta(x) = \frac{1}{7}x + 8$, $x \in I$. Find the Hermite interpolation polynomial for the set*

$$x_0 = -24, \quad x_1 = -4, \quad x_2 = 16,$$

$$y_0 = 1, \quad y_1 = -3, \quad y_2 = -1,$$

$$z_0 = -1, \quad z_1 = 1, \quad z_2 = -1.$$

Problem 4.6.8 *Let $I = \mathbb{R}$, $\beta(x) = \frac{1}{9}x + 3$, $x \in I$. Let also, $f : I \to \mathbb{R}$ be defined by*

$$f(x) = \frac{x+1}{x^2+1} + x^3, \quad x \in I.$$

Let also,

$$x_0 = -4, \quad x_1 = 0, \quad x_2 = 8.$$

Find the Hermite interpolation polynomial for the function f.

Problem 4.6.9 *Let $I = \mathbb{R}$, $\beta(x) = \frac{1}{2}x + 9$, $x \in I$. Let also,*

$$a = -3, \quad b = 29, \quad x_0 = -3, \quad x_1 = 0, \quad x_2 = 8,$$

$f : I \to \mathbb{R}$ be defined by

$$f(t) = e_1(t) + t^3 + 3t^2 + t - 10, \quad t \in I.$$

Find the β-Hermite interpolation polynomial for the function f.

4.7 Notes and References

In this chapter, we introduce Lagrange, β-Lagrange, Hermite and β-Hermite interpolation polynomials. We show that in the general case Lagrange and β-Lagrange interpolation polynomials are different, as well as Hermite and β-Hermite interpolation polynomials are different. From these interpolations they are explored approximations for the β-derivative of continuously β-differentiable functions.

5

Numerical β-Integration

The main purpose of this chapter is to evaluate the Cauchy β-integral and applying the results in the previous chapter on polynomial interpolation we derive formulae for numerical β-integration. We shall explain how one can estimate the associated approximation error.

Let $I \subseteq \mathbb{R}$ and $\beta : I \to \mathbb{R}$ be first kind or second kind general quantum operator.

5.1 Newton-Cotes Formula

Suppose that $f : I \to \mathbb{R}$ is continuous and $a, b \in I$, $a < b$. We will evaluate the integral

$$\int_a^b f(x) d_\beta x.$$

Let $n \in \mathbb{N}_0$, $x_j \in [a,b]$, $j \in \{0,1,\ldots,n\}$, be distinct. With $p_n \in \mathscr{P}_n$ we will denote the Lagrange interpolation polynomial for the function f. Then

$$p_n(x) = \sum_{k=0}^{n} L_k(x) f(x_k).$$

Then

$$\begin{aligned}\int_a^b f(x) d_\beta x &\approx \int_a^b p_n(x) d_\beta x \\ &= \sum_{k=0}^{n} \left(\int_a^b L_k(x) d_\beta x \right) f(x_k).\end{aligned}$$

Let

$$w_k = \int_a^b L_k(x) d_\beta x, \quad k \in \{0,1,\ldots,n\}.$$

Then

$$\int_a^b f(x) d_\beta x \approx \sum_{k=0}^{n} w_k f(x_k). \tag{5.1}$$

Definition 5.1.1 *The values w_k, $k \in \{0,1,\ldots,n\}$, are said to be quadrature weights, while the interpolation points x_j, $j \in \{0,1,\ldots,n\}$, are called the quadrature points.*

DOI: 10.1201/9781003472131-5

Definition 5.1.2 *The numerical quadrature rule* (5.1) *is said to be the Newton-Cotes formula.*

In the case $n = 1$, we take

$$a = x_0, \quad b = x_1.$$

Then

$$\begin{aligned} p_1(x) &= L_0(x)f(a) + L_1(x)f(b) \\ &= \frac{x-b}{a-b}f(a) + \frac{x-a}{b-a}f(b) \\ &= \frac{1}{b-a}\left((b-x)f(a) + (x-a)f(b)\right). \end{aligned}$$

Hence,

$$\begin{aligned} \int_a^b p_1(x)d_\beta x &= \frac{1}{b-a}\left(f(a)\int_a^b (b-x)d_\beta x + f(b)\int_a^b (x-a)d_\beta x\right) \\ &= \frac{1}{b-a}\left(f(a)\int_b^a (x-b)d_\beta x + f(b)\int_a^b (x-a)d_\beta x\right) \\ &= \frac{1}{b-a}\left(f(a)h_2(a,b) + f(b)h_2(b,a)\right), \end{aligned}$$

and

$$\int_a^b f(x)d_\beta x \approx \frac{1}{b-a}\left(f(a)h_2(a,b) + f(b)h_2(b,a)\right). \tag{5.2}$$

Definition 5.1.3 *The numerical integration formula* (5.2) *will be called the trapezium rule.*

Example 5.1.4 *Let $I = \mathbb{R}$, $\beta(t) = 2t$, $t \in I$, $a = 1$, $b = 8$, $f : I \to \mathbb{R}$ is defined by*

$$f(t) = \frac{t-1}{t+1}, \quad t \in I.$$

We will evaluate

$$\int_1^8 f(t)d_\beta t.$$

Let

$$f_1(t) = \frac{1}{3}t^2 - t,$$

$$f_2(t) = \frac{1}{3}t^2 - 8t, \quad t \in I.$$

Then

$$\begin{aligned}
D_\beta f_1(t) &= \frac{1}{3}(\beta(t)+t)-1 \\
&= \frac{1}{3}(2t+t)-1 \\
&= t-1, \\
D_\beta f_2(t) &= \frac{1}{3}(\beta(t)+t)-8 \\
&= \frac{1}{3}(2t+t)-8 \\
&= t-8, \\
h_2(a,b) &= \int_8^1 (t-8)d_\beta t \\
&= \int_8^1 D_\beta f_2(t)d_\beta t \\
&= f_2(1)-f_2(8) \\
&= \left(\frac{1}{3}-8\right)-\left(\frac{64}{3}-64\right) \\
&= -\frac{23}{3}-\frac{64}{3}+64 \\
&= -\frac{87}{3}+64 \\
&= \frac{-87+192}{3} \\
&= \frac{105}{3} \\
&= 35,
\end{aligned}$$

$$\begin{aligned}
h_2(b,a) &= \int_1^8 (t-1)d_\beta t \\
&= \int_1^8 D_\beta f_1(t)d_\beta t \\
&= f_1(8)-f_1(1) \\
&= \left(\frac{64}{3}-8\right)-\left(\frac{1}{3}-1\right) \\
&= \frac{64}{3}-8-\frac{1}{3}+1 \\
&= 9-7 \\
&= 2, \\
f(1) &= 0, \\
f(8) &= \frac{7}{9}.
\end{aligned}$$

Thus,

$$\begin{aligned}
\int_1^8 \frac{t-1}{t+1}d_\beta t &\approx \frac{1}{7}\left(\frac{7}{9}\cdot 2\right) \\
&= \frac{2}{9}.
\end{aligned}$$

Example 5.1.5 *Let $I=\mathbb{R}$, $\beta(t)=3t$, $t\in I$, $f:I\to\mathbb{R}$ be defined by*

$$f(t)=\frac{t^2+t+1}{t^3+t+1}(t-1)+e_{1,\beta}(t), \quad t\in I.$$

We will evaluate the integral

$$\int_1^9 f(t)d_\beta t.$$

Here

$$a=1, \quad b=9.$$

Let

$$f_1(t) = \frac{1}{4}t^2-t,$$

$$f_2(t) = \frac{1}{4}t^2 - 9t, \quad t \in I.$$

Then

$$\begin{aligned}
D_\beta f_1(t) &= \frac{1}{4}(\beta(t)+t) - 1 \\
&= \frac{1}{4}(3t+t) - 1 \\
&= t-1, \\
D_\beta f_2(t) &= \frac{1}{4}(\beta(t)+t) - 9 \\
&= \frac{1}{4}(3t+t) - 9 \\
&= t-9, \\
h_2(9,1) &= \int_1^9 (t-1) d_\beta t \\
&= \int_1^9 D_\beta f_1(t) d_\beta t \\
&= f_1(9) - f_1(1) \\
&= \frac{81}{4} - 9 - \frac{1}{4} + 1 \\
&= 20 - 8 \\
&= 12, \\
h_2(1,9) &= \int_9^1 (t-9) d_\beta t \\
&= \int_9^1 D_\beta f_2(t) d_\beta t \\
&= f_2(1) - f_2(9) \\
&= \left(\frac{1}{4} - 9\right) - \left(\frac{81}{4} - 81\right)
\end{aligned}$$

$$
\begin{aligned}
&= \frac{1}{4}-\frac{81}{4}-9+81 \\
&= -20+72 \\
&= 52, \\
f(1) &= 1, \\
f(9) &= \frac{81+9+1}{729+9+1}(9-1)+e_{1,\beta}(9) \\
&= \frac{728}{739}+e_{1,\beta}(9).
\end{aligned}
$$

Now, using the trapezium rule, we get

$$
\begin{aligned}
\int_1^9 f(t)d_\beta t &\approx \frac{1}{8}\left(52+12\cdot\left(\frac{728}{739}+e_{1,\beta}(9)\right)\right) \\
&= \frac{1}{2}\left(13+3\cdot\left(\frac{728}{739}+e_{1,\beta}(9)\right)\right).
\end{aligned}
$$

Exercise 5.1.6 *Let $I=\mathbb{R}$, $\beta(t)=\frac{2}{3}t+4$, $t\in I$. Using the trapezium rule, evaluate the integral*

$$
\int_1^8\left(\frac{t+1}{t^2+3t+4}+e_{2,\beta}(t)\right)d_\beta t.
$$

Definition 5.1.7 *For $t,s\in I$, define the monomials*

$$
\begin{aligned}
H_0(t,s) &= 1, \\
H_{k+1}(t,s) &= \int_s^t h_k(\beta(\tau),s)d_\beta\tau, \quad k\in\mathbb{N}_0.
\end{aligned}
$$

Lemma 5.1.8 *Let $t,s\in I$ and $t_1,t_2\in I$ are between s and t. Then*

$$
\begin{aligned}
\int_s^t(y-t_1)(y-t_2)d_\beta y &= h_2(t,t_2)h_1(t,t_1)-h_2(s,t_1)h_1(s,t_2) \\
&\quad +H_3(s,t_1)-H_3(t_2,t_1)-H_3(t,t_2).
\end{aligned}
$$

Proof 5.1.9 *We have*

$$
\int_s^t(y-t_1)(y-t_2)d_\beta y = \int_s^{t_1}(y-t_1)(y-t_2)d_\beta y+\int_{t_1}^{t_2}(y-t_1)(y-t_2)d_\beta y
$$

$$
\begin{aligned}
& + \int_{t_2}^{t} (y-t_1)(y-t_2)d_\beta y \\
= & - \int_{t_1}^{s} (y-t_1)(y-t_2)d_\beta y + \int_{t_1}^{t_2} D_\beta h_2(y,t_1)(y-t_2)d_\beta y \\
& + \int_{t_2}^{t} D_\beta h_2(y,t_2)(y-t_1)d_\beta y \\
= & - \int_{t_1}^{s} D_\beta h_2(y,t_1)(y-t_2)d_\beta y + h_2(y,t_1)(y-t_2)\Big|_{y=t_1}^{y=t_2} \\
& - \int_{t_1}^{t_2} h_2(\beta(y),t_1)d_\beta y + h_2(y,t_2)(y-t_1)\Big|_{y=t_1}^{y=t} \\
& - \int_{t_2}^{t} h_2(\beta(y),t_2)d_\beta y \\
= & -h_2(y,t_1)(y-t_2)\Big|_{y=t_1}^{y=s} + \int_{t_1}^{s} h_2(\beta(y),t_1)d_\beta y \\
& -H_3(t_2,t_1) + h_2(t,t_2)(t-t_1) - H_3(t,t_2) \\
= & -h_2(s,t_1)(s-t_2) + H_3(s,t_1) - H_3(t_2,t_1) \\
& +h_2(t,t_2)h_1(t,t_1) - H_3(t,t_2).
\end{aligned}
$$

This completes the proof.

Now, take three points

$$a = x_0 < x_1 < x_2 = b.$$

For $x \in [a,b]$, using Lemma 5.1.8, we get

$$
\begin{aligned}
\int_a^b L_0(x)d_\beta x &= \frac{1}{(x_0-x_1)(x_0-x_2)} \int_{x_0}^{x_2} (x-x_1)(x-x_2)d_\beta x \\
&= \frac{1}{(x_0-x_1)(x_0-x_2)} \Big(h_2(x_2,x_2)h_1(x_2,x_1) - h_2(x_0,x_1)h_1(x_0,x_2) \\
&\quad +H_3(x_0,x_1) - H_3(x_2,x_1) - H_3(x_2,x_2) \Big) \\
&= \frac{1}{(x_0-x_1)(x_0-x_2)} \left(-h_2(x_0,x_1)h_1(x_0,x_2) + H_3(x_0,x_1) - H_3(x_2,x_1)\right), \\
\int_a^b L_1(x)d_\beta x &= \frac{1}{(x_1-x_0)(x_1-x_2)} \int_{x_0}^{x_2} (x-x_0)(x-x_2)d_\beta x
\end{aligned}
$$

$$= \frac{1}{(x_1-x_0)(x_1-x_2)}\Big(h_2(x_2,x_2)h_1(x_2,x_0)-h_2(x_0,x_0)h_1(x_0,x_2)$$

$$+H_3(x_0,x_0)-H_3(x_2,x_0)-H_3(x_2,x_2)\Big)$$

$$= -\frac{H_3(x_2,x_0)}{(x_1-x_0)(x_1-x_2)},$$

$$\int_a^b L_2(x)d_\beta x = \frac{1}{(x_2-x_0)(x_2-x_1)}\int_{x_0}^{x_2}(x-x_0)(x-x_1)d_\beta x$$

$$= \frac{1}{(x_2-x_0)(x_2-x_1)}\Big(h_2(x_2,x_1)h_1(x_2,x_0)-h_2(x_0,x_0)h_1(x_0,x_1)$$

$$+H_3(x_0,x_0)-H_3(x_1,x_0)-H_3(x_2,x_1)\Big)$$

$$= \frac{1}{(x_2-x_0)(x_2-x_1)}\left(h_2(x_2,x_1)h_1(x_2,x_0)-H_3(x_1,x_0)-H_3(x_2,x_1)\right).$$

Hence and (5.1), we find

$$\begin{aligned}\int_a^b f(x)d_\beta x \approx\ & \frac{f(x_0)}{(x_0-x_1)(x_0-x_2)}\\ & \times(-h_2(x_0,x_1)h_1(x_0,x_2)+H_3(x_0,x_1)-H_3(x_2,x_1))\\ & -f(x_1)\frac{H_3(x_2,x_0)}{(x_1-x_0)(x_1-x_2)}\\ & +\frac{f(x_2)}{(x_2-x_0)(x_2-x_1)}\\ & \times(h_2(x_2,x_1)h_1(x_2,x_0)-H_3(x_1,x_0)-H_3(x_2,x_1)).\end{aligned} \tag{5.3}$$

Definition 5.1.10 *The numerical integration formula* (5.3) *is said to be the Simpson rule.*

Exercise 5.1.11 *Let $I=\mathbb{R}$, $\beta(t)=\frac{1}{2}t+1$, $t\in I$. Using the Simpson rule, evaluate the integral*

$$\int_{-6}^{9}\frac{d_\beta t}{t^2+t+1}.$$

5.2 β-Newton-Cotes Formula

In this section we will evaluate β-Riemann integral using the β-Lagrange interpolation polynomial. Suppose that $f : I \to \mathbb{R}$ is continuous, $n \in \mathbb{N}_0$, $a, b \in I$, $\beta(a) < b$, $x_j \in I$, $j \in \{0, 1, \dots, n\}$, are distinct points. Then

$$p_{\beta n}(x) = \sum_{k=0}^{n} L_{\beta k}(x) f(x_k), \quad x \in [a, b],$$

and

$$\int_a^b p_{\beta n}(x) d_\beta x = \sum_{k=0}^{n} \left(\int_a^b L_{\beta k}(x) d_\beta x \right) f(x_k).$$

Hence,

$$\begin{aligned} \int_a^b f(x) d_\beta x &\approx \int_a^b p_{\beta n}(x) d_\beta x \\ &= \sum_{k=0}^{n} \left(\int_a^b L_{\beta k}(x) d_\beta x \right) f(x_k). \end{aligned}$$

Denote

$$w_{\beta k} = \int_a^b L_{\beta k}(x) d_\beta x, \quad k \in \{0, 1, \dots, n\}.$$

Then

$$\int_a^b f(x) d_\beta x \approx \sum_{k=0}^{n} w_{\beta k} f(x_k). \tag{5.4}$$

Definition 5.2.1 *The values $w_{\beta k}$, $k \in \{0, 1, \dots, n\}$, will be called the β-quadrature weights and the β-interpolation points x_j, $j \in \{0, 1, \dots, n\}$, will be called the β-quadrature points.*

Definition 5.2.2 *The numerical quadrature formula* (5.4) *is said to be β-Newton-Cotes formula.*

For $n = 1$, we take

$$a = x_0 \leq \beta(x_0) < x_1 \leq \beta(x_1) = b.$$

Then

$$p_{\beta 1}(x) = L_{\beta 0}(x) f(x_0) + L_{\beta 1}(x) f(x_1)$$

$$\begin{aligned}
&= \frac{\beta(x)-\beta(x_1)}{\beta(x_0)-\beta(x_1)} f(x_0) + \frac{\beta(x)-\beta(x_0)}{\beta(x_1)-\beta(x_0)} f(x_1) \\
&= \frac{\beta(x)-b}{\beta(a)-b} f(a) + \frac{\beta(x)-\beta(a)}{b-\beta(a)} f(x_1) \\
&= \frac{1}{b-\beta(a)} \left(-(\beta(x)-b)f(a) + (\beta(x)-\beta(a))f(x_1)\right)
\end{aligned}$$

and

$$\int_a^b p_{\beta 1}(x) d_\beta x$$

$$\begin{aligned}
&= \frac{1}{b-\beta(a)} \left(-f(a) \int_a^b (\beta(x)-b) d_\beta x + f(x_1) \int_a^b (\beta(x)-\beta(a)) d_\beta x \right) \\
&= \frac{1}{b-\beta(a)} \left(f(a) \int_b^a (\beta(x)-b) d_\beta x + f(x_1) \int_a^{\beta(a)} (\beta(x)-\beta(a)) d_\beta x \right. \\
&\quad \left. + f(x_1) \int_{\beta(a)}^b (\beta(x)-\beta(a)) d_\beta x \right) \\
&= \frac{1}{b-\beta(a)} \left(f(a) g_2(a,b) + f(x_1)(\beta(a)-a)(\beta(a)-\beta(a)) \right. \\
&\quad \left. + f(x_1) g_2(b,\beta(a)) \right) \\
&= \frac{1}{b-\beta(a)} \left(f(a) g_2(a,b) + f(x_1) g_2(b,\beta(a)) \right).
\end{aligned}$$

Consequently we get

$$\int_a^b f(x) d_\beta x \approx \frac{1}{b-\beta(a)} \left(f(a) g_2(a,b) + f(x_1) g_2(b,\beta(a)) \right). \tag{5.5}$$

Definition 5.2.3 *The numerical integration formula* (5.5) *will be called the* β*-trapezium rule.*

Example 5.2.4 *Let* $I = \left\{ \frac{1}{12}, \frac{1}{4}, \frac{1}{3}, \frac{1}{2}, 1 \right\}$,

$$a = \frac{1}{4} = x_0, \quad x_1 = \frac{1}{2}, \quad b = 1,$$

$\beta : I \to \mathbb{R}$ *be defined by*

$$\beta\left(\frac{1}{12}\right) = \frac{1}{12},$$

$$\beta\left(\frac{1}{4}\right) = \frac{1}{3},$$

$$\beta\left(\frac{1}{3}\right) = \frac{1}{2},$$

$$\beta\left(\frac{1}{2}\right) = 1,$$

$$\beta(1) = 2,$$

the function $f: I \to \mathbb{R}$ be defined by

$$f(t) = \frac{t-1}{t+1}, \quad t \in I.$$

We will evaluate

$$\int_{\frac{1}{4}}^{1} f(t) d_\beta t.$$

We have

$$\begin{aligned}
\beta\left(\frac{1}{4}\right) - \frac{1}{4} &= \frac{1}{3} - \frac{1}{4} \\
&= \frac{1}{12}, \\
\beta\left(\frac{1}{3}\right) - \frac{1}{3} &= \frac{1}{2} - \frac{1}{3} \\
&= \frac{1}{6}, \\
\beta\left(\frac{1}{2}\right) - \frac{1}{2} &= \beta\left(\frac{1}{2}\right) - \frac{1}{2} \\
&= 1 - \frac{1}{2} \\
&= \frac{1}{2}, \\
g_2(a,b) &= \int_1^{\frac{1}{4}} (\beta(t) - 1) d_\beta t \\
&= -\int_{\frac{1}{4}}^{1} (\beta(t) - 1) d_\beta t
\end{aligned}$$

$$\begin{aligned}
&= -\left(\beta\left(\frac{1}{4}\right)-\frac{1}{4}\right)\left(\beta\left(\frac{1}{4}\right)-1\right)\\
&\quad -\left(\beta\left(\frac{1}{3}\right)-\frac{1}{3}\right)\left(\beta\left(\frac{1}{3}\right)-1\right)\\
&\quad -\left(\beta\left(\frac{1}{2}\right)-\frac{1}{2}\right)\left(\beta\left(\frac{1}{2}\right)-1\right)\\
&= -\frac{1}{12}\left(\frac{1}{3}-1\right)-\frac{1}{6}\left(\frac{1}{2}-1\right)-\frac{1}{2}(1-1)\\
&= -\frac{1}{12}\left(-\frac{2}{3}\right)-\frac{1}{6}\left(-\frac{1}{2}\right)\\
&= \frac{1}{18}+\frac{1}{12}\\
&= \frac{5}{36},\\
g_2(b,\beta(a)) &= \int_{\frac{1}{3}}^{1}\left(\beta(t)-\frac{1}{3}\right)d_\beta t\\
&= \left(\beta\left(\frac{1}{3}\right)-\frac{1}{3}\right)\left(\beta\left(\frac{1}{3}\right)-\frac{1}{3}\right)\\
&\quad +\left(\beta\left(\frac{1}{2}\right)-\frac{1}{2}\right)\left(\beta\left(\frac{1}{2}\right)-\frac{1}{3}\right)\\
&= \frac{1}{6}\left(\frac{1}{2}-\frac{1}{3}\right)+\frac{1}{2}\left(1-\frac{1}{3}\right)\\
&= \frac{1}{36}+\frac{1}{3}\\
&= \frac{13}{36},\\
f(a) &= f\left(\frac{1}{4}\right)\\
&= \frac{\frac{1}{4}-1}{\frac{1}{4}+1}\\
&= -\frac{3}{5},
\end{aligned}$$

$$
\begin{aligned}
f(x_1) &= f\left(\frac{1}{2}\right) \\
&= \frac{\frac{1}{2}-1}{\frac{1}{2}+1} \\
&= -\frac{1}{3}.
\end{aligned}
$$

Hence,

$$
\begin{aligned}
\int_{\frac{1}{4}}^{1} \frac{t-1}{t+1} d_\beta t &= \frac{1}{1-\beta\left(\frac{1}{4}\right)}\left(f\left(\frac{1}{4}\right) g_2\left(\frac{1}{4},1\right) + f\left(\frac{1}{2}\right) g_2\left(1,\frac{1}{3}\right)\right) \\
&= \frac{1}{1-\frac{1}{3}}\left(-\frac{3}{5}\cdot\frac{5}{36} - \frac{1}{3}\cdot\frac{13}{36}\right) \\
&= \frac{3}{2}\left(-\frac{1}{12} - \frac{13}{108}\right) \\
&= \frac{3}{2}\left(-\frac{9+13}{108}\right) \\
&= -\frac{11}{36}.
\end{aligned}
$$

Exercise 5.2.5 *Let $I=\mathbb{R}$, $\beta(t)=\frac{10}{11}t+1$, $t \in I$, and*

$$
a=-3=x_0, \quad x_1=-\frac{5}{6}, \quad b=2,
$$

$f: I \to \mathbb{R}$ be defined by

$$
f(t)=t^4+t^3+\frac{t+1}{t^2+1}, \quad t \in I.
$$

Using the β-trapezium rule, evaluate the integral

$$
\int_{-2}^{2} f(t) d_\beta t.
$$

Definition 5.2.6 *For $s,t \in I$, define the polynomials*

$$
\begin{aligned}
G_0(t,s) &= 1, \\
G_{k+1}(t,s) &= \int_s^t g_k(\tau,s) d_\beta \tau, \quad k \in \mathbb{N}_0.
\end{aligned}
$$

Lemma 5.2.7 *Let $t,s \in I$ and $t_1, t_2, \beta(t_1), \beta(t_2)$ are between s and t. Then*

$$\begin{aligned}\int_s^t (\beta(y)-\beta(t_1))(\beta(y)-\beta(t_2))d_\beta y &= -g_2(s,\beta(t_1))g_1(s,\beta(t_2))\\ &\quad -g_1(t,\beta(t_1))g_2(t,\beta(t_2))+(\beta(t_2)-t_2)g_2(\beta(t_2),\beta(t_1))\\ &\quad +G_3(s,\beta(t_1))-G_3(\beta(t_2),\beta(t_1))-G_3(t,\beta(t_2)).\end{aligned}$$

Proof 5.2.8 *We have*

$$\begin{aligned}&\int_s^t (\beta(y)-\beta(t_1))(\beta(y)-\beta(t_2))d_\beta y\\ &= \int_s^{\beta(t_1)} (\beta(y)-\beta(t_1))(\beta(y)-\beta(t_2))d_\beta y\\ &\quad +\int_{\beta(t_1)}^{\beta(t_2)} (\beta(y)-\beta(t_1))(\beta(y)-\beta(t_2))d_\beta y\\ &\quad +\int_{\beta(t_2)}^{t} (\beta(y)-\beta(t_1))(\beta(y)-\beta(t_2))d_\beta y\\ &= -\int_{\beta(t_1)}^{s} D_\beta g_2(y,\beta(t_1))(\beta(y)-\beta(t_2))d_\beta y\\ &\quad +\int_{\beta(t_1)}^{\beta(t_2)} D_\beta g_2(y,\beta(t_1))(\beta(y)-\beta(t_2))d_\beta y\\ &\quad +\int_{\beta(t_2)}^{t} (\beta(y)-\beta(t_1))D_\beta g_2(y,\beta(t_2))d_\beta y\\ &= -g_2(y,\beta(t_1))(y-\beta(t_2))\Big|_{y=\beta(t_1)}^{y=s}+\int_{\beta(t_1)}^{s} g_2(y,\beta(t_1))d_\beta y\\ &\quad +g_2(y,\beta(t_1))(y-t_2)\Big|_{y=\beta(t_1)}^{y=\beta(t_2)}-\int_{\beta(t_1)}^{\beta(t_2)} g_2(y,\beta(t_1))d_\beta y\\ &\quad +(y-\beta(t_1))g_2(y,\beta(t_2))\Big|_{y=\beta(t_2)}^{y=t}-\int_{\beta(t_2)}^{t} g_2(y,\beta(t_2))d_\beta y\\ &= -g_2(s,\beta(t_1))g_1(s,\beta(t_2))\\ &\quad -g_1(t,\beta(t_1))g_2(t,\beta(t_2))+(\beta(t_2)-t_2)g_2(\beta(t_2),\beta(t_1))\\ &\quad +G_3(s,\beta(t_1))-G_3(\beta(t_2),\beta(t_1))-G_3(t,\beta(t_2)).\end{aligned}$$

This completes the proof.

Let now,

$$a = x_0 < x_1 \leq \beta(x_1) < x_2 \leq \beta(x_2) = b.$$

Then, using Lemma 5.2.7, we get

$$\int_{x_0}^{\beta(x_2)} L_{\beta 0}(y) d_\beta y$$

$$\begin{aligned}
= & \frac{1}{(\beta(x_0)-\beta(x_1))(\beta(x_0)-\beta(x_2))}\Big(-g_2(x_0,\beta(x_1), g_1(x_0,\beta(x_2)) \\
& -g_1(\beta(x_2),\beta(x_1))g_2(\beta(x_2),\beta(x_2)) \\
& +(\beta(x_2)-x_2)g_2(\beta(x_2),\beta(x_1))+G_3(x_0,\beta(x_1)) \\
& -G_3(\beta(x_2),\beta(x_1))-G_3(\beta(x_2),\beta(x_2))\Big) \\
= & \frac{1}{(\beta(x_0)-\beta(x_1))(\beta(x_0)-\beta(x_2))}\Big(-g_2(x_0,\beta(x_1))g_1(x_0,\beta(x_2)) \\
& +(\beta(x_2)-x_2)g_2(\beta(x_2),\beta(x_1))+G_3(x_0,\beta(x_1))-G_3(\beta(x_2),\beta(x_1))\Big),
\end{aligned}$$

$$\int_{x_0}^{\beta(x_2)} L_{\beta 1}(y) d_\beta y$$

$$\begin{aligned}
= & \frac{1}{(\beta(x_1)-\beta(x_0))(\beta(x_1)-\beta(x_2))}\Big(-g_2(x_0,\beta(x_0))g_1(x_0,\beta(x_2)) \\
& -g_1(\beta(x_2),\beta(x_0))g_2(\beta(x_2),\beta(x_2)) \\
& +(\beta(x_2)-x_2)g_2(\beta(x_2),\beta(x_0))+G_3(x_0,\beta(x_0)) \\
& -G_3(\beta(x_2),\beta(x_0))-G_3(\beta(x_2),\beta(x_2))\Big) \\
= & \frac{1}{(\beta(x_1)-\beta(x_0))(\beta(x_1)-\beta(x_2))}\Big((\beta(x_2)-x_2)g_2(\beta(x_2),\beta(x_0)) \\
& -G_3(\beta(x_2),\beta(x_0))\Big),
\end{aligned}$$

$$\int_{x_0}^{\beta(x_2)} L_{\beta 2}(y) d_\beta y$$

$$= \frac{1}{(\beta(x_2)-\beta(x_0))(\beta(x_2)-\beta(x_1))}\Big(-g_2(x_0,\beta(x_2))g_1(x_0,\beta(x_1))$$

$$-g_1(\beta(x_2),\beta(x_0))g_2(\beta(x_2),\beta(x_1))$$

$$+(\beta(x_2)-x_2)g_2(\beta(x_2),\beta(x_0))+G_3(x_0,\beta(x_0))$$

$$-G_3(\beta(x_1),\beta(x_0))-G_3(\beta(x_2),\beta(x_1))\Big)$$

$$= \frac{1}{(\beta(x_2)-\beta(x_0))(\beta(x_2)-\beta(x_1))}\Big(-g_1(\beta(x_2),\beta(x_0))g_2(\beta(x_2),\beta(x_1))$$

$$+(\beta(x_2)-x_2)g_2(\beta(x_2),\beta(x_1))$$

$$-H_3(\beta(x_1),\beta(x_2))-G_3(\beta(x_2),\beta(x_1))\Big).$$

Hence and (5.4), we get

$$\int_a^b f(x)d_\beta x \tag{5.6}$$

$$\approx \frac{f(x_0)}{(\beta(x_0)-\beta(x_1))(\beta(x_0)-\beta(x_2))}\Big(-g_2(x_0,\beta(x_1))g_1(x_0,\beta(x_2))$$

$$+(\beta(x_2)-x_2)g_2(\beta(x_2),\beta(x_1))+G_3(x_0,\beta(x_1))-G_3(\beta(x_2),\beta(x_1))\Big)$$

$$+\frac{f(x_1)}{(\beta(x_1)-\beta(x_0))(\beta(x_1)-\beta(x_2))}\Big((\beta(x_2)-x_2)g_2(\beta(x_2),\beta(x_0))$$

$$-G_3(\beta(x_2),\beta(x_0))\Big)$$

$$+\frac{f(x_2)}{(\beta(x_2)-\beta(x_0))(\beta(x_2)-\beta(x_1))}\Big(-g_1(\beta(x_2),\beta(x_0))g_2(\beta(x_2),\beta(x_1))$$

$$+(\beta(x_2)-x_2)g_2(\beta(x_2),\beta(x_1))$$

$$-G_3(\beta(x_1),\beta(x_2))-G_3(\beta(x_2),\beta(x_1))\Big).$$

Definition 5.2.9 *The numerical integration formula* (5.6) *will be called the β-Simpson rule.*

Exercise 5.2.10 *Let* $I = \mathbb{R}$, $\beta(t) = \frac{5}{6}t + 12$, $t \in I$,

$$a = 2 = x_0, \quad x_1 = \frac{8}{3}, \quad x_2 = \frac{15}{4}, \quad b = 4,$$

$f : I \to \mathbb{R}$ *be defined by*

$$f(t) = t^2 + \frac{1+t+t^2}{1+t^4+t^6}, \quad t \in I.$$

Using the β*-Simpson rule, evaluate the integral*

$$\int_a^b f(t) d_\beta t.$$

5.3 Error Estimates

Our aim in this section is to evaluate the size of the error in the numerical integration (5.1). The error in (5.1) is defined by

$$E_n(f) = \int_a^b f(x) d_\beta x - \sum_{k=0}^{n} w_k f(x_k).$$

Theorem 5.3.1 *Let* $n \in \mathbb{N}$, $a, b \in I$, $a < b$. *Suppose that* $f : [a,b] \to \mathbb{R}$, $f \in \mathscr{C}_\beta^{n+1}([a,b])$. *Then*

$$E_n(f) = \int_a^b \frac{D_\beta^{n+1} f(\xi)}{\Pi_{n+1}^{n+1}(\xi)} \pi_{n+1}(x) d_\beta x$$

or

$$\int_a^b G_{min,n+1}(\xi, x) d_\beta x \le E_n(f) \le \int_a^b G_{max,n+1}(\xi, x) d_\beta x,$$

where

$$G_{max,n+1}(\xi,x) = \begin{cases} F_{max,n+1}(\xi)\pi_{n+1}(x) & \text{if} \quad \pi_{n+1}(x) > 0 \\ F_{min,n+1}(\xi)\pi_{n+1}(x) & \text{if} \quad \pi_{n+1}(x) \le 0, \end{cases}$$

$$G_{min,n+1}(\xi,x) = \begin{cases} F_{min,n+1}(\xi)\pi_{n+1}(x) & \text{if} \quad \pi_{n+1}(x) > 0 \\ F_{max,n+1}(\xi)\pi_{n+1}(x) & \text{if} \quad \pi_{n+1}(x) \le 0. \end{cases}$$

Proof 5.3.2 *We have that*

$$E_n(f) = \int_a^b (f(x) - p_n(x)) \, d_\beta x,$$

where $p_n \in \mathscr{P}_n$ *is the Lagrange interpolation polynomial for the function* f. *Hence and Theorem 4.1.13, we get the desired result. This completes the proof.*

5.4 β-Error Estimates

Now, we will evaluate the size of the error in the numerical integration (5.4). The β-error in (5.4) we define as follows.

$$E_{\beta n}(f) = \int_a^b f(x)d_\beta x - \sum_{k=0}^{n} w_{\beta k} f(x_k).$$

Theorem 5.4.1 *Let $n \in \mathbb{N}$, $a, b \in I$, $a < b$. Suppose that $f : [a,b] \to \mathbb{R}$, $f \in \mathscr{C}_\beta^{n+1}([a,b])$. Then*

$$E_{\beta n}(f) = \int_a^b \frac{D_\beta^{n+1} f(\xi)}{\Pi_{\beta n+1}^{n+1}(\xi)} \pi_{\beta n+1}(x) d_\beta x$$

or

$$\int_a^b G_{\beta min,n+1}(\xi,x)d_\beta x \le E_{\beta n}(f) \le \int_a^b G_{\beta max,n+1}(\xi,x)d_\beta x,$$

where

$$G_{\beta max,n+1}(\xi,x) = \begin{cases} F_{\beta max,n+1}(\xi)\pi_{\beta n+1}(x) & \text{if} \quad \pi_{\beta n+1}(x) > 0 \\ F_{\beta min,n+1}(\xi)\pi_{\beta n+1}(x) & \text{if} \quad \pi_{\beta n+1}(x) \le 0, \end{cases}$$

$$G_{\beta min,n+1}(\xi,x) = \begin{cases} F_{\beta min,n+1}(\xi)\pi_{\beta n+1}(x) & \text{if} \quad \pi_{\beta n+1}(x) > 0 \\ F_{\beta max,n+1}(\xi)\pi_{\beta n+1}(x) & \text{if} \quad \pi_{\beta n+1}(x) \le 0. \end{cases}$$

Proof 5.4.2 *Observe that*

$$E_{\beta n}(f) = \int_a^b \left(f(x) - p_{\beta n}(x)\right) d_\beta x,$$

where $p_{\beta n} \in \mathscr{P}_n^\beta$ is the β-Lagrange interpolation polynomial for the function f. Hence and Theorem 4.2.17, we get the expected β-error estimate. This completes the proof.

5.5 Composite Quadrature Rules

Suppose that $a, b \in I$, $a < b$, $f : [a,b] \to \mathbb{R}$ is a given continuous function. Let

$$a = x_0 < x_1 < \ldots < x_n = b.$$

Then

$$\int_a^b f(x)d_\beta x = \sum_{j=0}^{n-1} \int_{x_j}^{x_{j-1}} f(x)d_\beta x.$$

Now, applying the trapezium rule, we find

$$\int_{x_j}^{x_{j+1}} f(x)d_\beta x \approx \frac{1}{x_{j+1}-x_j}\left(f(x_j)h_2(x_j,x_{j+1}) + f(x_{j+1})h_2(x_{j+1},x_j)\right),$$

$j \in \{0,1,\dots,n\}$. Hence,

$$\int_a^b f(x)d_\beta x \approx \sum_{j=0}^{n-1} \frac{1}{x_{j+1}-x_j}\left(f(x_j)h_2(x_j,x_{j+1}) + f(x_{j+1})h_2(x_{j+1},x_j)\right). \tag{5.7}$$

Definition 5.5.1 *The numerical integration formula* (5.7) *is said to be the composite trapezium rule.*

Example 5.5.2 *Let $I = \mathbb{R}$, $\beta(t) = 2t$, $t \in I$, $f : I \to \mathbb{R}$ is defined by*

$$f(t) = \frac{t+1}{t^2+t+1}, \quad t \in I.$$

Let also,

$$1 = x_0 < 4 = x_1 < 16 = x_2 = b.$$

Here $h_1(t,s) = t-s$, $t,s \in I$. For a fixed $s \in I$, denote

$$f_1(t) = \frac{1}{3}t^2 - st, \quad t \in I.$$

Then

$$\begin{aligned} D_\beta f_1(t) &= \frac{1}{3}(\beta(t)+t) - s \\ &= \frac{1}{3}(2t+t) - s \\ &= t-s \\ &= h_1(t,s), \quad t \in I, \end{aligned}$$

and

$$h_2(t,s) = \int_s^t h_1(\tau,s)d_\beta \tau$$

$$
\begin{aligned}
&= \int_s^t (\tau - s) d_\beta \tau \\
&= \int_s^t D_\beta f_1(\tau) d_\beta \tau \\
&= f_1(\tau)\Big|_{\tau=s}^{\tau=t} \\
&= \frac{1}{3}t^2 - st - \frac{1}{3}s^2 + s^2 \\
&= \frac{t^2 + 2s^2 - 3st}{3}, \quad t \in I.
\end{aligned}
$$

Hence,

$$
\begin{aligned}
h_2(x_1, x_0) &= h_2(4, 1) \\
&= \frac{16 - 12 + 2}{3} \\
&= 2, \\
h_2(x_0, x_1) &= h_2(1, 4) \\
&= \frac{1 + 32 - 12}{3} \\
&= 7, \\
h_2(x_2, x_1) &= h_2(16, 4) \\
&= \frac{256 - 192 + 32}{3} \\
&= \frac{256 - 160}{3} \\
&= \frac{96}{3} \\
&= 32, \\
h_2(x_1, x_2) &= h_2(4, 16)
\end{aligned}
$$

$$
\begin{aligned}
&= \frac{16-192+512}{3} \\
&= \frac{16+320}{3} \\
&= 112, \\
f(1) &= \frac{2}{3}, \\
f(4) &= \frac{5}{21}, \\
f(16) &= \frac{17}{273}.
\end{aligned}
$$

Thus,

$$
\begin{aligned}
\int_1^{16} \frac{t+1}{t^2+t+1} d_\beta t &\approx \frac{1}{x_1-x_0}(f(x_0)h_2(x_0,x_1)+f(x_1)h_2(x_1,x_0)) \\
&\quad + \frac{1}{x_2-x_1}(f(x_1)h_2(x_1,x_2)+f(x_2)h_2(x_2,x_1)) \\
&= \frac{1}{3}\left(\frac{2}{3}\cdot 7+\frac{5}{21}\cdot 2\right) \\
&\quad + \frac{1}{12}\left(\frac{5}{21}\cdot 112+\frac{17}{273}\cdot 32\right) \\
&= \frac{14}{9}+\frac{10}{63}+\frac{280}{126}+\frac{136}{819} \\
&= \frac{300}{126}+\frac{1410}{819} \\
&= \frac{50}{21}+\frac{1410}{819} \\
&= \frac{13650+9870}{5733} \\
&= \frac{23520}{5733}.
\end{aligned}
$$

Exercise 5.5.3 *Let $I = \mathbb{R}$, $\beta(t) = \frac{1}{3}t + 2$, $t \in I$. Using the composite trapezium rule, evaluate the integral*

$$\int_{-1}^{\frac{1}{4}} \left(\frac{t^2+1}{t^4+1} + \sin_1(t,0) \right) d_\beta t.$$

Now, suppose that

$$a = x_0 < x_1 < x_2 < \ldots < x_{2m} = b.$$

Then

$$\int_a^b f(x) d_\beta x = \sum_{j=1}^{m} \int_{x_{2j-2}}^{x_{2j}} f(x) d_\beta x.$$

Hence, using the Simpson rule, we find

$$\int_a^b f(x) d_\beta x \tag{5.8}$$

$$\begin{aligned}\approx \sum_{j=1}^{m} \Bigg(& \frac{f(x_{2j-2})}{(x_{2j-2}-x_{2j-1})(x_{2j-2}-x_{2j})} \Big(-h_2(x_{2j-2},x_{2j-1})h_1(x_{2j-2},x_{2j}) \\ & + H_3(x_{2j-2},x_{2j-1}) - H_3(x_{2j},x_{2j-1}) \Big) \\ & - f(x_{2j-1}) \frac{H_3(x_{2j},x_{2j-2})}{(x_{2j-1}-x_{2j-2})(x_{2j-1}-x_{2j})} \\ & + \frac{f(x_{2j})}{(x_{2j}-x_{2j-2})(x_{2j}-x_{2j-1})} \Big(h_2(x_{2j},x_{2j-1})h_1(x_{2j},x_{2j-2}) \\ & - H_3(x_{2j-1},x_{2j-2}) - H_3(x_{2j},x_{2j-1}) \Big) \Bigg).\end{aligned}$$

Definition 5.5.4 *The numerical integration formula* (5.8) *is said to be the composite Simpson rule.*

Exercise 5.5.5 *Let $I = \mathbb{R}$, $\beta(t) = 3t$, $t \in I$,*

$$a = 1 = x_0, \quad x_1 = 9, \quad x_2 = 81,$$

$f : I \to \mathbb{R}$ is defined by

$$f(t) = 1 + t - \frac{3t}{1+t^4} + e_2(t,1), \quad t \in I.$$

Using the composite Simpson rule, evaluate the integral

$$\int_a^b f(t) d_\beta t.$$

5.6 β-Composite Quadrature Rules

Suppose that $a, b \in I$, $a < b$, $n \in \mathbb{N}$, $f : [a,b] \to \mathbb{R}$ is continuous, $x_j \in \mathbb{N}$, $j \in \{0, 1, \ldots, n\}$, be distinct. Then, we have

$$a = x_0 \leq \beta(x_0) < x_1 \leq \beta(x_1) < \ldots < x_n \leq \beta(x_n) = b.$$

Then

$$\int_a^b f(x)d_\beta x = \sum_{j=0}^{n-1} \int_{x_j}^{\beta(x_{j+1})} f(x)d_\beta x.$$

We apply the β-trapezium rule and we obtain

$$\int_a^b f(x)d_\beta x \tag{5.9}$$

$$\approx \sum_{j=0}^{n-1} \frac{1}{\beta(x_{j+1}) - \beta(x_j)} \left(f(x_j)g_2(x_j, \beta(x_{j+1})) + f(x_{j+1})g_2(\beta(x_{j+1}), \beta(x_j))\right).$$

Definition 5.6.1 *The numerical integral formula* (5.9) *is said to be the β-composite trapezium rule.*

Example 5.6.2 *Let*

$$I = \left\{-1, 0, \frac{1}{8}, \frac{1}{6}, \frac{1}{4}, \frac{1}{2}, 1\right\}$$

and $\beta : I \to \mathbb{R}$ be defined as follows

$$\begin{aligned}
\beta(-1) &= -1, \\
\beta(0) &= \frac{1}{8}, \\
\beta\left(\frac{1}{8}\right) &= \frac{1}{6}, \\
\beta\left(\frac{1}{6}\right) &= \frac{1}{4}, \\
\beta\left(\frac{1}{4}\right) &= \frac{1}{2}, \\
\beta\left(\frac{1}{2}\right) &= 1,
\end{aligned}$$

$$\beta(1) \quad = \quad 2.$$

Let also,

$$a = x_0 = 0, \quad x_1 = \frac{1}{6}, \quad x_2 = \frac{1}{2}, \quad b = 1.$$

We have

$$x_0 = 0 < \beta(x_0) = \frac{1}{8} < x_1 = \frac{1}{6} < \beta(x_1) = \frac{1}{4} < x_2 = \frac{1}{2} < \beta(x_2) = 1 = b.$$

Let $f(t) = t$, $t \in I$. We will evaluate the integral

$$\int_a^b f(t)d_\beta t$$

using the β-composite trapezium rule. We have

$$g_2(x_0, \beta(x_1)) \quad = \quad \int_{\beta(x_1)}^{x_0} (\beta(t) - \beta(x_1))d_\beta t$$

$$\begin{aligned}
&= \int_{\frac{1}{4}}^{\frac{1}{8}} \left(\beta(t) - \frac{1}{4}\right) d_\beta t \\
&= -\int_{\frac{1}{8}}^{\frac{1}{4}} \left(\beta(t) - \frac{1}{4}\right) d_\beta t \\
&= -\left(\beta\left(\frac{1}{8}\right) - \frac{1}{8}\right)\left(\beta\left(\frac{1}{8}\right) - \frac{1}{4}\right) - \left(\beta\left(\frac{1}{6}\right) - \frac{1}{6}\right)\left(\beta\left(\frac{1}{6}\right) - \frac{1}{4}\right) \\
&= -\left(\frac{1}{6} - \frac{1}{8}\right)\left(\frac{1}{6} - \frac{1}{4}\right) - \left(\frac{1}{4} - \frac{1}{6}\right)\left(\frac{1}{4} - \frac{1}{4}\right) \\
&= -\frac{1}{24}\left(-\frac{1}{12}\right) \\
&= \frac{1}{288},
\end{aligned}$$

$$g_2(\beta(x_1), \beta(x_0)) \quad = \quad \int_{\beta(x_0)}^{\beta(x_1)} (\beta(t) - \beta(x_0))d_\beta t$$

$$\begin{aligned}
&= \int_{\frac{1}{8}}^{\frac{1}{4}} \left(\beta(t) - \frac{1}{8}\right) d_\beta t \\
&= \left(\beta\left(\frac{1}{8}\right) - \frac{1}{8}\right)\left(\beta\left(\frac{1}{8}\right) - \frac{1}{8}\right) + \left(\beta\left(\frac{1}{6}\right) - \frac{1}{6}\right)\left(\beta\left(\frac{1}{6}\right) - \frac{1}{8}\right)
\end{aligned}$$

$$
\begin{aligned}
&= \left(\frac{1}{6}-\frac{1}{8}\right)\left(\frac{1}{6}-\frac{1}{8}\right)+\left(\frac{1}{4}-\frac{1}{6}\right)\left(\frac{1}{4}-\frac{1}{8}\right)\\
&= \frac{1}{576}+\frac{1}{12}\left(\frac{1}{8}\right)\\
&= \frac{1}{576}+\frac{1}{96}\\
&= \frac{7}{576},
\end{aligned}
$$

$$
g_2(x_1,\beta(x_2)) = \int_{\beta(x_2)}^{x_1}(\beta(t)-\beta(x_2))d_\beta t
$$

$$
\begin{aligned}
&= \int_1^{\frac{1}{6}}(\beta(t)-1)d_\beta t\\
&= -\int_{\frac{1}{6}}^{1}(\beta(t)-1)d_\beta t\\
&= -\left(\beta\left(\frac{1}{6}\right)-\frac{1}{6}\right)\left(\beta\left(\frac{1}{6}\right)-1\right)-\left(\beta\left(\frac{1}{4}\right)-\frac{1}{4}\right)\left(\beta\left(\frac{1}{4}\right)-1\right)\\
&\quad -\left(\beta\left(\frac{1}{2}\right)-\frac{1}{2}\right)\left(\beta\left(\frac{1}{2}\right)-1\right)\\
&= -\left(\frac{1}{4}-\frac{1}{6}\right)\left(\frac{1}{4}-1\right)-\left(\frac{1}{2}-\frac{1}{4}\right)\left(\frac{1}{2}-1\right)\\
&\quad -\left(1-\frac{1}{2}\right)(1-1)\\
&= -\frac{1}{12}\left(-\frac{3}{4}\right)-\frac{1}{4}\left(-\frac{1}{2}\right)\\
&= \frac{1}{16}+\frac{1}{8}\\
&= \frac{3}{16},
\end{aligned}
$$

$$
\begin{aligned}
g_2(\beta(x_2),\beta(x_1)) &= \int_{\beta(x_1)}^{\beta(x_2)}(\beta(t)-\beta(x_1))d_\beta t\\
&= \int_{\frac{1}{4}}^{1}\left(\beta(t)-\frac{1}{4}\right)d_\beta t
\end{aligned}
$$

$$\begin{aligned}
&= \left(\beta\left(\frac{1}{4}\right)-\frac{1}{4}\right)\left(\beta\left(\frac{1}{4}\right)-\frac{1}{4}\right)\\
&\quad +\left(\beta\left(\frac{1}{2}\right)-\frac{1}{2}\right)\left(\beta\left(\frac{1}{2}\right)-\frac{1}{4}\right)\\
&= \left(\frac{1}{2}-\frac{1}{4}\right)\left(\frac{1}{2}-\frac{1}{4}\right)+\left(1-\frac{1}{2}\right)\left(1-\frac{1}{4}\right)\\
&= \frac{1}{16}+\frac{3}{8}\\
&= \frac{7}{16},
\end{aligned}$$

$$\begin{aligned}
f(0) &= 0,\\
f\left(\frac{1}{6}\right) &= \frac{1}{6},\\
f\left(\frac{1}{2}\right) &= \frac{1}{2}.
\end{aligned}$$

Hence and the β-composite rule, we get

$$\begin{aligned}
\int_0^1 t d_\beta t &\approx \frac{1}{\beta(x_1)-\beta(x_0)}\left(f(x_0)g_2(x_0,\beta(x_1))+f(x_1)g_2(\beta(x_1),\beta(x_0))\right)\\
&\quad +\frac{1}{\beta(x_2)-\beta(x_1)}\left(f(x_1)g_2(x_1,\beta(x_2))+f(x_2)g_2(\beta(x_2),\beta(x_1))\right)\\
&= \frac{1}{\frac{1}{4}}\left(\frac{1}{6}\cdot\frac{7}{576}\right)+\frac{1}{1-\frac{1}{4}}\left(\frac{1}{6}\cdot\frac{3}{16}+\frac{1}{2}\cdot\frac{7}{16}\right)\\
&= \frac{14}{1728}+\frac{4}{3}\left(\frac{3}{96}+\frac{7}{32}\right)\\
&= \frac{14}{1728}+\frac{4}{3}\left(\frac{1}{4}\right)\\
&= \frac{14}{1728}+\frac{1}{3}\\
&= \frac{590}{1728}\\
&= \frac{295}{864}.
\end{aligned}$$

Exercise 5.6.3 *Let $I = \mathbb{R}$, $\beta(t) = \frac{7}{8}t + 2$, $t \in I$, and*

$$a = -2 = x_0, \quad x_1 = -1, \quad x_2 = -\frac{1}{8}, \quad b = 0,$$

$f : I \to \mathbb{R}$ be defined by

$$f(t) = \frac{t^2 - t + 1}{t^4 + t^2 + 1} + t, \quad t \in I.$$

Using the β-composite trapezium rule, evaluate the integral

$$\int_a^b f(t) d_\beta t.$$

Let $m \in \mathbb{N}$ and

$$a = x_0 \le \beta(x_0) < x_1 \le \beta(x_1) < \ldots < x_{2m} \le \beta(x_{2m}) = b.$$

Then

$$\int_a^b f(x) d_\beta x = \sum_{j=1}^m \int_{x_{2j-2}}^{\beta(x_{2j})} f(x) d_\beta x.$$

Now, applying the β-Simpson rule, we arrive at

$$\begin{aligned}
\int_a^b f(x) d_\beta x \approx & \sum_{j=1}^m \left(\frac{f(x_{2j-2})}{(\beta(x_{2j-2}) - \beta(x_{2j-1}))(\beta(x_{2j-2}) - \beta(x_{2j}))} \right. \\
& \times \Big(-g_2(x_{2j-2}, \beta(x_{2j-1})) g_1(x_{2j-2}, \beta(x_{2j})) \\
& + (\beta(x_{2j}) - x_{2j}) g_2(\beta(x_{2j}), \beta(x_{2j-1})) \\
& + G_3(x_{2j-2}, \beta(x_{2j-1})) - G_3(\beta(x_{2j}), \beta(x_{2j-1})) \Big) \\
& + \frac{f(x_{2j-1})}{(\beta(x_{2j-1}) - \beta(x_{2j-2}))(\beta(x_{2j-1}) - \beta(x_{2j}))} \\
& \times \Big((\beta(x_{2j}) - x_{2j}) g_2(\beta(x_{2j}), \beta(x_{2j-2})) - G_3(\beta(x_{2j}), \beta(x_{2j-2})) \Big) \\
& + \frac{f(x_{2j})}{(\beta(x_{2j}) - \beta(x_{2j-2}))(\beta(x_{2j}) - \beta(x_{2j-1}))} \\
& \times \Big(-g_1(\beta(x_{2j}), \beta(x_{2j-2})) g_2(\beta(x_{2j}), \beta(x_{2j-1}))
\end{aligned} \tag{5.10}$$

$$+(\beta(x_{2j})-x_{2j})g_2(\beta(x_{2j}),\beta(x_{2j-1}))$$

$$-G_3(\beta(x_{2j-1}),\beta(x_{2j}))-G_3(\beta(x_{2j}),\beta(x_{2j-1}))\Big).\Big).$$

Definition 5.6.4 *The numerical integration formula* (5.10) *will be called the β-composite Simpson rule.*

Exercise 5.6.5 *Let $I=\mathbb{R}$, $\beta(t)=\frac{1}{3}t+2$*

$$a=x_0=0,\quad x_1=4,\quad x_2=8,\quad x_3=12,\quad x_4=16=b,$$

$f:I\to\mathbb{R}$ be defined by

$$f(t)=1+t+t^2+e_2(t,0),\quad t\in I.$$

Using the β-composite Simpson rule, evaluate the integral

$$\int_a^b f(t)d_\beta t.$$

5.7 The Euler-Maclauren Expansion

Theorem 5.7.1 *Let $a,b\in I$, $a<b$, $k,m\in\mathbb{N}$, $x_j\in I$, $j\in\{0,1,\ldots,m\}$,*

$$a=x_0<x_1<\ldots<x_m=b,$$

$f\in\mathscr{C}_\beta^{2k}([a,b])$ and $T(m)$ be the result of approximation the integral $I=\int_a^b f(x)d_\beta x$ by the composite trapezium rule with m subintervals $[x_{j-1},x_j]$, $j\in\{1,\ldots,m\}$. Then

$$\begin{aligned} I-T(m) &= \sum_{j=1}^{m}\Bigg(\frac{1}{h_1(x_j,x_{j-1})}\sum_{l=1}^{2k-1}\Bigg((-1)^l\Bigg(H_{l+1}(x_j,x_{j-1})D_\beta^l f(x_j) \\ &\quad +H_{l+1}(x_{j-1},x_j)D_\beta^l f(x_{j-1})\Bigg) \\ &\quad +\int_{x_{j-1}}^{x_j}\left(H_{2k+1}(\beta(\tau),x_{j-1})-H_{2k+1}(\beta(\tau),x_j)\right)D_\beta^{2k}f(\tau)d_\beta\tau\Bigg)\Bigg). \end{aligned} \tag{5.11}$$

Definition 5.7.2 *The formula* (5.11) *is said to be the Euler-Maclauren expansion.*

Proof 5.7.3 *Fix $j \in \{1,\ldots,m\}$. Then, using the trapezium rule, we find*

$$\int_{x_{j-1}}^{x_j} f(x)d_\beta x - \frac{1}{h_1(x_j,x_{j-1})}\left(f(x_{j-1})h_2(x_{j-1},x_j)+f(x_j)h_2(x_j,x_{j-1})\right)$$

$$= \frac{1}{h_1(x_j,x_{j-1})}\left(\int_{x_{j-1}}^{x_j} h_1(\tau,x_{j-1})f(\tau)d_\beta\tau - \int_{x_{j-1}}^{x_j} h_1(\tau,x_j)f(\tau)d_\beta\tau\right)$$

$$-\frac{1}{h_1(x_j,x_{j-1})}(f(x_{j-1})h_2(x_{j+1},x_j)+f(x_j)h_2(x_j,x_{j-1}))$$

$$= \frac{1}{h_1(x_j,x_{j-1})}\left(\int_{x_{j-1}}^{x_j} D_\beta h_2(\tau,x_{j-1})f(\tau)d_\beta\tau\right.$$

$$\left.-\int_{x_{j-1}}^{x_j} D_\beta h_2(\tau,x_j)f(\tau)d_\beta\tau\right)$$

$$-\frac{1}{h_1(x_j,x_{j-1})}(f(x_{j-1})h_2(x_{j+1},x_j)+f(x_j)h_2(x_j,x_{j-1}))$$

$$= \frac{1}{h_1(x_j,x_{j-1})}(f(x_{j-1})h_2(x_{j+1},x_j)+f(x_j)h_2(x_j,x_{j-1}))$$

$$-\frac{1}{h_1(x_j,x_{j-1})}\int_{x_{j-1}}^{x_j}\left(h_2(\beta(\tau),x_{j-1})-h_2(\beta(\tau),x_j)\right)D_\beta f(\tau)d_\beta\tau$$

$$-\frac{1}{h_1(x_j,x_{j-1})}(f(x_{j-1})h_2(x_{j+1},x_j)+f(x_j)h_2(x_j,x_{j-1}))$$

$$= -\frac{1}{h_1(x_j,x_{j-1})}\int_{x_{j-1}}^{x_j}\left(D_\beta H_3(\tau,x_{j-1})-D_\beta H_3(\tau,x_j)\right)D_\beta f(\tau)d_\beta\tau$$

$$= -\frac{1}{h_1(x_j,x_{j-1})}\left(H_3(x_j,x_{j-1})D_\beta f(x_j)+H_3(x_{j-1},x_j)D_\beta f(x_{j-1})\right)$$

$$+\frac{1}{h_1(x_j,x_{j-1})}\int_{x_{j-1}}^{x_j}\left(D_\beta H_4(\tau,x_{j-1})-D_\beta H_4(\tau,x_j)\right)D_\beta^2 f(\tau)d_\beta\tau$$

$$= -\frac{1}{h_1(x_j,x_{j-1})}\left(H_3(x_j,x_{j-1})D_\beta f(x_j)+H_3(x_{j-1},x_j)D_\beta f(x_{j-1})\right)$$

$$+\frac{1}{h_1(x_j,x_{j-1})}\left(H_4(x_j,x_{j-1})D_\beta^2 f(x_j)+H_4(x_{j-1},x_j)D_\beta^2 f(x_{j-1})\right)$$

$$-\frac{1}{h_1(x_j,x_{j-1})}\int_{x_{j-1}}^{x_j}\left(D_\beta H_5(\tau,x_{j-1})-D_\beta H_5(\tau,x_j)\right)D_\beta^3 f(\tau)d_\beta\tau$$

$$= \cdots$$

$$= \frac{1}{h_1(x_j,x_{j-1})}\sum_{l=1}^{2k-1}\left((-1)^l\left(H_{l+1}(x_j,x_{j-1})D^l_\beta f(x_j)\right.\right.$$

$$\left. +H_{l+1}(x_{j-1},x_j)D^l_\beta f(x_{j-1})\right)$$

$$\left. +\int_{x_{j-1}}^{x_j}\left(H_{2k+1}(\beta(\tau),x_{j-1})-H_{2k+1}(\beta(\tau),x_j)\right)D^{2k}_\beta f(\tau)d_\beta\tau\right).$$

The summation over all the subintervals $[x_{j-1},x_j]$, $j\in\{1,\ldots,m\}$. gives the required result. This completes the proof.

5.8 The β-Euler-Maclauren Expansion

Theorem 5.8.1 *Let $a,b\in I$, $a<b$, $k,m\in\mathbb{N}$, $x_j\in I$, $j\in\{0,1,\ldots,m\}$,*

$$a=x_0\le\beta(x_0)<x_1\le\beta(x_1)<\ldots<x_m\le\beta(x_m)=b,$$

$f\in\mathscr{C}^{2k}_\beta([a,b])$ and $T_\beta(m)$ be the result of approximation the integral $I=\int_a^b f(x)d_\beta x$ by the β-composite trapezium rule with m subintervals $[x_{j-1},x_j]$, $j\in\{1,\ldots,m\}$. Then

$$I-T_\beta(m) = \sum_{j=1}^m\frac{(\beta(x_j)-x_j)}{\beta(x_j)-\beta(x_{j-1})}D_\beta f(x_j) \tag{5.12}$$

$$+\sum_{j=1}^m\left(\frac{1}{g_1(\beta(x_j),\beta(x_{j-1}))}\sum_{l=1}^{2k-1}\left((-1)^l\right.\right.$$

$$\times\left(g_{l+1}(\beta(x_j),\beta(x_{j-1}))D^l_\beta f(\beta(x_j))\right.$$

$$\left. +g_{l+1}(x_{j-1},\beta(x_j))D^l_\beta f(x_{j-1})\right)$$

$$\left.\left. +\int_{x_{j-1}}^{\beta(x_j)}\left(g_{2k+1}(\beta(\tau),\beta(x_{j-1}))-g_{2k+1}(\beta(\tau),\beta(x_j))\right)D^{2k}_\beta f(\tau)d_\beta\tau\right)\right).$$

Definition 5.8.2 *The formula* (5.12) *is said to be the β-Euler-Maclauren expansion.*

Proof 5.8.3 *Take $j\in\{1,\ldots,m\}$ arbitrarily. Applying the β-trapezium rule, we obtain*

$$\int_{x_{j-1}}^{\beta(x_j)}f(x)d_\beta x-\frac{1}{g_1(\beta(x_j),\beta(x_{j-1}))}\left(f(x_{j-1})g_2(x_{j-1},\beta(x_j))\right.$$

$$
\begin{aligned}
& +f(x_j)g_2(\beta(x_j),\beta(x_{j-1}))\Big) \\
= \; & \frac{1}{g_1(\beta(x_j),\beta(x_{j-1}))}\left(\int_{x_{j-1}}^{\beta(x_j)} g_1(\beta(\tau),\beta(x_{j-1}))f(\tau)d_\beta\tau\right. \\
& \left.-\int_{x_{j-1}}^{\beta(x_j)} g_1(\beta(\tau),\beta(x_j))f(\tau)d_\beta\tau\right) \\
& -\frac{1}{g_1(\beta(x_j),\beta(x_{j-1}))} \\
& \times(f(x_{j-1})g_2(x_{j-1},\beta(x_j))+f(x_j)g_2(\beta(x_j),\beta(x_{j-1}))) \\
= \; & \frac{1}{g_1(\beta(x_j),\beta(x_{j-1}))}\left(\int_{x_{j-1}}^{\beta(x_j)} D_\beta g_2(\tau,\beta(x_{j-1}))f(\tau)d_\beta\tau\right. \\
& \left.-\int_{x_{j-1}}^{\beta(x_j)} D_\beta g_2(\tau,\beta(x_j))f(\tau)d_\beta\tau\right) \\
& -\frac{1}{g_1(\beta(x_j),\beta(x_{j-1}))} \\
& \times(f(x_{j-1})g_2(x_{j-1},\beta(x_j))+f(x_j)g_2(\beta(x_j),\beta(x_{j-1}))) \\
= \; & \frac{1}{g_1(\beta(x_j),\beta(x_{j-1}))} \\
& \times(f(x_{j-1})g_2(x_{j-1},\beta(x_j))+f(\beta(x_j))g_2(\beta(x_j),\beta(x_{j-1}))) \\
& -\frac{1}{g_1(\beta(x_j),\beta(x_{j-1}))}\int_{x_{j-1}}^{\beta(x_j)} \\
& \times\left(g_2(\beta(\tau),\beta(x_{j-1}))-g_2(\beta(\tau),\beta(x_j))\right)D_\beta f(\tau)d_\beta\tau \\
& -\frac{1}{g_1(\beta(x_j),\beta(x_{j-1}))} \\
& \times(f(x_{j-1})g_2(x_{j-1},\beta(x_j))+f(x_j)g_2(\beta(x_j),\beta(x_{j-1}))) \\
= \; & \frac{(\beta(x_j)-x_j)}{\beta(x_j)-\beta(x_{j-1})}D_\beta f(x_j) \\
& -\frac{1}{g_1(\beta(x_j),\beta(x_{j-1}))}
\end{aligned}
$$

$$
\begin{aligned}
&\times \int_{x_{j-1}}^{\beta(x_j)} \left(D_\beta g_3(\tau,\beta(x_{j-1})) - D_\beta g_3(\tau,\beta(x_j))\right) D_\beta f(\tau) d_\beta \tau \\
= \quad & \frac{\beta(x_j)-x_j}{\beta(x_j)-\beta(x_{j-1})} D_\beta f(x_j) \\
&- \frac{1}{g_1(\beta(x_j),\beta(x_{j-1}))} \\
&\times \left(g_3(\beta(x_j),\beta(x_{j-1}))D_\beta f(\beta(x_j)) + g_3(x_{j-1},\beta(x_j))D_\beta f(x_{j-1})\right) \\
&+ \frac{1}{g_1(\beta(x_j),\beta(x_{j-1}))} \\
&\times \int_{x_{j-1}}^{\beta(x_j)} \left(D_\beta g_4(\tau,\beta(x_{j-1})) - D_\beta g_4(\tau,\beta(x_j))\right) D_\beta^2 f(\tau) d_\beta \tau \\
= \quad & \frac{\beta(x_j)-x_j}{\beta(x_j)-\beta(x_{j-1})} D_\beta f(x_j) \\
&- \frac{1}{g_1(\beta(x_j),\beta(x_{j-1}))} \\
&\times \left(g_3(\beta(x_j),\beta(x_{j-1}))D_\beta f(\beta(x_j)) + g_3(x_{j-1},\beta(x_j))D_\beta f(x_{j-1})\right) \\
&+ \frac{1}{g_1(\beta(x_j),\beta(x_{j-1}))} \\
&\times \Big(g_4(\beta(x_j),\beta(x_{j-1}))D_\beta^2 f(\beta(x_j)) + g_4(x_{j-1},\beta(x_j))D_\beta^2 f(x_{j-1})\Big) \\
&- \frac{1}{g_1(\beta(x_j),\beta(x_{j-1}))} \\
&\times \int_{x_{j-1}}^{\beta(x_j)} \left(D_\beta g_5(\tau,\beta(x_{j-1})) - D_\beta g_5(\tau,\beta(x_j))\right) D_\beta^3 f(\tau) d_\beta \tau \\
= \quad & \cdots \\
= \quad & \frac{\beta(x_j)-x_j}{\beta(x_j)-\beta(x_{j-1})} D_\beta f(x_j) \\
&+ \frac{1}{g_1(\beta(x_j),\beta(x_{j-1}))} \sum_{l=1}^{2k-1} \Bigg((-1)^l \Big(g_{l+1}(\beta(x_j),\beta(x_{j-1}))D_\beta^l f(x_j)
\end{aligned}
$$

$$+g_{l+1}(x_{j-1},\beta(x_j))D^l_\beta f(x_{j-1})\Big)$$

$$+\int_{x_{j-1}}^{\beta(x_j)}\left(g_{2k+1}(\beta(\tau),\beta(x_{j-1}))-g_{2k+1}(\beta(\tau),\beta(x_j))\right)D^k_\beta f(\tau)d_\beta\tau\Big).$$

The summation over all the subintervals $[x_{j-1},x_j]$, $j\in\{1,\ldots,m\}$, gives the desired result. This completes the proof.

5.9 Construction of Gauss Quadrature Rules

Let $a,b\in I$, $a<b$, $f:[a,b]\to\mathbb{R}$, $w:[a,b]\to[0,\infty)$, $f\in\mathscr{C}^1_\beta([a,b])$, $w\in\mathscr{C}_\beta([a,b])$. We wish construct a quadrature formula for the approximate evaluation of the integral

$$\int_a^b w(x)f(x)d_\beta x.$$

Suppose that $n\in\mathbb{N}$, $x_j\in I$, $j\in\{0,1,\ldots,n\}$, and

$$a=x_0\le\beta(x_0)<x_1\le\beta(x_1)<\ldots<x_n\le\beta(x_n)=b.$$

Define the polynomials

$$\begin{aligned} P_j(x) &= \left(1-\frac{D_\beta M_j(x_j)L_j(x_j)+M_j(\beta(x_j))D_\beta L_j(x_j)}{M_j(\beta(x_j))L_j(\beta(x_j))}(x-x_j)\right)\\ &\quad\times M_j(x)L_j(x),\\ K_j(x) &= \frac{x-x_j}{M_j(\beta(x_j))L_j(\beta(x_j))}M_j(x)L_j(x),\quad j\in\{0,1,\ldots,n\},\quad x\in[a,b]. \end{aligned}$$

where the polynomials M_j, L_j, $j\in\{0,1,\ldots,n\}$, are defined as in Chapter 4. Then the Hermite interpolation polynomial of degree $2n+1$ for the function f is given by the expression.

$$p_{2n+1}(x)=\sum_{j=0}^n\left(P_j(x)f(x_j)+K_j(x)D_\beta f(x_j)\right),\quad x\in[a,b].$$

Thus,

$$\begin{aligned} \int_a^b w(x)f(x)d_\beta x &\approx \int_a^b w(x)p_{2n+1}(x)d_\beta x\\ &= \int_a^b w(x)\left(\sum_{j=0}^n\left(P_j(x)f(x_j)+K_j(x)D_\beta f(x_j)\right)\right)d_\beta x \end{aligned}$$

$$= \sum_{j=0}^{n}\left(\int_a^b w(x)P_j(x)d_\beta x\right)f(x_j)+\sum_{j=0}^{n}\left(\int_a^b w(x)K_j(x)d_\beta x\right)D_\beta f(x_j).$$

Let

$$W_j = \int_a^b w(x)P_j(x)d_\beta x,$$

$$V_j = \int_a^b w(x)K_j(x)d_\beta x, \quad j\in\{0,1,\ldots,n\}.$$

Therefore

$$\int_a^b w(x)f(x)d_\beta x \approx \sum_{j=0}^{n}\left(W_jf(x_j)+V_jD_\beta f(x_j)\right). \tag{5.13}$$

Now, we take the function w so that $V_j\equiv 0$, $j\in\{0,1,\ldots,n\}$. Such function exists.

Example 5.9.1 *Let*

$$I=\left\{-1,0,\frac{1}{8},\frac{1}{6},\frac{1}{5},\frac{1}{4},\frac{1}{3},\frac{1}{2},1\right\},$$

$\beta : I\to\mathbb{R}$ *be defined by*

$$\beta(-1) = -1,$$

$$\beta(0) = \frac{1}{8},$$

$$\beta\left(\frac{1}{8}\right) = \frac{1}{6},$$

$$\beta\left(\frac{1}{6}\right) = \frac{1}{4},$$

$$\beta\left(\frac{1}{4}\right) = \frac{1}{3},$$

$$\beta\left(\frac{1}{3}\right) = \frac{1}{2},$$

$$\beta\left(\frac{1}{2}\right) = 1,$$

$$\beta(1) = 2,$$

and

$$a=x_0=0,\quad x_1=\frac{1}{6},\quad x_2=\frac{1}{4},\quad x_3=\frac{1}{2},\quad b=\beta(x_3)=1,$$

$$\begin{aligned} w(x) &= \left(x-\frac{1}{8}\right)\left(x-\frac{1}{5}\right)\left(x-\frac{1}{3}\right) \\ &\quad \times sign\left(\left(x-\frac{1}{8}\right)\left(x-\frac{1}{5}\right)\left(x-\frac{1}{3}\right)\right), \quad x \in [0,1]. \end{aligned}$$

Note that

$$\begin{aligned} K_j(x_k) &= 0, \quad j,k \in \{0,1,2,3\}, \\ w\left(\frac{1}{8}\right) &= w\left(\frac{1}{5}\right) = w\left(\frac{1}{3}\right) = 0. \end{aligned}$$

Thus,

$$\begin{aligned} \int_0^1 w(x)K_j(x)d_\beta x &= (\beta(0)-0)w(0)K_j(0) \\ &\quad + \left(\beta\left(\frac{1}{8}\right)-\frac{1}{8}\right)w\left(\frac{1}{8}\right)K_j\left(\frac{1}{8}\right) \\ &\quad + \left(\beta\left(\frac{1}{6}\right)-\frac{1}{6}\right)w\left(\frac{1}{6}\right)K_j\left(\frac{1}{6}\right) \\ &\quad + \left(\beta\left(\frac{1}{4}\right)-\frac{1}{4}\right)w\left(\frac{1}{4}\right)K_j\left(\frac{1}{4}\right) \\ &\quad + \left(\beta\left(\frac{1}{3}\right)-\frac{1}{3}\right)w\left(\frac{1}{3}\right)K_j\left(\frac{1}{3}\right) \\ &\quad + \left(\beta\left(\frac{1}{2}\right)-\frac{1}{2}\right)w\left(\frac{1}{2}\right)K_j\left(\frac{1}{2}\right) \\ &= 0, \quad j \in \{0,1,2,3\}. \end{aligned}$$

Then, the formula (5.13) takes the form

$$\int_a^b w(x)f(x)d_\beta x \approx \sum_{j=0}^{n} W_j f(x_j). \tag{5.14}$$

Definition 5.9.2 *The numerical integration formula* (5.14) *is said to be the Gauss quadrature rule, the quantities* W_j*,* $j \in \{0,1,\ldots,n\}$*, are said to be quadrature weights.*

Example 5.9.3 *Let* $I = \left\{-1, 0, \frac{1}{4}, \frac{1}{3}, 1\right\}$ *and* $\beta : I \to \mathbb{R}$ *be defined by*

$$\beta(-1) = -1,$$

$$\begin{aligned}\beta(0) &= \frac{1}{4},\\ \beta\left(\frac{1}{4}\right) &= \frac{1}{3},\\ \beta\left(\frac{1}{3}\right) &= 1,\\ \beta(1) &= 2,\end{aligned}$$

and

$$\begin{aligned}a &= x_0 = 0, \quad x_1 = \frac{1}{3}, \quad \beta(x_1) = 1 = b,\\ w(x) &= 1 - x,\\ f(x) &= x, \quad x \in [0,1].\end{aligned}$$

We will evaluate

$$\int_a^b f(x) d_\beta x,$$

using the Gauss quadrature rule. We have

$$\begin{aligned}L_0(x) &= \frac{x - x_1}{x_0 - x_1}\\ &= -3x + 1,\\ L_0(x_0) &= 1,\\ D_\beta L_0(x) &= -3,\\ L_0(\beta(x_0)) &= L_0\left(\frac{1}{4}\right)\\ &= -\frac{3}{4} + 1\\ &= \frac{1}{4},\\ L_1(x) &= \frac{x - x_0}{x_1 - x_0}\end{aligned}$$

$$
\begin{aligned}
&= 3x, \\
L_1(x_1) &= 1, \\
D_\beta L_1(x) &= 3, \\
L_1(\beta(x_1)) &= L_1(1) \\
&= 3, \\
M_0(x) &= \frac{x-\beta(x_1)}{x_0-\beta(x_1)} \\
&= -x+1, \\
D_\beta M_0(x) &= -1, \\
M_0(\beta(x_0)) &= -\frac{1}{4}+1 \\
&= \frac{3}{4}, \\
M_1(x) &= \frac{x-\beta(x_0)}{x_1-\beta(x_0)} \\
&= \frac{x-\frac{1}{4}}{\frac{1}{3}-\frac{1}{4}} \\
&= \frac{\frac{4x-1}{4}}{\frac{1}{12}} \\
&= 12x-3, \\
D_\beta M_1(x) &= 12, \\
M_1(\beta(x_1)) &= 12-3 \\
&= 9,
\end{aligned}
$$

$$\begin{aligned}
P_0(x) &= \left(1-\frac{D_\beta M_0(x_0)L_0(x_0)+M_0(\beta(x_0))D_\beta L_0(x_0)}{M_0(\beta(x_0))L_0(\beta(x_0))}(x-x_0)\right)\\
&\quad \times M_0(x)L_0(x)\\
&= \left(1-\frac{-1+\frac{3}{4}(-3)}{\frac{3}{4}\cdot\frac{1}{4}}x\right)(1-x)(1-3x)\\
&= \left(1-\frac{-\frac{13}{4}}{\frac{3}{16}}x\right)(1-x)(1-3x)\\
&= \left(1-\frac{52}{3}x\right)(1-x)(1-3x),\\
P_1(x) &= \left(1-\frac{D_\beta M_1(x_1)L_1(x_1)+M_1(\beta(x_1))D_\beta L_1(x_1)}{M_1(\beta(x_1))L_1(\beta(x_1))}(x-x_1)\right)\\
&\quad \times M_1(x)L_1(x)\\
&= \left(1-\frac{12+9\cdot 3}{9\cdot 3}\left(x-\frac{1}{3}\right)\right)(12x-3)(3x)\\
&= \left(1-\frac{13}{9}\left(x-\frac{1}{3}\right)\right)(12x-3)(3x)\\
&= 3x(12x-3)-13x\left(x-\frac{1}{3}\right)(4x-1), \quad x\in[0,1],\\
P_1(0) &= 0,\\
P_1\left(\frac{1}{4}\right) &= 0,\\
P\left(\frac{1}{3}\right) &= 1,\\
w(0) &= 1,\\
w\left(\frac{1}{4}\right) &= \frac{3}{4},\\
w\left(\frac{1}{3}\right) &= \frac{2}{3},
\end{aligned}$$

$$\begin{aligned}\beta\left(\frac{1}{4}\right)-\frac{1}{4} &= \frac{1}{3}-\frac{1}{4}\\ &= \frac{1}{12},\\ \beta\left(\frac{1}{3}\right)-\frac{1}{3} &= 1-\frac{1}{3}\\ &= \frac{2}{3}.\end{aligned}$$

Hence,

$$\begin{aligned}\int_0^1 (1-x)x d_\beta x &\approx f(x_0)\int_0^1 w(x)P_0(x)d_\beta x\\ &\quad +f(x_1)\int_0^1 w(x)P_1(x)d_\beta x\\ &= \frac{1}{3}\int_0^1 w(x)P_1(x)d_\beta x\\ &= \frac{1}{3}(\beta(0)-0)w(0)P_1(0)\\ &\quad +\frac{1}{3}\left(\beta\left(\frac{1}{4}\right)-\frac{1}{4}\right)w\left(\frac{1}{4}\right)O_1\left(\frac{1}{4}\right)\\ &\quad +\frac{1}{3}\left(\beta\left(\frac{1}{3}\right)-\frac{1}{3}\right)w\left(\frac{1}{3}\right)O_1\left(\frac{1}{3}\right)\\ &= \frac{1}{3}\cdot\frac{2}{3}\cdot\frac{2}{3}\\ &= \frac{4}{27}.\end{aligned}$$

Exercise 5.9.4 *Let* $I=\mathbb{R}$, $\beta(t)=\frac{1}{7}t+2$, $t\in I$,

$$\begin{aligned}a &= x_0=-1,\quad x_1=-\frac{1}{2},\quad x_2=\frac{1}{3},\quad b=\frac{2}{3},\\ w(x) &= \left|x^3-\frac{4}{9}x\right||,\\ f(x) &= \frac{1+x}{1+x+x^2},\quad x\in I.\end{aligned}$$

Using the Gauss quadrature rule, evaluate

1.
$$\int_a^b w(x)f(x)d_\beta x.$$

2.
$$\int_a^b (w(x))^2 f(x)d_\beta x.$$

3.
$$\int_a^b (w(x))^3 f(x)d_\beta x.$$

5.10 Error Estimation for Gauss Quadrature Rules

Let a, b, f, w, x_j, $j \in \{0,1,\dots,n\}$, be as in Section 5.9. By Section 4.3, we have

$$G_{min,2n+2}(\xi) \le \frac{f(x)-p_{2n+1}(x)}{\pi_{n+1}(x)\zeta_{n+1}(x)} \le G_{max,2n+2}(\xi)$$

for some $\xi = \xi(x) \in (a,b)$, $x \in [a,b]$. Observe that

$$\text{sign}(\pi_{n+1}(x)) = \text{sign}(\zeta_{n+1}(x)), \quad x \in [a,b].$$

Then

$$\begin{aligned} G_{min,2n+2}(\xi)\pi_{n+1}(x)\zeta_{n+1}(x) &\le f(x)-p_{2n+1}(x) \\ &\le G_{max,2n+2}(\xi)\pi_{n+1}(x)\zeta_{n+1}(x), \end{aligned} \tag{5.15}$$

$x \in [a,b]$. Next,

$$\int_a^b w(x)f(x)d_\beta x - \sum_{j=0}^n W_j f(x_j) = \int_a^b w(x)(f(x)-p_{2n+1}(x))d_\beta x.$$

Now, applying (5.15), we arrive at

$$\begin{aligned} &\int_a^b w(x)G_{min,2n+2}(\xi)\pi_{n+1}(x)\zeta_{n+1}(x)d_\beta x \\ &\le \int_a^b w(x)f(x)d_\beta x - \sum_{j=0}^n W_j f(x_j) \\ &\le \int_a^b w(x)G_{max,2n+2}(\xi)\pi_{n+1}(x)\zeta_{n+1}(x)d_\beta x. \end{aligned}$$

5.11 β-Gauss Quadrature Rules

Throughout this section, suppose that β is β-differentiable. Let $a,b \in I$, $a < b$, $f : [a,b] \to \mathbb{R}$, $w : [a,b] \to [0,\infty)$, $f \in \mathscr{C}^1_\beta([a,b])$, $w \in \mathscr{C}_\beta([a,b])$. We wish construct a quadrature formula for the approximate evaluation of the integral

$$\int_a^b w(x)f(x)d_\beta x.$$

Suppose that $n \in \mathbb{N}$, $x_j \in I$, $j \in \{0,1,\ldots,n\}$, and

$$a = x_0 \le \beta(x_0) < x_1 \le \beta(x_1) < \ldots < x_n \le \beta(x_n) = b.$$

Define the polynomials

$$\begin{aligned} P_{\beta j}(x) &= \left(1 - \frac{D_\beta M_j(x_j) + M_j(\beta(x_j))D_\beta L_j(x_j)}{D_\beta\beta(x_j)M_j(\beta(x_j))L_j(\beta(x_j))}(\beta(x) - \beta(x_j))\right) \\ &\quad \times M_j(x)L_j(x), \\ K_{\beta j}(x) &= \frac{\beta(x) - \beta(x_j)}{D_\beta\beta(x_j)M_j(\beta(x_j))L_j(\beta(x_j))} M_j(x)L_j(x), \end{aligned}$$

$x \in [a,b]$, $j \in \{0,1,\ldots,n\}$, where the polynomials L_j and M_j are defined as in Chapter 4. Then the β-Hermite interpolation polynomial of degree $2n+1$ for the function f is given by the following way.

$$p_{\beta 2n+1}(x) = \sum_{j=0}^{n} \left(P_{\beta j}(x)f(x_j) + K_{\beta j}(x)D_\beta f(x_j)\right), \quad x \in [a,b].$$

Thus,

$$\begin{aligned} \int_a^b w(x)f(x)d_\beta x &\approx \int_a^b w(x)p_{\beta 2n+1}(x)d_\beta x \\ &= \int_a^b w(x)\left(\sum_{j=0}^{n}\left(P_{\beta j}(x)f(x_j) + K_{\beta j}(x)D_\beta f(x_j)\right)\right)d_\beta x \\ &= \sum_{j=0}^{n}\left(\int_a^b w(x)P_{\beta j}(x)d_\beta x\right)f(x_j) \\ &\quad + \sum_{j=0}^{n}\left(\int_a^b w(x)K_{\beta j}(x)d_\beta x\right)D_\beta f(x_j). \end{aligned}$$

Set

$$W_{\beta j} = \int_a^b w(x)P_{\beta j}(x)d_\beta x,$$

$$V_{\beta j} = \int_a^b w(x)K_{\beta j}(x)d_\beta x, \quad j \in \{0,1,\ldots,n\}.$$

Therefore

$$\int_a^b w(x)f(x)d_\beta x \approx \sum_{j=0}^{n} \left(W_{\beta j}f(x_j) + V_{\beta j}D_\beta f(x_j)\right). \tag{5.16}$$

Now, we take the function w so that $V_{\beta j} \equiv 0$, $j \in \{0,1,\ldots,n\}$. Thus, (5.16) takes the form

$$\int_a^b w(x)f(x)d_\beta x \approx \sum_{j=0}^{n} W_{\beta j}f(x_j). \tag{5.17}$$

Definition 5.11.1 *The numerical integration formula* (5.17) *is said to be the β-Gauss quadrature rule and the quantities $W_{\beta j}$, $j \in \{0,1,\ldots,n\}$, are said to be the β-quadrature weights.*

Exercise 5.11.2 *Let $I = \mathbb{R}$, $\beta(t) = \frac{1}{2}t + 4$, $t \in I$,*

$$a = x_0 = 0, \quad x_1 = \frac{5}{6}, \quad x_2 = 3, \quad x_3 = 11, \quad b = 12,$$

$$f(t) = t + t^2, \quad w(t) = \left(t - \frac{1}{3}\right)^2 (t-1)^2(t-7)^2(t-12)^2, \quad t \in I.$$

Using the β-Gauss quadrature rule, evaluate the integrals

1. $$\int_a^b w(t)f(t)d_\beta t.$$

2. $$\int_a^b (w(t))^2 f(t)d_\beta t.$$

5.12 Error Estimation for β-Gauss Quadrature Rules

Let β, a, b, f, w, x_j, $j \in \{0,1,\ldots,n\}$, be as in the previous section. By Section 4.4, we have

$$G_{min,2n+2}(\xi) \le \frac{f(x) - p_{\beta 2n+1}(x)}{\pi_{n+1}(x)\zeta_{n+1}(x)} \le G_{max,2n+2}(\xi)$$

for some $\xi = \xi(x) \in (a,b)$, $x \in [a,b]$. Since

$$\operatorname{sign}(\pi_{n+1}(x)) = \operatorname{sign}(\zeta_{n+1}(x)), \quad x \in [a,b],$$

we find

$$\begin{aligned} G_{min,2n+2}(\xi)\pi_{n+1}(x)\zeta_{n+1}(x) &\leq f(x) - p_{\beta 2n+1}(x) \\ &\leq G_{max,2n+2}(\xi)\pi_{n+1}(x)\zeta_{n+1}(x), \end{aligned}$$

$x \in [a,b]$, and

$$\int_a^b w(x)f(x)d_\beta x - \sum_{j=0}^{n} W_j f(\beta x_j) = \int_a^b w(x)(f(x) - p_{\beta 2n+1}(x))d_\beta x,$$

whereupon we arrive at

$$\begin{aligned} \int_a^b w(x)G_{min,2n+2}(\xi)\pi_{n+1}(x)\zeta_{n+1}(x)d_\beta x &\leq \int_a^b w(x)f(x)d_\beta x - \sum_{j=0}^{n} W_j f(x_j) \\ &\leq \int_a^b w(x)G_{max,2n+2}(\xi)\pi_{n+1}(x)\zeta_{n+1}(x)d_\beta x. \end{aligned}$$

5.13 Advanced Practical Problems

Problem 5.13.1 *Let $I = \mathbb{R}$, $\beta(t) = \frac{1}{3}t + 2$, $t \in I$. Using the trapezium rule, evaluate the integral*

$$\int_{\frac{1}{256}}^{1} \frac{t^2+t+1}{t^4 - 3t^3 + 7t^2 + 4t + 1} d_\beta t.$$

Problem 5.13.2 *Let $I = \mathbb{R}$, $\beta(t) = 3t$. Using the Simpson rule, evaluate the integral*

$$\int_1^{27} \frac{t^5+t+1}{t^2+3t+10} d_\beta t.$$

Problem 5.13.3 *Let $I = \mathbb{R}$, $\beta(t) = \frac{1}{8}t$, $t \in I$. Using the composite trapezium rule, evaluate the integral*

$$\int_1^{81} \left(\frac{t-1}{t^4+t+1} + \cos_{1,\beta}(t,1) \right) d_\beta t.$$

Problem 5.13.4 *Let* $I = \mathbb{R}$, $\beta(t) = \frac{3}{4}t + 2$, $t \in I$,

$$a = -1 = x_0, \quad x_1 = \frac{1}{2}, \quad b = 3,$$

$f : I \to \mathbb{R}$ *be defined by*

$$f(t) = \cosh_{1,\beta}(t,-1) + \sinh_{2,\beta}(t,-1) + t^3, \quad t \in I.$$

Using the β*-trapezium rule, evaluate the integral*

$$\int_{-1}^{3} f(t)d_\beta t.$$

Problem 5.13.5 *Let* $I = \mathbb{R}$, $\beta(t) = \frac{1}{3}t$, $t \in I$,

$$a = -3 = x_0, \quad x_1 = \frac{1}{2}, \quad x_2 = \frac{4}{3}, \quad b = 2,$$

$f : I \to \mathbb{R}$ *be defined by*

$$f(t) = \sin_{2,\beta}(t,0) + e_{1,\beta}(t,0) - \frac{t}{2+t+t^2}, \quad t \in I.$$

Using the β*-Simpson rule, evaluate the integral*

$$\int_a^b f(t)d_\beta t.$$

Problem 5.13.6 *Let* $I = \mathbb{R}$, $\beta(t) = \frac{1}{8}t$, $t \in I$,

$$a = 1 = x_0, \quad x_1 = 8, \quad x_2 = 64,$$

$f : I \to \mathbb{R}$ *be defined by*

$$f(t) = \frac{1+t+t^2}{1+t+t^2+t^3+t^4+t^5+t^6} + \cosh_1(t,3), \quad t \in I.$$

Using the composite Simpson rule, evaluate the integral

$$\int_a^b f(t)d_\beta t.$$

Problem 5.13.7 *Let* $I = \mathbb{R}$, $\beta(t) = \frac{1}{11}t + 2$, $t \in I$,

$$a = x_0 = 1, \quad x_1 = \frac{4}{3}, \quad x_2 = \frac{5}{3}, \quad x_3 = \frac{7}{3}, \quad x_4 = 4, \quad b = 4,$$

$f : I \to \mathbb{R}$ be defined by

$$f(t) = t^4 + t^2 + \frac{1}{1+t} + e_1(t,1), \quad t \in I.$$

Using the β-composite trapezium rule, evaluate the integral

$$\int_a^b f(t) d_\beta t.$$

Problem 5.13.8 *Let $I = \mathbb{R}$, $\beta(t) = 4t$, $t \in I$,*

$$a = 1 = x_0, \quad x_1 = 9, \quad x_2 = 81, \quad x_3 = 729, \quad x_4 = 6561,$$

$f : I \to \mathbb{R}$ be defined by

$$f(t) = \frac{1+t}{1+t^6} + \sin_1(t,1) + \cosh_2(t,1), \quad t \in I.$$

Using the β-composite Simpson rule, evaluate the integral

$$\int_a^b f(t) d_\beta t.$$

Problem 5.13.9 *Let $I = \mathbb{R}$, $\beta(t) = 2t$, $t \in I$,*

$$a = x_0 = 0, \quad x_1 = \frac{1}{9}, \quad x_2 = \frac{1}{7}, \quad x_3 = \frac{1}{5}, \quad x_4 = \frac{1}{3}, \quad x_5 = \frac{5}{6}, \quad b = 1,$$

$$\begin{aligned} w(x) &= \left(x - \frac{1}{10}\right)^2 \left(x - \frac{1}{8}\right)^2 \left(x - \frac{1}{6}\right)^2 \left(x - \frac{1}{4}\right)^2 \\ &\quad \times \left(x - \frac{1}{10}\right)^2 \left(x - \frac{2}{3}\right)^2 (x-1)^2, \\ f(x) &= 1 + x + e_1(x,0), \quad x \in [0,1]. \end{aligned}$$

Using the Gauss quadrature rule, evaluate

1. $$\int_a^b w(x) f(x) d_\beta x.$$

2. $$\int_a^b (w(x))^2 f(x) d_\beta x.$$

3. $$\int_a^b (w(x))^3 f(x) d_\beta x.$$

Problem 5.13.10 *Let* $I = \mathbb{R}$, $\beta(t) = \frac{1}{3}t + 7$, $t \in I$,

$$a = x_0 = -1, \quad x_1 = 0, \quad x_2 = 1, \quad b = 7,$$

$$f(t) = 1 + t + t^2, \quad w(t) = \left(t + \frac{1}{4}\right)^2 \left(t - \frac{1}{8}\right)^2 (t-7)^2, \quad t \in I.$$

Using the β*-Gauss quadrature rule, evaluate the integral*

$$\int_a^b (w(t))^2 \left(f(t) + (f(t))^3\right) d_\beta t.$$

5.14 Notes and References

In this chapter, using the Lagrange, β-Lagrange, Hermite and β-Hermite interpolation polynomials, we derive formulae for numeraical β-integration. We deduct the trapezium and β-trapezium rules, Newton-Cotes and β-Newton-Cotes formulas and we give some error estimates. In the chapter they are constructed Gauss and β-Gauss quadrature rules. They are considered Euler-Maclauren and β-Euler-Maclauren expansions.

6

Piecewise Polynomial Approximation

The focus of our discussions in the previous chapters has been the question of approximation of a given continuous function f, defined on an interval $[a,b]$ by a polynomial on that interval either through Lagrange interpolation, β-Lagrange interpolation, Hermite interpolation or β-Hermite interpolation polynomials. Each of these constructions was global in nature, in the sense that the approximation was defined by the same analytical expression on the whole interval $[a,b]$. In this chapter, we will present an alternative and more flexible way for approximation of a function f and this approach is based on the dividing the interval $[a,b]$ into a number of subintervals and to look for a piecewise approximation by polynomials of low degree. Such piecewise-polynomial approximations are called splines, and the endpoints of the subintervals are known as the knots.

Let $I \subseteq \mathbb{R}$ and β be first kind or second kind general quantum operator.

6.1 Linear Interpolating Splines

Let $a,b \in I$, $a < b$, and $m \in \mathbb{N}$, $m \geq 2$.

Definition 6.1.1 *Suppose that $f \in \mathscr{C}([a,b])$, $K = \{x_0, x_1, \ldots, x_m\}$ be a subset of $[a,b]$ such that*

$$a = x_0 < x_1 < \ldots < x_m = b.$$

The linear spline s_L, interpolating f at the points x_j, $j \in \{0,1,\ldots,m\}$, is defined by

$$s_L(x) = \frac{x_j - x}{x_j - x_{j-1}} f(x_{j-1}) + \frac{x - x_{j-1}}{x_j - x_{j-1}} f(x_j), \quad x \in [x_{j-1}, x_j],$$

$j \in \{1,\ldots,m\}$. The points x_j, $j \in \{0,1,\ldots,m\}$, are the knots of the spline, and K is said to be the set of knots.

DOI: 10.1201/9781003472131-6

By Definition 6.1.1, it follows that

$$s_L(x_{j-1}) = f(x_{j-1}), \quad s_L(x_j) = f(x_j), \quad j \in \{1, \ldots, m\}.$$

Below, denote

$$\begin{aligned} r_{j+1} &= x_{j+1} - x_j, \quad j \in \{0, 1, \ldots, m-1\}, \\ r &= \max_{j \in \{1, \ldots, m\}} r_j. \end{aligned}$$

Theorem 6.1.2 *Let $f \in \mathscr{C}^1_\beta([a,b])$. Then*

$$\max_{x \in [a,b]} |s_L(x) - f(x)| \leq 2r \max_{x \in [a,b]} \left| D_\beta f(x) \right|.$$

Proof 6.1.3 *Let $j \in \{1, \ldots, m\}$ be arbitrarily chosen and fixed. We have*

$$\begin{aligned} |s_L(x) - f(x)| &= \left| \frac{x_j - x}{x_j - x_{j-1}} f(x_{j-1}) + \frac{x - x_{j-1}}{x_j - x_{j-1}} f(x_j) - f(x) \right| \\ &= \left| \frac{x_j - x}{x_j - x_{j_1}} (f(x_{j-1}) - f(x)) + \frac{x - x_{j-1}}{x_j - x_{j-1}} (f(x_j) - f(x)) \right| \\ &\leq \frac{x_j - x}{x_j - x_{j-1}} |f(x_{j-1}) - f(x)| + \frac{x - x_{j-1}}{x_j - x_{j-1}} |f(x_j) - f(x)|, \end{aligned}$$

$x \in [a,b]$. By the Mean Value Theorem, it follows that

$$\begin{aligned} |f(x_{j+1}) - f(x)| &\leq \max_{x \in [x_{j-1}, x_j]} \left| D_\beta f(x) \right| (x - x_{j-1}) \\ &\leq \max_{x \in [a,b]} \left| D_\beta f(x) \right| (x - x_{j-1}), \quad x \in [x_{j-1}, x_j], \end{aligned}$$

and

$$\begin{aligned} |f(x_j) - f(x)| &\leq \max_{x \in [x_{j-1}, x_j]} \left| D_\beta f(x) \right| (x_j - x) \\ &\leq \max_{x \in [a,b]} \left| D_\beta f(x) \right| (x_j - x), \quad x \in [x_{j-1}, x_j]. \end{aligned}$$

Therefore

$$\begin{aligned} |s_L(x) - f(x)| \leq & \frac{(x_j - x)^2}{r_j} \max_{x \in [a,b]} \left| D_\beta f(x) \right| \\ & + \frac{(x - x_{j-1})^2}{r_j} \max_{x \in [a,b]} \left| D_\beta f^{D_\beta}(x) \right| \end{aligned}$$

$$\leq 2r_j \max_{x\in[a,b]} |f(x)|$$

$$\leq 2r \max_{x\in[a,b]} \left|D_\beta f(x)\right|, \quad x \in [x_{j-1}, x_j].$$

Since $j \in \{1,\ldots,m\}$, was arbitrarily chosen, we obtain

$$\sup_{x\in[a,b]} |s_L(x) - f(x)| \leq 2r \max_{x\in[a,b]} \left|D_\beta f(x)\right|.$$

This completes the proof.

Example 6.1.4 *Let*

$$I = \left\{-1, 0, \frac{1}{8}, \frac{1}{7}, \frac{1}{6}, \frac{1}{5}, \frac{1}{4}, \frac{1}{3}, \frac{1}{2}, 1, 2\right\}$$

and

$$\begin{aligned}
\beta(-1) &= -1, \\
\beta(0) &= \frac{1}{8}, \\
\beta\left(\frac{1}{8}\right) &= \frac{1}{7}, \\
\beta\left(\frac{1}{7}\right) &= \frac{1}{6}, \\
\beta\left(\frac{1}{6}\right) &= \frac{1}{5}, \\
\beta\left(\frac{1}{5}\right) &= \frac{1}{4}, \\
\beta\left(\frac{1}{4}\right) &= \frac{1}{3}, \\
\beta\left(\frac{1}{3}\right) &= \frac{1}{2}, \\
\beta\left(\frac{1}{2}\right) &= 1, \\
\beta(1) &= 2,
\end{aligned}$$

and

$$\begin{aligned} a &= x_0 = 0, \quad x_1 = \frac{1}{8}, \quad x_2 = \frac{1}{7}, \quad x_3 = \frac{1}{4}, \quad x_4 = b = 1, \\ f(t) &= t^2 + t + 1, \quad t \in I. \end{aligned}$$

We have

$$\begin{aligned} r_1 &= x_1 - x_0 \\ &= \frac{1}{8}, \\ r_2 &= x_2 - x_1 \\ &= \frac{1}{56}, \\ r_3 &= x_3 - x_2 \\ &= \frac{3}{28}, \\ r_4 &= x_4 - x_3 \\ &= \frac{3}{4}. \end{aligned}$$

Hence, $r = \frac{3}{4}$. *We have*

$$D_\beta f(t) = \beta(t) + t + 1, \quad t \in I.$$

From here,

$$\begin{aligned} \max_{t \in [0,1]} D_\beta f(t) &= \beta(1) + 1 + 1 \\ &= 2 + 2 \\ &= 4. \end{aligned}$$

Thus,

$$\begin{aligned} \sup_{x \in [0,1]} |s_L(x) - f(x)| &\leq 2 \cdot \frac{3}{4} \cdot 4 \\ &= 6. \end{aligned}$$

Exercise 6.1.5 *Let* $I = \mathbb{R}$, $\beta(t) = t+1$, $t \in I$, *and*

$$\begin{aligned} a &= x_0 = 0, \quad x_1 = 2, \quad x_2 = 4, \quad x_3 = 6, \quad x_4 = 8, \quad x_5 = 10 = b, \\ f(t) &= t^3 + t, \quad t \in I. \end{aligned}$$

Prove that

$$\sup_{x \in [0,10]} |s_L(x) - f(x)| \leq 1328.$$

Remark 6.1.6 *Fix* $j \in \{1, \ldots, m\}$ *arbitrarily. Suppose that* $f \in \mathscr{C}^2_\beta([a,b])$. *Then, by Theorem 4.1.13, we have*

$$f(x) - s_L(x) = \frac{D^2_\beta f(\xi)}{\Pi^2_2(\xi)} \pi_2(x), \quad x \in [x_{j-1}, x_j],$$

or

$$F_{min,2}(\xi) \leq \frac{f(x) - s_L(x)}{\pi_2(x)} \leq F_{max,2}(\xi), \quad x \in [x_{j-1}, x_j].$$

Now, suppose that s_L is a linear spline with knots x_j, $j \in \{0, 1, \ldots, m\}$. We can express s_L as a linear combination of "suitable basis functions" ϕ_j, $j \in \{0, 1, \ldots, m\}$, as follows.

$$s_L(x) = \sum_{j=0}^{m} \phi_j(x) f(x_j), \quad x \in [a,b].$$

We require each function ϕ_j, $j \in \{0, 1, \ldots, m\}$, is itself a linear spline that vanishes at every knot except x_j, and $\phi_j(x_j) = 1$.

Definition 6.1.7 *The functions* ϕ_j, $j \in \{0, 1, \ldots, m\}$, *are said to be the linear basis splines or hat functions.*

Definition 6.1.8 *The formal definition of* ϕ_j, $j \in \{, 1, \ldots, m\}$, *is as follows.*

$$\phi_j(x) = \begin{cases} 0 \quad \textit{if} \quad x \leq x_{j-1} \\ \dfrac{x - x_{j-1}}{x_j - x_{j-1}} \quad \textit{if} \quad x_{j-1} \leq x \leq x_j \\ \dfrac{x_{j+1} - x}{x_{j+1} - x_j} \quad \textit{if} \quad x_j \leq x \leq x_{j+1} \\ 0 \quad \textit{if} \quad x_{j+1} \leq x, \end{cases}$$

for $j \in \{1, \ldots, m-1\}$, *and*

$$\phi_0(x) = \begin{cases} \dfrac{x_1 - x}{x_1 - x_0} & \text{if} \quad a = x_0 \le x \le x_1 \\ 0 \quad \text{if} \quad x_1 \le x, \end{cases}$$

and

$$\phi_m(x) = \begin{cases} 0 \quad \text{if} \quad x \le x_{m-1} \\ \dfrac{x - x_{m-1}}{x_m - x_{m-1}} & \text{if} \quad x_{m-1} \le x \le x_m = b. \end{cases}$$

6.2 Linear Interpolating β-Splines

Let $a, b \in I$, $a < b$, and $m \in \mathbb{N}$, $m \ge 2$.

Definition 6.2.1 *Suppose that* $f \in \mathscr{C}([a,b])$, $K = \{x_0, x_1, \ldots, x_m\}$ *be a subset of* $[a,b]$ *such that*

$$a = x_0 \le \beta(x_0) < x_1 \le \beta(x_1) < \ldots < x_m \le \beta(x_m) = b.$$

The linear β*-spline* $s_\beta L$*, interpolating* f *at the points* x_j*,* $j \in \{0, 1, \ldots, m\}$*, is defined by*

$$s_{\beta L}(x) = \frac{\beta(x_j) - \beta(x)}{\beta(x_j) - \beta(x_{j-1})} f(x_{j-1}) + \frac{\beta(x) - \beta(x_{j-1})}{\beta(x_j) - \beta(x_{j-1})} f(x_j), \quad x \in [x_{j-1}, x_j],$$

$j \in \{1, \ldots, m\}$*. The points* x_j*,* $j \in \{0, 1, \ldots, m\}$*, are the knots of the* β*-spline, and* K *is said to be the set of knots.*

By Definition 6.2.1, it follows that

$$s_{\beta L}(x_{j-1}) = f(x_{j-1}), \quad s_{\beta L}(x_j) = f(x_j), \quad j \in \{1, \ldots, m\}.$$

Below, denote

$$\begin{aligned} r_{j+1} &= x_{j+1} - x_j, \quad j \in \{0, 1, \ldots, m-1\}, \\ r &= \max_{j \in \{1, \ldots, m\}} r_j. \end{aligned}$$

Theorem 6.2.2 *Let* $f \in \mathscr{C}^1_\beta([a,b])$*. Then*

$$\max_{x \in [a,b]} |s_{\beta L}(x) - f(x)| \le 2r \max_{x \in [a,b]} |D_\beta f(x)|.$$

Proof 6.2.3 *Take $j \in \{1,\ldots,m\}$ arbitrarily. Then*

$$\begin{aligned}
|s_{\beta L}(x)-f(x)| &= \left|\frac{\beta(x_j)-\beta(x)}{\beta(x_j)-\beta(x_{j-1})}f(x_{j-1})+\frac{\beta(x)-\beta(x_{j-1})}{\beta(x_j)-\beta(x_{j-1})}f(x_j)-f(x)\right| \\
&= \left|\frac{\beta(x_j)-\beta(x)}{\beta(x_j)-\beta(x_{j_1})}(f(x_{j-1})-f(x))+\frac{\beta(x)-\beta(x_{j-1})}{\beta(x_j)-\beta(x_{j-1})}(f(x_j)-f(x))\right| \\
&\le \frac{\beta(x_j)-\beta(x)}{\beta(x_j)-\beta(x_{j-1})}|f(x_{j-1})-f(x)|+\frac{\beta(x)-\beta(x_{j-1})}{\beta(x_j)-\beta(x_{j-1})}|f(x_j)-f(x)|,
\end{aligned}$$

$x \in [a,b]$. Now, applying the Mean Value Theorem, we get

$$\begin{aligned}
|f(x_{j+1})-f(x)| &\le \max_{x\in[x_{j-1},x_j]}\left|D_\beta f(x)\right|(x-x_{j-1}) \\
&\le \max_{x\in[a,b]}\left|D_\beta f(x)\right|(x-x_{j-1}), \quad x\in[x_{j-1},x_j],
\end{aligned}$$

and

$$\begin{aligned}
|f(x_j)-f(x)| &\le \max_{x\in[x_{j-1},x_j]}\left|D_\beta f(x)\right|(x_j-x) \\
&\le \max_{x\in[a,b]}\left|D_\beta f(x)\right|(x_j-x), \quad x\in[x_{j-1},x_j].
\end{aligned}$$

Therefore

$$\begin{aligned}
|s_{\beta L}(x)-f(x)| &\le r_j\max_{x\in[a,b]}\left|D_\beta f(x)\right|+r_j\max_{x\in[a,b]}\left|D_\beta f(x)\right| \\
&= 2r_j\max_{x\in[a,b]}\left|D_\beta f(x)\right| \\
&\le 2r\max_{x\in[a,b]}\left|D_\beta f(x)\right|, \quad x\in[x_{j-1},x_j].
\end{aligned}$$

Using that $j \in \{1,\ldots,m\}$ was arbitrarily chosen, we arrive at

$$\sup_{x\in[a,b]}|s_{\beta L}(x)-f(x)| \le 2r\max_{x\in[a,b]}\left|D_\beta f(x)\right|.$$

This completes the proof.

Example 6.2.4 *Let*

$$I = \left\{-1,0,\frac{1}{6},\frac{5}{6},1,\frac{4}{3},\frac{11}{6},2,\frac{5}{2},\frac{8}{3},3\right\}$$

and $\beta : I \to I$ be defined as follows

$$\beta(-1) = -1,$$

$$\begin{aligned}
\beta(0) &= \frac{1}{6}, \\
\beta\left(\frac{1}{6}\right) &= \frac{5}{6}, \\
\beta\left(\frac{5}{6}\right) &= 1, \\
\beta\left(\frac{4}{3}\right) &= \frac{11}{6}, \\
\beta\left(\frac{11}{6}\right) &= 2, \\
\beta(2) &= \frac{5}{2}, \\
\beta\left(\frac{5}{2}\right) &= \frac{8}{3}, \\
\beta\left(\frac{8}{3}\right) &= 3,
\end{aligned}$$

and

$$\begin{aligned}
a &= x_0 = 0, \quad x_1 = \frac{5}{6}, \quad x_2 = \frac{4}{3}, \quad x_3 = 2, \quad x_4 = \frac{8}{3}, \quad b = 3, \\
f(t^) &= \frac{1}{t+1}, \quad t \in I.
\end{aligned}$$

We have

$$x_0 < \beta(x_0) < x_1 < \beta(x_1) < x_2 < \beta(x_2) < x_3 < \beta(x_3) < x_4 < \beta(x_4) < \beta(x_4) = b.$$

Next, we have

$$\begin{aligned}
D_\beta f(t) &= -\frac{1}{(t+1)(\beta(t)+1)}, \\
|D_\beta f(t)| &= \frac{1}{(t+1)(\beta(t)+1)}, \quad t \in I.
\end{aligned}$$

Thus,

$$\max_{t\in[a,b]} |D_\beta f(t)| = \frac{1}{\beta(0)+1}$$

$$= \frac{1}{\frac{1}{6}+1}$$

$$= \frac{6}{7}.$$

Next,

$$\begin{aligned}
r_1 &= x_1 - x_0 \\
&= \frac{5}{6}, \\
r_2 &= x_2 - x_1 \\
&= \frac{4}{3} - \frac{5}{6} \\
&= \frac{1}{2}, \\
r_3 &= x_3 - x_2 \\
&= 2 - \frac{4}{3} \\
&= \frac{2}{3}, \\
r_4 &= x_4 - x_3 \\
&= \frac{8}{3} - 2 \\
&= \frac{2}{3}.
\end{aligned}$$

Consequently $r = \frac{2}{3}$ *and*

$$\begin{aligned}
|s_{\beta L}(x) - f(x)| &\leq 2 \cdot \frac{2}{3} \cdot \frac{6}{7} \\
&= \frac{8}{7}.
\end{aligned}$$

Exercise 6.2.5 *Let* $I = \mathbb{R}$, $\beta(t) = 6t + \frac{1}{8}$, $t \in I$. *Evaluate*

$$|s_{\beta L}(x) - f(x)|, \quad x \in [a,b].$$

Remark 6.2.6 *Fix $j \in \{1,\dots,m\}$ arbitrarily. Suppose that $f \in \mathscr{C}^2_\beta([a,b])$. Then, by Theorem 4.2.17, we have*

$$f(x) - s_{\beta L}(x) = \frac{D^2_\beta f(\xi)}{\Pi^2_{\beta 2}(\xi)} \pi_{\beta 2}(x), \quad x \in [x_{j-1}, x_j],$$

or

$$F_{\beta min,2}(\xi) \le \frac{f(x) - s_{\beta L}(x)}{\pi_{\beta 2}(x)} \le F_{\beta max,2}(\xi), \quad x \in [x_{j-1}, x_j].$$

Now, suppose that $s_{\beta L}$ is a linear spline with knots x_j, $j \in \{0,1,\dots,m\}$. We can express $s_{\beta L}$ as a linear combination of "suitable β-basis functions" $\phi_{\beta j}$, $j \in \{0,1,\dots,m\}$, as follows.

$$s_{\beta L}(x) = \sum_{j=0}^{m} \phi_{\beta j}(x) f(x_j), \quad x \in [a,b].$$

We require each function $\phi_{\beta j}$, $j \in \{0,1,\dots,m\}$, is itself a linear spline that vanishes at every knot except x_j, and $\phi_{\beta j}(x_j) = 1$.

Definition 6.2.7 *The functions $\phi_{\beta j}$, $j \in \{0,1,\dots,m\}$, are said to be the linear basis β-splines or β-hat functions.*

Definition 6.2.8 *The formal definition of $\phi_{\beta j}$, $j \in \{,1,\dots,m\}$, is as follows.*

$$\phi_{\beta j}(x) = \begin{cases} 0 \quad \text{if} \quad x \le x_{j-1} \\[2ex] \dfrac{\beta(x) - \beta(x_{j-1})}{\beta(x_j) - \beta(x_{j-1})} \quad \text{if} \quad x_{j-1} \le x \le x_j \\[2ex] \dfrac{\beta(x_{j+1}) - \beta(x)}{\beta(x_{j+1}) - \beta(x_j)} \quad \text{if} \quad x_j \le x \le x_{j+1} \\[2ex] 0 \quad \text{if} \quad x_{j+1} \le x, \end{cases}$$

for $j \in \{1,\dots,m-1\}$, and

$$\phi_{\beta 0}(x) = \begin{cases} \dfrac{\beta(x_1) - \beta(x)}{\beta(x_1) - \beta(x_0)} \quad \text{if} \quad a = x_0 \le x \le x_1 \\[2ex] 0 \quad \text{if} \quad x_1 \le x, \end{cases}$$

and

$$\phi_{\beta m}(x) = \begin{cases} 0 \quad \text{if} \quad x \le x_{m-1} \\[2ex] \dfrac{\beta(x) - \beta(x_{m-1})}{\beta(x_m) - \beta(x_{m-1})} \quad \text{if} \quad x_{m-1} \le x \le x_m = b. \end{cases}$$

6.3 Cubic Splines

Suppose that $a, b \in I$, $a < b$,

$$K = \{a = x_0 \leq \beta(x_0) < x_1 \leq \beta(x_1) < \ldots < x_m \leq \beta(x_m) = b\},$$

$m \in \mathbb{N}$, $m \geq 3$, and $f \in \mathscr{C}([a,b])$.

Definition 6.3.1 *The natural cubic spline is denoted by s_2 and it satisfies the end conditions*

$$D_\beta^2 s_2(x_0) = D_\beta^2 s_2(x_m) = 0.$$

Now, we will construct the natural cubic spline. Let

$$a_j = D_\beta^2 s_2(x_j), \quad j \in \{0, 1, \ldots, m\}.$$

We have $a_0 = a_m = 0$. In addition, suppose that

$$D_\beta s_2(x_j-) = D_\beta s_2(x_j+), \quad j \in \{1, \ldots, m-1\},$$

and

$$D_\beta^2 s_2(x) = \frac{x_j - x}{x_j - x_{j-1}} a_{j-1} + \frac{x - x_{j-}}{x_j - x_{j-1}} a_j, \quad x \in [x_{j-1}, x_j], \quad j \in \{1, \ldots, m\}. \quad (6.1)$$

Fix $j \in \{1, \ldots, m-1\}$. We integrate the equality (6.1) from x to x_j, $x \in [x_{j-1}, x_j]$, and we find

$$\int_x^{x_j} D_\beta^2 s_2(t) D_\beta t = \frac{a_{j-1}}{x_j - x_{j-1}} \int_x^{x_j} (x_j - t) d_\beta t + \frac{a_j}{x_j - x_{j-1}} \int_x^{x_j} (t - x_{j-1}) d_\beta t$$

$$= \frac{a_{j-1}}{x_j - x_{j-1}} \int_{x_j}^{x} (t - x_j) d_\beta t + \frac{a_j}{x_j - x_{j-1}} \int_x^{x_j} (t - x_j) d_\beta t$$

$$+ \frac{a_j}{x_j - x_{j-1}} \int_x^{x_j} (x_j - x_{j-1}) d_\beta t$$

$$= \frac{a_{j-1}}{h_1(x_j, x_{j-1})} h_2(x, x_j) - \frac{a_j}{h_1(x_j, x_{j-1})} h_2(x, x_j) + a_j h_1(x_j, x),$$

$x \in [x_{j-1}, x_j]$, or

$$D_\beta s(x_j-) - D_\beta s(x) = \frac{a_{j-1}}{h_1(x_j, x_{j-1})} h_2(x, x_j) - \frac{a_j}{h_1(x_j, x_{j-1})} h_2(x, x_j)$$

$$+ a_j h_1(x_j, x), \quad x \in [x_{j-1}, x_j],$$

or

$$D_\beta s(x) = D_\beta s(x_j-) - \frac{a_{j-1}}{h_1(x_j, x_{j-1})} h_2(x, x_j) + \frac{a_j}{h_1(x_j, x_{j-1})} h_2(x, x_j)$$

$$-a_j h_1(x_j,x), \quad x \in [x_{j-1},x_j].$$

Now, we integrate the last equality from x to x_j and we find

$$\begin{aligned}\int_x^{x_j} D_\beta s(t)d_\beta t &= D_\beta s(x_j-)h_1(x_j,x) + \frac{a_{j-1}}{h_1(x_j,x_{j-1})}\int_{x_j}^x h_2(t,x_j)d_\beta t\\ &\quad -\frac{a_j}{h_1(x_j,x_{j-1})}\int_{x_j}^x h_2(t,x_j)d_\beta t - a_j\int_{x_j}^x (t-x_j)d_\beta t\\ &= s^{D_\beta}(x_j-)h_1(x_j,x) + a_{j-1}\frac{h_3(x,x_j)}{h_1(x_j,x_{j-1})}\\ &\quad -a_j\frac{h_3(x,x_j)}{h_1(x_j,x_{j-1})} - a_j h_2(x,x_j), \quad x \in [x_{j-1},x_j],\end{aligned}$$

and

$$\begin{aligned}s_2(x_j)-s_2(x) &= D_\beta s_2(x_j-)h_1(x_j,x) + a_{j-1}\frac{h_3(x,x_j)}{h_1(x_j,x_{j-1})}\\ &\quad a_j\left(\frac{h_3(x,x_j)}{h_1(x_j,x_{j-1})} + h_2(x,x_j)\right), \quad x \in [x_{j-1},x_j].\end{aligned} \tag{6.2}$$

In particular,

$$\begin{aligned}f(x_j)-f(x_{j-1}) &= D_\beta s_2(x_j-)h_1(x_j,x_{j-1}) + a_{j-1}\frac{h_3(x_{j-1},x_j)}{h_1(x_j,x_{j-1})}\\ &\quad -a_j\left(\frac{h_3(x_{j-1},x_j)}{h_1(x_j,x_{j-1})} + h_2(x_{j-1},x_j)\right)\end{aligned}$$

and

$$\begin{aligned}D_\beta s_2(x_j-) &= \frac{f(x_j)-f(x_{j-1})}{h_1(x_j,x_{j-1})} - a_{j-1}\frac{h_3(x_{j-1},x_j)}{(h_1(x_j,x_{j-1}))^2}\\ &\quad +a_j\left(\frac{h_3(x_{j-1},x_j)}{(h_1(x_j,x_{j-1}))^2} + \frac{h_2(x_{j-1},x_j)}{h_1(x_j,x_{j-1})}\right).\end{aligned} \tag{6.3}$$

Now, we consider the equality

$$D_\beta^2 s_2(x) = \frac{x_{j+1}-x}{x_{j+1}-x}a_j + \frac{x-x_j}{x_{j+1}-x_j}a_{j+1}, \quad x \in [x_j,x_{j+1}],$$

which we integrate from x_j to x, $x \in [x_j,x_{j+1}]$, and we find

$$D_\beta s_2(x) - D_\beta s_2(x_j-) = \frac{a_j}{x_{j+1}-x_j}\int_{x_j}^x (x_{j+1}-t)d_\beta t$$

$$+\frac{a_{j+1}}{x_{j+1}-x_j}\int_{x_j}^{x}(t-x_j)d_\beta t$$

$$\begin{aligned}
&= \frac{a_j}{h_1(x_{j+1},x_j)}\int_{x_j}^{x}(x_{j+1}-x_j)d_\beta t-\frac{a_j}{h_1(x_{j+1},x_j)}\int_{x_j}^{x}(t-x_j)d_\beta t\\
&\quad +\frac{a_{j+1}}{h_1(x_{j+1},x_j)}h_2(x,x_j)\\
&= a_jh_1(x,x_j)-\frac{a_j}{h_1(x_{j+1},x_j)}h_2(x,x_j)+\frac{a_{j+1}h_1(x_{j+1},x_j)}{h}{}_2(x,x_j),
\end{aligned}$$

$x\in[x_j,x_{j+1}]$, and

$$\begin{aligned}
D_\beta s_2(x) &= D_\beta s_2(x_j+)+a_jh_1(x,x_j)-\frac{a-j}{h_1(x_{j+1},x_j)}h_2(x,x_j)\\
&\quad +\frac{a_{j+1}}{h_1(x_{j+1},x_j)}h_2(x,x_j),\quad x\in[x_j,x_{j+1}].
\end{aligned}$$

We integrate the last equality from x_j to x and we obtain

$$\begin{aligned}
s_2(x)-s_2(x_j) &= D_\beta s_2(x_j+)h_1(x,x_j)+a_j\int_{x_j}^{x}h_1(t,x_j)d_\beta t\\
&\quad -\frac{a_j}{h_1(x_{j+1},x_j)}\int_{x_j}^{x}h_2(t,x_j)d_\beta t+\frac{a_{j+1}}{h_1(x_{j+1},x_j)}\int_{x_j}^{x}h_2(t,x_j)d_\beta t\\
&= D_\beta s_2(x_j+)h_1(x,x_j)+a_jh_2(x,x_j)-a_j\frac{h_3(x,x_j)}{h_1(x_{j+1},x_j)}\\
&\quad +a_{j+1}\frac{h_3(x,x_j)}{h_1(x_{j+1},x_j)},\quad x\in[x_j,x_{j+1}].
\end{aligned}$$

In particular,

$$\begin{aligned}
f(x_{j+1})-f(x_j) &= D_\beta s_2(x_j+)h_1(x_{j+1},x_j)\\
&\quad +a_j\left(h_2(x_{j+1},x_j)-\frac{h_3(x_{j+1},x_j)}{h_1(x_{j+1},x_j)}\right)\\
&\quad +a_{j+1}\frac{h_3(x_{j+1},x_j)}{h_1(x_{j+1},x_j)}.
\end{aligned}$$

Thus,

$$D_\beta s_2(x_j+) = \frac{f(x_{j+1})-f(x_j)}{h_1(x_{j+1},x_j)}+a_j\left(\frac{h_2(x_{j+1},x_j)}{h_1(x_{j+1},x_j)}+\frac{h_3(x_{j+1},x_j)}{(h_1(x_{j+1},x_j))^2}\right)$$

$$-a_{j+1}\frac{h_3(x_{j+1},x_j)}{(h_1(x_{j+1},x_j))^2}.$$

By the last equality and by (6.3), we arrive at

$$\begin{aligned}
&\frac{f(x_j)-f(x_{j-1})}{h_1(x_j,x_{j-1})}-a_{j-1}\frac{h_3(x_{j-1},x_j)}{(h_1(x_j,x_{j-1}))^2}\\
&\quad+a_j\left(\frac{h_3(x_{j-1},x_j)}{(h_1(x_j,x_{j-1}))^2}+\frac{h_2(x_{j-1},x_j)}{h_1(x_j,x_{j-1})}\right)\\
&= \frac{f(x_{j+1})-f(x_j)}{h_1(x_{j+1},x_j)}+a_j\left(\frac{h_2(x_{j+1},x_j)}{h_1(x_{j+1},x_j)}+\frac{h_3(x_{j+1},x_j)}{(h_1(x_{j+1},x_j))^2}\right)\\
&\quad-a_{j+1}\frac{h_3(x_{j+1},x_j)}{(h_1(x_{j+1},x_j))^2},
\end{aligned}$$

whereupon

$$\begin{aligned}
&-a_{j-1}\frac{h_3(x_{j+1},x_j)}{(h_1(x_j,x_{j-1}))^2}+a_j\left(\frac{h_3(x_{j-1},x_j)}{(h_1(x_j,x_{j-1}))^2}+\frac{h_2(x_{j+1},x_j)}{h_1(x_j,x_{j-1})}\right. \qquad (6.4)\\
&\quad+\frac{h_2(x_{j+1},x_j)}{h_1(x_{j+1},x_j)}\\
&\quad\left.-\frac{h_3(x_{j+1},x_j)}{(h_1(x_{j+1},x_j))^2}\right)+a_{j+1}\frac{h_3(x_{j+1},x_j)}{(h_1(x_{j+1},x_j))^2}\\
&= \frac{f(x_{j+1})-f(x_j)}{h_1(x_{j+1},x_j)}-\frac{f(x_j)-f(x_{j-1})}{h_1(x_j,x_{j-1})},\quad j\in\{1,\ldots,m-1\}.
\end{aligned}$$

Note that

$$\begin{aligned}
h_2(x,x_j) &= \int_{x_j}^{x}(t-x_j)d_\beta t\\
&= \int_{x}^{x_j}(x_j-t)d_\beta t\\
&\leq (x_j-x)^2,\quad x\in[x_{j-1},x_j],\\
h_3(x_{j-1},x_j) &= \int_{x_j}^{x_{j-1}} h_2(t,x_j)d_\beta t\\
&= -\int_{x_{j-1}}^{x_j} h_2(t,x_j)d_\beta t
\end{aligned}$$

$$
\begin{aligned}
&\geq -\int_{x_{j-1}}^{x_j}(x_j-t)^2 d_\beta t\\
&> -(x_j-x_{j-1})\int_{x_{j-1}}^{x_j}(x_j-t)d_\beta t\\
&= -h_1(x_j,x_{j-1})\int_{x_j}^{x_{j-1}}(t-x_j)d_\beta t\\
&= -h_1(x_j,x_{j-1})h_2(x_{j-1},x_j),
\end{aligned}
$$

and

$$
\frac{h_3(x_{j-1},x_j)}{(h_1(x_j,x_{j-1}))^2} > -\frac{h_2(x_{j-1},x_j)}{h_1(x_j,x_{j-1})}
$$

or

$$
\frac{h_3(x_{j-1},x_j)}{h_1(x_j,x_{j-1})} + \frac{h_2(x_{j-1},x_j)}{h_1(x_j,x_{j-1})} > 0. \tag{6.5}
$$

Next,

$$
\begin{aligned}
h_3(x_{j+1},x_j) &= \int_{x_j}^{x_{j+1}} h_2(t,x_j)d_\beta t\\
&\leq \frac{1}{2}\int_{x_j}^{x_{j+1}}(t-x_j)^2 d_\beta t\\
&\leq \frac{h_1(x_{j+1},x_j)}{2}\int_{x_j}^{x_{j+1}}(t-x_j)d_\beta t\\
&= \frac{1}{2}h_1(x_{j+1},x_j)h_2(x_{j+1},x_j),
\end{aligned}
$$

whereupon

$$
2\frac{h_3(x_{j+1},x_j)}{(h_1(x_{j+1},x_j))^2} \leq \frac{h_2(x_{j+1},x_j)}{h_1(x_{j+1},x_j)}. \tag{6.6}
$$

Hence and (6.5), we obtain

$$
\frac{h_3(x_{j+1},x_j)}{(h_1(x_j,x_{j-1}))^2} + \frac{h_2(x_{j+1},x_j)}{h_1(x_j,x_{j-1})} + \frac{h_2(x_{j+1},x_j)}{h_1(x_{j+1},x_j)} - \frac{h_3(x_{j+1},x_j)}{(h_1(x_j,x_{j-1}))^2} \geq 0. \tag{6.7}
$$

Let

$$
l(t) = (t-x_j)^2, \quad t \in [x_{j-1},x_j].
$$

Then

$$
\begin{aligned}
h_2(x,x_j) &= \int_{x_j}^{x}(t-x_j)d_\beta t\\
&= -\int_{x}^{x_j}(t-x_j)d_\beta t
\end{aligned}
$$

$$
\begin{aligned}
&= -\frac{1}{2}\int_x^{x_j}((t-x_j)+(t-x_j))d_\beta t \\
&\leq -\frac{1}{2}\int_x^{x_j}((t-x_j)+(\beta(t)-x_j))d_\beta t \\
&= -\frac{1}{2}\int_x^{x_j} D_\beta l(t)d_\beta t \\
&= -\frac{1}{2}(l(x_j)-l(x) \\
&= \frac{1}{2}(x-x_j)^2, \quad x\in[x_{j-1},x_j].
\end{aligned}
$$

Hence,

$$
\begin{aligned}
h_3(x_{j-1},x_j) &= \int_{x_j}^{x_{j-1}} h_2(t,x_j)d_\beta t \\
&= -\int_{x_{j-1}}^{x_j} h_2(t,x_j)d_\beta t \\
&\geq -\frac{1}{2}\int_{x_{j-1}}^{x_j}(t-x_j)^2 d_\beta t \\
&= -\frac{1}{2}\int_{x_{j-1}}^{x_j}(x_j-t)(x_j-t)d_\beta t \\
&\geq -\frac{1}{2}h_1(x_j,x_{j-1})\int_{x_{j-1}}^{x_j}(x_j-t)d_\beta t \\
&= -\frac{1}{2}h_1(x_j,x_{j-1})\int_{x_j}^{x_{j-1}}(t-x_j)d_\beta t \\
&= -\frac{1}{2}h_1(x_j,x_{j-1})h_2(x_{j-1},x_j),
\end{aligned}
$$

whereupon

$$
2\frac{h_3(x_{j-1},x_j)}{(h_1(x_j,x_{j-1}))^2} > -\frac{h_2(x_{j-1},x_j)}{h_1(x_j,x_{j-1})}. \tag{6.8}
$$

Next,

$$
\begin{aligned}
h_3(x_{j+1},x_j) &= \int_{x_j}^{x_{j+1}} h_2(t,x_j)d_\beta t \\
&\leq \frac{1}{2}\int_{x_j}^{x_{j+1}}(t-x_j)^2 d_\beta t \\
&< \frac{1}{2}h_1(x_{j+1},x_j)\int_{x_j}^{x_{j+1}}(t-x_j)d_\beta t
\end{aligned}
$$

$$= \frac{1}{2}h_1(x_{j+1},x_j)h_2(x_{j+1},x_j),$$

from where

$$2\frac{h_3(x_{j+1},x_j)}{(h_1(x_{j+1},x_j))^2} < \frac{h_2(x_{j+1},x_j)}{h_1(x_{j+1},x_j)}. \tag{6.9}$$

Now, applying (6.5), (6.6), (6.7) and (6.9), we obtain

$$\begin{aligned} &\left| \frac{h_3(x_{j-1},x_j)}{(h_1(x_j,x_{j-1}))^2} + \frac{h_2(x_{j+1},x_j)}{h_1(x_j,x_{j-1})} + \frac{h_2(x_{j+1},x_j)}{h_1(x_{j+1},x_j)} - \frac{h_3(x_{j+1},x_j)}{(h_1(x_{j+1},x_j))^2} \right| \\ &> \left| -\frac{h_3(x_{j-1},x_j)}{(h_1(x_j,x_{j-1}))^2} \right| + \left| \frac{h_3(x_{j+1},x_j)}{(h_1(x_{j+1},x_j))^2} \right| \iff \\ &\frac{h_3(x_{j-1},x_j)}{(h_1(x_j,x_{j-1}))^2} + \frac{h_2(x_{j+1},x_j)}{h_1(x_j,x_{j-1})} + \frac{h_2(x_{j+1},x_j)}{h_1(x_{j+1},x_j)} - \frac{h_3(x_{j+1},x_j)}{(h_1(x_{j+1},x_j))^2} \\ &> -\frac{h_3(x_{j-1},x_j)}{(h_1(x_j,x_{j-1}))^2} + \frac{h_3(x_{j+1},x_j)}{(h_1(x_{j+1},x_j))^2} \iff \end{aligned}$$

$$2\frac{h_3(x_{j-1},x_j)}{(h_1(x_j,x_{j-1}))^2} + \frac{h_2(x_{j+1},x_j)}{h_1(x_j,x_{j-1})} + \frac{h_2(x_{j+1},x_j)}{h_1(x_{j+1},x_j)} - 2\frac{h_3(x_{j+1},x_j)}{(h_1(x_{j+1},x_j))^2} > 0.$$

Therefore, the system (6.4) has a unique solution a_j, $j \in \{0,1,\ldots,m\}$. By (6.2), we get

$$\begin{aligned} s_2(x) &= f(x_j) - D_\beta s_2(x_j-)h_1(x_j,x) - a_{j-1}\frac{h_3(x,x_j)}{h_1(x_j,x_{j-1})} \\ &\quad + a_j\left(\frac{h_3(x,x_j)}{h_1(x_j,x_{j-1})} + h_2(x,x_j)\right), \quad x \in [x_{j-1},x_j], \quad j \in \{1,\ldots,m\}. \end{aligned}$$

Now, applying (6.3), we arrive at

$$\begin{aligned} s_2(x) &= f(x_j) - \frac{f(x_j)-f(x_{j-1})}{h_1(x_j,x_{j-1})}h_1(x_j,x) \\ &\quad + a_{j-1}\left(\frac{h_3(x_{j-1},x_j)}{(h_1(x_j,x_{j-1}))^2}h_1(x_j,x) - \frac{h_3(x,x_j)}{h_1(x_j,x_{j-1})}\right) \\ &\quad + a_j\left(-\frac{h_3(x_{j-1},x_j)}{(h_1(x_j,x_{j-1}))^2}h_1(x_j,x) - \frac{h_2(x_{j-1},x_j)}{h_1(x_j,x_{j-1})}h_1(x_j,x)\right. \\ &\quad \left. + \frac{h_3(x,x_j)}{h_1(x_j,x_{j-1})} + h_2(x,x_j)\right), \quad x \in [x_{j-1},x_j], \quad j \in \{1,\ldots,m\}. \end{aligned}$$

Example 6.3.2 *Let* $I = \{-1, 0, 1, 2, 3, 4\}$, $\beta : I \to I$ *be defined by*

$$\begin{aligned} \beta(-1) &= -1, \\ \beta(0) &= 1, \\ \beta(1) &= 2, \\ \beta(3) &= 4, \\ \beta(4) &= 4, \end{aligned}$$

and

$$f(t) = \frac{1+t}{1+t^2}, \quad t \in I.$$

Take

$$a = x_0 = 0, \quad x_1 = 2, \quad x_2 = b = 4.$$

Then, by (6.4) *and using that* $a_0 = 0$, $a_2 = 0$, *we get*

$$\begin{aligned} a_1\left(\frac{h_3(x_0,x_1)}{(h_1(x_1,x_0))^2} + \frac{h_2(x_0,x_1)}{h_1(x_1,x_0)} + \frac{h_2(x_2,x_1)}{h_1(x_2,x_1)} - \frac{h_3(x_2,x_1)}{(h_1(x_2,x_1))^2}\right) \\ = \frac{f(x_2)-f(x_1)}{h_1(x_2,x_1)} - \frac{f(x_1)-f(x_0)}{h_1(x_1,x_0)} \end{aligned}$$

or

$$\begin{aligned} a_1\left(\frac{h_3(0,2)}{(h_1(2,0))^2} + \frac{h_2(0,2)}{h_1(2,0)} + \frac{h_2(4,2)}{h_1(4,2)} - \frac{h_3(4,2)}{(h_1(4,2))^2}\right) \\ = \frac{f(4)-f(2)}{h_1(4,2)} - \frac{f(2)-f(0)}{h_1(2,0)}. \end{aligned}$$

Now, we will compute $h_i(\cdot,\cdot)$, $i \in \{1,2,3\}$. *We have*

$$\begin{aligned} h_1(2,0) &= 2, \\ h_2(0,2) &= \int_2^0 h_1(\tau,2)d_\beta\tau \\ &= -\int_0^2 (\tau-2)d_\beta\tau \end{aligned}$$

$$
\begin{aligned}
&= -(\beta(0)-)(-2)-(\beta(1)-1)(1-2) \\
&= 2+1 \\
&= 3, \\
h_3(0,2) &= \int_2^0 h_2(\tau,2)d_\beta\tau \\
&= -\int_0^2 h_2(\tau,2)d_\beta\tau \\
&= -(\beta(0)-0)h_2(0,2)-(\beta(1)-1)h_2(1,2) \\
&= -3-\int_2^1 h_1(\tau,2)d_\beta\tau \\
&= -3+\int_1^2 (\tau-2)d_\beta\tau \\
&= -3+(\beta(1)-1)(1-2) \\
&= -3-1 \\
&= -4, \\
h_3(2,4) &= \int_4^2 h_2(\tau,4)d_\beta\tau \\
&= -\int_2^4 h_2(\tau,4)d_\beta\tau \\
&= -(\beta(2)-2)h_2(2,4)-(\beta(3)-3)h_2(3,4) \\
&= -\int_4^2 h_1(\tau,4)d_\beta\tau-\int_4^3 h_1(\tau,4)d_\beta\tau \\
&= \int_2^4 (\tau-4)d_\beta\tau+\int_3^4 (\tau-4)d_\beta\tau \\
&= (\beta(2)-2)(2-4)+(\beta(3)-3)(3-4)+(\beta(3)-3)(3-4)
\end{aligned}
$$

$$\begin{aligned}
&= -2-1-1\\
&= -4,\\
h_1(4,2) &= 4-2\\
&= 2,\\
h_2(4,2) &= \int_2^4 h_1(\tau,2)d_\beta\tau\\
&= \int_2^4 (\tau-2)d_\beta\tau\\
&= (\beta(2)-2)(2-2)+(\beta(3)-3)(3-2)\\
&= 1,\\
h_3(4,2) &= \int_2^4 h_2(\tau,2)d_\beta\tau\\
&= (\beta(2)-2)h_2(2,2)+(\beta(3)-3)h_2(3,2)\\
&= h_2(3,2)\\
&= \int_2^3 h_1(\tau,2)d_\beta\tau\\
&= (\beta(2)-2)h_1(2,2)\\
&= 0,\\
f(0) &= 1,\\
f(2) &= \frac{3}{5},\\
f(4) &= \frac{5}{17}.
\end{aligned}$$

Hence, for a_1 we get the following equation

$$a_1\left(-\frac{4}{2^2}+\frac{3}{2}+\frac{1}{2}-0\right)=\frac{\frac{5}{17}-\frac{3}{5}}{2}-\frac{\frac{3}{5}-1}{2}$$

or

$$\begin{aligned} a_1 = &= a_1(-1+2) \\ &= \frac{-\frac{26}{85}}{2}+\frac{1}{5} \\ &= -\frac{13}{85}+\frac{1}{5} \\ &= \frac{4}{85}. \end{aligned}$$

Let

$$\begin{aligned} l_1(t) &= \frac{t^2}{2}-\frac{5}{2}t, \\ l_2(t) &= \frac{1}{6}t^3-\frac{3}{2}t^2+\frac{13}{3}t, \\ l_3(t) &= \frac{t^2-9t}{2}, \\ l_4(t) &= \frac{1}{6}t^3-5t^2+\frac{89}{6}t, \quad t\in I. \end{aligned}$$

Then

$$\begin{aligned} D_\beta l_1(t) &= \frac{\beta(t)+t}{2}-\frac{5}{2} \\ &= \frac{t+1+t}{2}-\frac{5}{2} \\ &= t-2, \\ D_\beta l_2(t) &= \frac{1}{6}((\beta(t))^2+t\beta(t)+t^2)-\frac{3}{2}(\beta(t)+t)+\frac{13}{3} \\ &= \frac{1}{6}((t+1)^2+t(t+1)+t^2)-\frac{3}{2}(t+1+t)+\frac{13}{3} \\ &= \frac{1}{6}(t^2+2t+1+t^2+t+t^2)-\frac{3}{2}(2t+1)+\frac{13}{3} \end{aligned}$$

$$
\begin{aligned}
&= \frac{t^2}{2}+\frac{t}{2}+\frac{1}{6}-3t-\frac{3}{2}+\frac{13}{3}\\
&= \frac{t^2-5t}{2}+\frac{1-9+26}{6}\\
&= \frac{t^2-5t}{2}+3,\\
D_\beta l_3(t) &= \frac{1}{2}(\beta(t)+t-9)\\
&= \frac{1}{2}(t+1+t-9)\\
&= t-4,\\
D_\beta l_4(t) &= \frac{1}{6}((\beta(t))^2+t\beta(t)+t^2)-5(\beta(t)+t)+\frac{89}{6}\\
&= \frac{1}{6}((t+1)^2+t(t+1)+t^2)-5(t+1+t)+\frac{89}{6}\\
&= \frac{1}{6}(3t^2+3t+1)-5(2t+1)+\frac{89}{6}\\
&= \frac{t^2}{2}-\frac{9}{2}t+15-5\\
&= \frac{t^2}{2}-\frac{9}{2}t+10, \quad t\in I.
\end{aligned}
$$

From here,

$$
\begin{aligned}
h_2(x,2) &= \int_2^x (\tau-2)d_\beta\tau\\
&= l_1(\tau)\Big|_{\tau=2}^{\tau=x}\\
&= l_1(x)-l_1(2)\\
&= \frac{x^2}{2}-\frac{5}{2}x-2+5\\
&= \frac{x^2-5x}{2}+3,
\end{aligned}
$$

$$
\begin{aligned}
h_3(x,2) &= \int_2^x h_2(\tau,2)d_\beta\tau \\
&= \int_2^x l_2^{D_\beta}(\tau)d_\beta\tau \\
&= l_2(x)-l_2(2) \\
&= \frac{x^3}{6}-\frac{3x^2}{2}+\frac{13}{3}x-\frac{8}{6}+6+\frac{26}{3} \\
&= \frac{x^3}{6}-\frac{3}{2}x^2w+\frac{13}{3}x+\frac{40}{3}, \quad x\in[0,2]. \\
h_2(x,4) &= \int_4^x (\tau-4)d_\beta\tau \\
&= {}'\int_4^x D_\beta l_3(\tau)d_\beta\tau \\
&= l_3(\tau)\Big|_{\tau=4}^{\tau=x} \\
&= \frac{x^2-9x}{2}-\frac{16-36}{2} \\
&= \frac{x^2-9x}{2}+10, \\
h_3(x,4) &= \int_4^x h_2(\tau,4)d_\beta\tau \\
&= \int_4^x D_\beta l_4(\tau)d_\beta\tau \\
&= l_4(x)-l_4(4) \\
&= \frac{1}{6}x^3-5x^2+\frac{89}{6}-\frac{64}{6}+80+\frac{178}{3} \\
&= \frac{x^3-30x^2+89-64+480+356}{6} \\
&= \frac{x^3-30x^2+861}{6}, \quad x\in[2,4].
\end{aligned}
$$

Therefore

$$\begin{aligned}
s_2(x) &= f(x_1) - \frac{f(x_1)-f(x_0)}{x_1-x_0} h_1(x_1,x) + a_1\Bigg(- \frac{h_3(x_0,x_1)}{(h_1(x_1,x_0))^2} h_1(x_1,x) \\
&\quad - \frac{h_2(x_0,x_1)}{h_1(x_1,x_0)} h_1(x_1,x) + \frac{h_3(x,x_1)}{h_1(x_1,x_0)} + h_2(x,x_1)\Bigg) \\
&= f(2) - \frac{f(2)-f(0)}{2} h_1(2,x) + \frac{4}{85}\Bigg(- \frac{h_3(0,2)}{(h_1(2,0))^2} h_1(2,x) \\
&\quad - \frac{h_2(0,2)}{h_1(2,0)} h_1(2,x) + \frac{h_3(x,2)}{h_1(2,0)} + h_2(x,2)\Bigg) \\
&= \frac{3}{5} - \frac{\frac{3}{5}-1}{2}(2-x) + \frac{4}{85}\Bigg(\frac{4}{4}(2-x) - \frac{3}{2}(2-x) \\
&\quad + \frac{1}{2}\left(\frac{x^3}{6} - \frac{3}{2}x^2 + \frac{13}{3}x + \frac{40}{3}\right) + \frac{x^2-5x}{2} + 3\Bigg) \\
&= \frac{3}{5} + \frac{1}{5}(2-x) + \frac{4}{85}\Bigg(- \frac{1}{2}(2-x) + \frac{x^3}{12} - \frac{3x^2}{4} + \frac{13x}{6} \\
&\quad + \frac{40}{6} + \frac{x^2-5x}{2} + 3\Bigg) \\
&= 1 - \frac{x}{5} \\
&\quad + \frac{4}{85} \cdot \frac{-12+6x+x^3-9x^2+26x+80+6x^2-30x+36}{12} \\
&= 1 - \frac{x}{5} + \frac{4}{85} \cdot \frac{x^3-3x^2+2x+104}{12} \\
&= \frac{255-51x+x^3-3x^2+2x+104}{255} \\
&= \frac{x^3-3x^2-49x+359}{255}, \quad x \in [0,2],
\end{aligned}$$

and

$$\begin{aligned}
s_2(x) &= f(x_2) - \frac{f(x) - f(x_1)}{h_1(x_2,x_1)} h_1(x_1,x) \\
&\quad + a_1\left(\frac{h_3(x_1,x_2)}{(h_1(x_2,x_1))^2} h_1(x_2,x_1) - \frac{h_3(x,x_2)}{h_1(x_2,x_1)}\right)
\end{aligned}$$

$$
\begin{aligned}
&= f(4) - \frac{f(4)-f(2)}{h_1(4,2)} h_1(4,x) \\
&\quad + \frac{4}{85}\left(\frac{h_3(2,4)}{(h_1(4,2))^2} h_1(4,x) - \frac{h_3(x,4)}{h_1(4,2)} \right) \\
&= \frac{5}{17} - \frac{\frac{5}{17}-\frac{3}{5}}{2}(4-x) \\
&\quad + \frac{4}{85}\left(-\frac{4}{4}(4-x) - \frac{x^3-30x^2+861}{12} \right) \\
&= \frac{5}{17} - \frac{25-81}{170}(4-x) \\
&\quad + \frac{4}{85}\left(\frac{-48+12x-x^3+30x^2-861}{12} \right) \\
&= \frac{50-56(4-x)}{170} - \frac{x^3-30x^2-12x+909}{255} \\
&= \frac{-174+56x}{170} - \frac{x^3-30x^2-12x+909}{255} \\
&= \frac{-522+168x-2x^3+60x^2+24x+1818}{510} \\
&= \frac{-2x^3+60x^2+192x+1296}{510} \\
&= \frac{-x^3+30x^2+96x+648}{255}, \quad x \in [2,4].
\end{aligned}
$$

Exercise 6.3.3 *Let* $I = \mathbb{R}$, $\beta(t) = \frac{1}{5}t + 1$, $t \in I$,

$$
f(t) = 2 + 3t + t^2 + \frac{1+t-t^2}{1+t+t^2}, \quad t \in I,
$$

$$
a = x_0 = 0, \quad x_1 = \frac{1}{6}, \quad x_2 = b = 1.
$$

Find $s_2(x)$, $x \in [x_{j-1}, x_j]$, $j \in \{1,2\}$.

6.4 Hermite Cubic Splines

Let $a, b \in I$, $a < b$,

$$K = \{a = x_0 \leq \beta(x_0) < x_1 \leq \beta(x_1) < \ldots < x_m \leq \beta(x_m) = b\},$$

$f \in \mathscr{C}^1_\beta([a,b])$.

Definition 6.4.1 *Define the Hermite cubic polynomial as a polynomial in the form*

$$s(x) = c_0 + c_1 h_1(x, x_{j-1}) + c_2 h_2(x, x_{j-1}) + c_3 h_3(x, x_{j-1}), \quad x \in [x_{j-1}, x_j], \tag{6.10}$$

$j \in \{1, \ldots, m\}$, *where* c_0, c_1, c_2 *and* c_3 *are constants so that*

$$s(x_j) = f(x_j), \quad D_\beta s(x_j) = D_\beta f(x_j), \quad j \in \{0, 1, \ldots, m\}. \tag{6.11}$$

Now, we will determine the constants c_0, c_1, c_2 and c_3. We have

$$\begin{aligned} s(x_{j-1}) &= c_0 \\ &= f(x_{j-1}) \end{aligned}$$

and

$$\begin{aligned} s(x_j) &= f(x_{j-1}) + c_1 h_1(x, x_{j-1}) + c_2 h_2(x, x_{j-1}) + c_3 h_3(x, x_{j-1}) \\ &= f(x_j), \end{aligned}$$

whereupon

$$c_1 h_1(x_j, x_{j-1}) + c_2 h_2(x_j, x_{j-1}) + c_3 h_3(x_j, x_{j-1}) = f(x_j) - f(x_{j-1}).$$

Next,

$$D_\beta s(x) = c_1 + c_2 h_1(x, x_{j-1}) + c_3 h_2(x, x_{j-1}), \quad x \in [x_{j-1}, x_j].$$

From here,

$$\begin{aligned} D_\beta s(x_{j-1}) &= c_1 \\ &= D_\beta f(x_{j-1}) \end{aligned}$$

and

$$D_\beta s(x_j) = D_\beta f(x_{j-1}) + c_2 h_1(x_j, x_{j-1}) + c_3 h_2(x_j, x_{j-1})$$

$$= \quad D_\beta f(x_j).$$

Thus, we get the system

$$\begin{aligned} c_2 h_2(x_j, x_{j_1}) + c_3 h_3(x_j, x_{j-1}) &= f(x_j) - f(x_{j-1}) \\ &\quad - D_\beta f(x_{j-1}) h_1(x_j, x_{j-1}) \\ c_2 h_1(x_j, x_{j-1}) + c_3 h_2(x_j, x_{j-1}) &= D_\beta f(x_j) - D_\beta f(x_{j-1}). \end{aligned}$$

By the last system, we find

$$\begin{aligned} c_2 &= \frac{\left(f(x_j) - f(x_{j-1}) - D_\beta f(x_{j-1}) h_1(x_j, x_{j-1})\right) h_2(x_j, x_{j-1})}{(h_2(x_j, x_{j-1}))^2 - h_1(x_j, x_{j-1}) h_3(x_j, x_{j-1})} \\ &\quad - \frac{\left(D_\beta f(x_j) - D_\beta f(x_{j-1})\right) h_3(x_j, x_{j-1})}{(h_2(x_j, x_{j-1}))^2 - h_1(x_j, x_{j-1}) h_3(x_j, x_{j-1})} \end{aligned}$$

and

$$\begin{aligned} c_3 &= \frac{\left(f(x_j) - f(x_{j-1}) - D_\beta f(x_{j-1}) h_1(x_j, x_{j-1})\right) h_2(x_j, x_{j-1}) h_1(x_j, x_{j-1})}{h_3(x_j, x_{j-1}) h_1(x_j, x_{j-1}) - (h_2(x_j, x_{j-1}))^2} \\ &\quad - \frac{\left(D_\beta f(x_j) - D_\beta f(x_{j-1})\right) h_2(x_j, x_{j-1})}{h_3(x_j, x_{j-1}) h_1(x_j, x_{j-1}) - (h_2(x_j, x_{j-1}))^2}. \end{aligned}$$

Consequently

$$\begin{aligned} s(x) &= f(x_{j-1}) + D_\beta f(x_{j-1}) h_1(x, x_{j-1}) \\ &\quad + \left(\frac{\left(f(x_j) - f(x_{j-1}) - D_\beta f(x_{j-1}) h_1(x_j, x_{j-1})\right) h_2(x_j, x_{j-1})}{(h_2(x_j, x_{j-1}))^2 - h_1(x_j, x_{j-1}) h_3(x_j, x_{j-1})} \right. \\ &\quad \left. - \frac{\left(D_\beta f(x_j) - D_\beta f(x_{j-1})\right) h_3(x_j, x_{j-1})}{(h_2(x_j, x_{j-1}))^2 - h_1(x_j, x_{j-1}) h_3(x_j, x_{j-1})} \right) h_2(x, x_{j-1}) \\ &\quad + \left(\frac{\left(f(x_j) - f(x_{j-1}) - D_\beta f(x_{j-1}) h_1(x_j, x_{j-1})\right) h_2(x_j, x_{j-1}) h_1(x_j, x_{j-1})}{h_3(x_j, x_{j-1}) h_1(x_j, x_{j-1}) - (h_2(x_j, x_{j-1}))^2} \right. \\ &\quad \left. - \frac{\left(D_\beta f(x_j) - D_\beta f(x_{j-1})\right) h_2(x_j, x_{j-1})}{h_3(x_j, x_{j-1}) h_1(x_j, x_{j-1}) - (h_2(x_j, x_{j-1}))^2} \right) h_3(x, x_{j-1}), \end{aligned}$$

$x \in [x_{j-1}, x_j]$, $j \in \{1, \ldots, m\}$.

Remark 6.4.2 *We can define the Hermite cubic polynomial in the following way*

$$s(x) = c_0 + c_1 g_1(x, x_{j-1}) + c_2 g_2(x, x_{j-1}) + c_3 g_3(x, x_{j-1}), \quad x \in [x_{j-1}, x_j],$$

$j \in \{1, \ldots, m\}$, *where the constants* c_0, c_1, c_2 *and* c_3 *will be determined by the condition* (6.11). *As above, one can deduct that*

$$\begin{aligned} s(x) &= f(x_{j-1}) + D_\beta f(x_{j-1}) g_1(x, x_{j-1}) \\ &\quad + \left(\frac{\left(f(x_j) - f(x_{j-1}) - D_\beta f(x_{j-1}) g_1(x_j, x_{j-1})\right) g_2(x_j, x_{j-1})}{(g_2(x_j, x_{j-1}))^2 - g_1(x_j, x_{j-1}) g_3(x_j, x_{j-1})} \right. \\ &\quad \left. - \frac{\left(D_\beta f(x_j) - D_\beta f(x_{j-1})\right) g_3(x_j, x_{j-1})}{(g_2(x_j, x_{j-1}))^2 - g_1(x_j, x_{j-1}) g_3(x_j, x_{j-1})} \right) g_2(x, x_{j-1}) \\ &\quad + \left(\frac{\left(f(x_j) - f(x_{j-1}) - D_\beta f(x_{j-1}) g_1(x_j, x_{j-1})\right) g_2(x_j, x_{j-1}) g_1(x_j, x_{j-1})}{g_3(x_j, x_{j-1}) g_1(x_j, x_{j-1}) - (g_2(x_j, x_{j-1}))^2} \right. \\ &\quad \left. - \frac{\left(D_\beta f(x_j) - D_\beta f(x_{j-1})\right) g_2(x_j, x_{j-1})}{g_3(x_j, x_{j-1}) g_1(x_j, x_{j-1}) - (g_2(x_j, x_{j-1}))^2} \right) g_3(x, x_{j-1}), \end{aligned}$$

$x \in [x_{j-1}, x_j]$, $j \in \{1, \ldots, m\}$.

Example 6.4.3 *Let* $I = \mathbb{R}$, $\beta(t) = 2t$, $t \in I$,

$$a = x_0 = 1, \quad x_1 = 4, \quad x_2 = 16 = b,$$

$$f(t) = \frac{1}{1+t}, \quad t \in I.$$

We will find $s(x)$, $x \in [1, 4]$. *We have*

$$\begin{aligned} f(x_0) &= \frac{1}{2}, \\ f(x_1) &= \frac{1}{5}, \\ D_\beta f(t) &= -\frac{1}{(1+t)(1+\beta(t))} \\ &= -\frac{1}{(1+t)(1+2t)}, \quad t \in I, \\ D_\beta f(1) &= -\frac{1}{6}, \\ D_\beta f(4) &= -\frac{1}{45}, \\ h_1(x, x_0) &= x - 1, \end{aligned}$$

$$\begin{aligned}
h_1(x_1,x_0) &= 3,\\
h_2(x,x_0) &= \int_1^x (t-1)d_\beta t\\
&= \left(\frac{1}{3}t^2 - t\right)\Big|_{t=1}^{t=x}\\
&= \frac{x^2}{3} - x - \left(\frac{1}{3} - 1\right)\\
&= \frac{x^2}{3} - x + \frac{2}{3},\\
h_2(x_1,x_0) &= \frac{16}{3} - 4 + \frac{2}{3}\\
&= 6 - 4\\
&= 2,\\
h_3(x,x_0) &= \int_1^x \left(\frac{t^2}{3} - t + \frac{2}{3}\right) d_\beta t\\
&= \left(\frac{t^3}{21} - \frac{t^2}{3} + \frac{2}{3}t\right)\Big|_{t=1}^{t=x}\\
&= \frac{x^3}{21} - \frac{x^2}{3} + \frac{2}{3}x - \frac{1}{21} + \frac{1}{3} - \frac{2}{3}\\
&= \frac{x^3}{21} - \frac{x^2}{3} + \frac{2}{3}x - \frac{8}{21},\\
h_3(x_1,x_0) &= \frac{64}{21} - \frac{16}{3} + \frac{8}{3} - \frac{8}{21}\\
&= \frac{56}{21} - \frac{8}{3}\\
&= 0, \quad x \in [1,4].
\end{aligned}$$

Hence,

$$\begin{aligned}
c_0 &= f(x_0)\\
&= \frac{1}{2},
\end{aligned}$$

$$
\begin{aligned}
c_1 &= D_\beta f(x_0) \\
&= -\frac{1}{6}, \\
c_2 &= \frac{\left(\frac{1}{5}-\frac{1}{2}+\frac{1}{6}\cdot 3\right)2}{4-3\cdot 0} - \frac{\left(-\frac{1}{45}+\frac{1}{6}\right)0}{4-3\cdot 0} \\
&= \frac{\frac{2}{5}}{4} \\
&= \frac{1}{10}, \\
c_3 &= \frac{\left(\frac{1}{5}-\frac{1}{2}+\frac{1}{6}\cdot 3\right)3}{3\cdot 0-2^2} - \frac{\left(-\frac{1}{45}+\frac{1}{6}\right)2}{3\cdot 0-2^2} \\
&= -\frac{\frac{3}{5}}{4} - \frac{\frac{13}{45}}{-4} \\
&= -\frac{3}{20}+\frac{13}{180} \\
&= -\frac{14}{180} \\
&= --\frac{7}{90}
\end{aligned}
$$

and

$$
\begin{aligned}
s(x) &= \frac{1}{2}-\frac{1}{6}(x-1)+\frac{1}{10}\left(\frac{x^2}{3}-x+\frac{2}{3}\right) \\
&\quad -\frac{7}{90}\left(\frac{x^3}{21}-\frac{x^2}{3}+\frac{2}{3}x-\frac{8}{21}\right) \\
&= \frac{1}{90}\left(45-15x+15+3x^2-9x+6-\frac{x^3}{3}+\frac{7}{3}x^2-\frac{14}{3}x+\frac{8}{3}\right) \\
&= \frac{1}{90}\left(-\frac{x^3}{3}=\frac{16}{3}x^2-\frac{86}{3}x+\frac{206}{3}\right) \\
&= -\frac{1}{270}\left(x^3-16x^2+86x-206\right), \quad x \in [1,4].
\end{aligned}
$$

Exercise 6.4.4 *Let* $I = \mathbb{R}$, $\beta(t) = \frac{1}{3}t + 2$, $t \in I$,

$$a = x_0 = \frac{1}{81}, \quad x_1 = \frac{1}{9}, \quad x_2 = b = 1,$$

$$f(t) = 1 + t - t^2 + \frac{1}{1+t^2}, \quad t \in I.$$

Find $s(x)$, $x \in \left[\frac{1}{81}, \frac{1}{9}\right]$.

6.5 Advanced Practical Problems

Problem 6.5.1 *Let* $I = \mathbb{R}$, $\beta(t) = 4t$, $t \in I$,

$$a = x_0 = 1, \quad x_1 = \frac{4}{3}, \quad x_2 = \frac{5}{2} = b,$$

$$f(t) = 2 + 3t + t^2, \quad t \in I.$$

Prove that

$$\sup_{x \in [a,b]} |s_L(x) - f(x)| \leq \frac{133}{6}.$$

Problem 6.5.2 *Let* $i = \mathbb{R}$, $\beta(t) = 2t + 3$, $t \in I$,

$$a = x_0 = 1, \quad x_1 = \frac{1}{4}, \quad x_2 = \frac{7}{6}, \quad x_3 = \frac{7}{4}, \quad x_4 = \frac{5}{2}, \quad x_5\frac{7}{2}, \quad b = 4,$$

$$f(t) = t + \frac{1}{t^3 + t^2 + t + 1}, \quad t \in I.$$

Evaluate

$$|s_{\beta L}(x) - f(x)|, \quad x \in [a,b].$$

Problem 6.5.3 *Let* $I = \mathbb{R}$, $\beta(t) = 2t$, $t \in I$,

$$f(t) = 1 + t + e_1(t,1) + \sin_1(t,1), \quad t \in I,$$

$$a = x_0 = 1, \quad x_1 = 4, \quad x_2 = b = 16.$$

Find $s_2(x)$, $x \in [x_{j-1}, x_j]$, $j \in \{1,2\}$.

Problem 6.5.4 *Let* $I = \mathbb{R}$, $\beta(t) = \frac{1}{2}t$, $t \in I$,

$$a = x_0 = \frac{1}{256}, \quad x_1 = \frac{1}{64}, \quad x_2 = \frac{1}{8}, \quad x_3 = 1,$$

$$f(t) = \frac{1+t}{1+t+t^2}, \quad t \in I.$$

Find $s(x)$, $x \in \left[\frac{1}{64}, \frac{1}{8}\right]$.

Problem 6.5.5 *Let* $a, b \in I$, $a < b$,

$$K = \{a = x_0 \leq \beta(x_0) < x_1 \leq \beta(x_1) < \ldots < x_m \leq \beta(x_m) = b\},$$

$f \in \mathscr{C}^1_\beta([a,b])$. *Find Hermite cubic spline in the form*

$$s(x) = c_0 + c_1 h_1(x, x_{j-1}) + c_2 g_2(x, x_{j-1}) + c_3 g_3(x, x_{j-1}), \quad x \in [x_{j-1}, x_j],$$

so that (6.11) *hold.*

Problem 6.5.6 *Let* $a, b \in I$, $a < b$,

$$K = \{a = x_0 \leq \beta(x_0) < x_1 \leq \beta(x_1) < \ldots < x_m \leq \beta(x_m) = b\},$$

$f \in \mathscr{C}^1_\beta([a,b])$. *Find Hermite cubic spline in the form*

$$s(x) = c_0 + c_1 h_1(x, x_{j-1}) + c_2 h_2(x, x_{j-1}) + c_3 g_3(x, x_{j-1}), \quad x \in [x_{j-1}, x_j],$$

so that (6.11) *hold.*

Problem 6.5.7 *Let* $a, b \in I$, $a < b$,

$$K = \{a = x_0 \leq \beta(x_0) < x_1 \leq \beta(x_1) < \ldots < x_m \leq \beta(x_m) = b\},$$

$f \in \mathscr{C}^1_\beta([a,b])$. *Find Hermite cubic spline in the form*

$$s(x) = c_0 + c_1 h_1(x, x_{j-1}) + c_2 g_2(x, x_{j-1}) + c_3 h_3(x, x_{j-1}), \quad x \in [x_{j-1}, x_j],$$

so that (6.11) *hold.*

6.6 Notes and References

In this chapter, we introduce piecewise approximations of a given function. They are defined linear interpolating splines, linear interpolating β-splines, cubic and Hermite splines. We deduct some of their properties. The results in this chapter are provided with suitable examples.

7

The Euler Method

In this chapter, we generalise the Euler method for β-differential equations. Let $I \subseteq \mathbb{R}$ and $\beta : I \to I$ be a quantum operator.

7.1 Analysing the Method

Consider the Cauchy problem

$$\begin{cases} D_\beta y(x) & = & f(x, y(x)) \quad x \in I \\ \\ y(x_0) & = & y_0, \end{cases} \tag{7.1}$$

where $f : I \times \mathbb{R} \to \mathbb{R}$, $y_0 \in \mathbb{R}$, $x_0 \in I$ are given, $y : I \to \mathbb{R}$ is unknown. Suppose that $x \in [x_0, \overline{x}]_I$ and take the points

$$x_0 < x_1 < \ldots < x_N = \overline{x}, \quad x_i \in I, \quad i = 1, \ldots, N,$$

so that

$$x_i = \begin{cases} \beta^{l_{i-1}}(x_{i-1}), & l_{i-1} \in \mathbb{N} \quad \text{if} \quad x_{i-1} \neq s_0, \\ \\ x_{i-1} + q_{i-1}, & q_{i-1} \in \mathbb{R} \quad \text{if} \quad x_{i-1} = s_0, \end{cases} \tag{7.2}$$

$i = 1, \ldots, N$. Define $\{y_i\}_{i=1}^N$ by

$$\begin{aligned} y_{i+1} - y_i &= (x_{i+1} - x_i) f(x_i, y_i) \\ &= \begin{cases} \left(\beta^{l_i}(x_i) - x_i\right) f(x_i, y_i) \quad \text{if} \quad x_i \neq s_0, \\ \\ q_i f(x_i, y_i) \quad \text{if} \quad x_i = s_0. \end{cases} \end{aligned} \tag{7.3}$$

7.2 Local Truncation Error

To obtain the local truncation error, we consider the cases when $x_0 = s_0$ and $x_0 \neq s_0$ separately.

DOI: 10.1201/9781003472131-7

1. $x_0 \neq s_0$.

   In a single step of the Euler method, the computed result is

$$y_1 = y_0 + \left(\beta^{l_0}(x_0) - x_0\right) f(x_0, y(x_0))$$

   and it differs from the exact answer $y(x_1) = y(\beta^{l_0}(x_0))$ by

$$\begin{aligned} y(x_1) - y_1 &= y\left(\beta^{l_0}(x_0)\right) - y(x_0) - \left(\beta^{l_0}(x_0) - x_0\right) f(x_0, y(x_0)) \\ &= y\left(\beta^{l_0}(x_0)\right) - y(x_0) - \left(\beta^{l_0}(x_0) - x_0\right) D_\beta y(x_0). \end{aligned}$$

   If $l_0 = 1$, then $y(x_1) = y_1$. Assuming y has continuous first-order and second-order derivatives, this can be written, using the quantum Taylor formula, in the form

$$\int_{x_0}^{\beta^{-1}(\beta^{l_0}(x_0))} h_1\left(\beta^{l_0}(x_0), \beta(\tau)\right) D_\beta^2 y(\tau) d_\beta \tau.$$

   Another way of writing the error, assuming that the third derivative $D_\beta^3 y$ also exists and is bounded, is

$$h_2\left(\beta^{l_0}(x_0), x_0\right) D_\beta^2 y(x_0) + O\left(h_3(\beta^{l_0}(x_0), x_0)\right).$$

2. x_0 is right-dense.

   In a single step of the Euler method, the computed result

$$y_1 = y_0 + q_0 f(x_0, y(x_0))$$

   differs from the exact solution $y(x_1)$ by

$$\begin{aligned} y(x_1) - y_1 &= y(x_0 + q_0) - y(x_0) - q_0 f(x_0, y(x_0)) \\ &= y(x_0 + q_0) - y(x_0) - q_0 D_\beta y(x_0). \end{aligned}$$

   Assuming y has continuous first and second derivatives, this can be written in the form

$$\int_{x_0}^{\beta^{-1}(x_0 + q_0)} h_1(x_0 + q_0, \beta(\tau)) D_\beta^2 y(\tau) d_\beta \tau.$$

   Another way of writing the error, assuming that the third derivative exists and is bounded, is

$$h_2(x_0 + q_0, x_0) D_\beta^2 y(x_0) + O\left(h_3(x_0 + q_0, x_0)\right).$$

Example 7.2.1 *Let $I = \mathbb{R}$, $\beta(t) = 2t$, and consider the IVP*

$$D_\beta y(x) = -\frac{1}{1+2x} y(x), \quad x > 1,$$

$$y(1) = \frac{1}{2}.$$

Take

$$x_0 = 1, \quad x_1 = 8, \quad x_2 = 16, \quad x_3 = 32.$$

We have $x_1 = \beta^3(x_0)$, i.e., $l_0 = 3$. Let

$$g_1(x) = \frac{x^2}{3} - x,$$

$$g_2(x) = \frac{x^3}{21} - \frac{x^2}{3} + \frac{2}{3}x,$$

$$y(x) = \frac{1}{1+x}, \quad x \in I.$$

Then

$$\begin{aligned} D_\beta g_1(x) &= \frac{\beta(x)+x}{3} - 1 \\ &= \frac{2x+x}{3} - 1 \\ &= x - 1, \\ D_\beta g_2(x) &= \frac{(\beta(x))^2 + x\beta(x) + x^2}{21} - \frac{\beta(x)+x}{3} + \frac{2}{3} \\ &= \frac{4x^2+2x^2+x^2}{21} - \frac{2x+x}{3} + \frac{2}{3} \\ &= \frac{x^2}{3} - x + \frac{2}{3}, \\ D_\beta y(x) &= -\frac{1}{(1+x)(1+\beta(x))} \\ &= -\frac{1}{(1+x)(1+2x)} \\ &= -\frac{1}{2x^2+3x+1}, \end{aligned}$$

$$\begin{aligned} D_\beta^2 y(x) &= \frac{2(\beta(x)+x)+3}{(2x^2+3x+1)(2(\beta(x))^2+3\beta(x)+1)} \\ &= \frac{6x+3}{(2x^2+3x+1)(8x^2+6x+1)}, \quad x\in I, \\ D_\beta^2 y(1) &= \frac{9}{6\cdot 15} \\ &= \frac{1}{10}. \end{aligned}$$

Hence,

$$\begin{aligned} D_\beta y(x) &= -\frac{1}{(1+x)(1+2x)} \\ &= -\frac{1}{1+x}y(x), \quad x>1, \\ y(1) &= \frac{1}{2}, \end{aligned}$$

i.e., y is a solution of the considered IVP. Next,

$$\begin{aligned} h_2(x,x_0) &= \int_1^x (t-1)d_\beta t \\ &= \int_1^x D_\beta g_1(t)d_\beta t \\ &= g_1(x)-g_1(1) \\ &= \frac{x^2}{3}-x-\left(\frac{1}{3}-1\right) \\ &= \frac{x^2}{3}-x+\frac{2}{3}, \\ h_2(\beta^{l_0}(x_0),x_0) &= h_2(8,1) \\ &= \frac{64}{3}-8+\frac{2}{3} \\ &= 22-8 \\ &= 14, \end{aligned}$$

$$
\begin{aligned}
h_3(x,x_0) &= \int_1^x h_2(t,x_0)d_\beta t \\
&= \int_1^x \left(\frac{t^2}{3} - t + \frac{2}{3}\right) d_\beta t \\
&= \int_1^x D_\beta g_2(t) d_\beta t \\
&= g_2(x) - g_1(1) \\
&= \frac{x^3}{21} - \frac{x^2}{3} + \frac{2}{3}x - \left(\frac{1}{21} - \frac{1}{3} + \frac{2}{3}\right) \\
&= \frac{x^3}{21} - \frac{x^2}{3} + \frac{2}{3}x - \left(\frac{1}{21} + \frac{1}{3}\right) \\
&= \frac{x^3}{21} - \frac{x^2}{3} + \frac{2}{3}x - \frac{8}{21}, \quad x \in I, \\
h_3(\beta^{l_0}(x_0),x_0) &= h_3(8,1) \\
&= \frac{512}{21} - \frac{64}{3} + \frac{16}{3} - \frac{8}{21} \\
&= \frac{504}{21} - \frac{48}{3} \\
&= 24 - 16 \\
&= 8.
\end{aligned}
$$

Therefore, the local truncation error is

$$
\begin{aligned}
h_2\left(\beta^{l_0}(x_0),x_0\right) D_\beta^2 y(x_0) + O\left(h_3(\beta^{l_0}(x_0),x_0)\right) &= \frac{14}{10} + O(8) \\
&= \frac{7}{5} + O(8).
\end{aligned}
$$

Exercise 7.2.2 *Let* $I = \mathbb{R}$, $\beta(t) = \frac{2}{3}t + 4$, $t \in I$,

$$
a = x_0 = 0, \quad x_1 = \frac{1}{6}, \quad x_2 = \frac{1}{4}, \quad x_3 = \frac{1}{2}, \quad x_4 = 1.
$$

Consider the IVP

$$
D_\beta y(x) = 1 + (y(x))^2, \quad x > 0,
$$

$$y(0) = 2.$$

Find the local truncation error.

7.3 Global Truncation Error

Let $\tilde{y}(x)$ denote the computed solution on the interval $[x_0, \bar{x}]$. That is, at step values $x_0, x_1, \ldots, x_N = \bar{x}$ defined by (7.2), $\tilde{y}$ is computed, using the equation (7.3). For "off-step" points, $\tilde{y}(x)$ is defined by linear interpolation, or what is equivalent, $\tilde{y}(x)$ is evaluated using a partial step from the most recently computed step values. That is, if $x \in [x_{k-1}, x_k]$, $k = 1, \ldots, N$, then

$$\tilde{y}(x) = y_{k-1} + h_1(x, x_{k-1}) f(x_{k-1}, y_{k-1}), \tag{7.4}$$

Define the maximum step size as

$$m = \max_{1 \le i \le N} \{x_i - x_{i-1}\}.$$

Also,

$$\alpha(x) = y(x) - \tilde{y}(x), \tag{7.5}$$

$$\beta_1(x) = f(x, y(x)) - f(x, \tilde{y}(x)). \tag{7.6}$$

Suppose that

$$|f(x,y) - f(x,z)| \le L|y - z| \quad \text{for all } x \in I \text{ and } y, z \in \mathbb{R},$$

where $L > 0$. From (7.5) and (7.6), we have

$$|\beta_1(x)| \le L|\alpha(x)|.$$

Define $E(x)$ so that the exact solution satisfies

$$\begin{aligned} y(x) &= y(x_{k-1}) + h_1(x, x_{k-1}) f(x_{k-1}, y(x_{k-1})) \\ &\quad + h_2(x, x_{k-1}) E(x), \quad x \in [x_{k-1}, x_k], \end{aligned} \tag{7.7}$$

and assume that $|E(x)| \le p$. Subtracting (7.4) from (7.7), we get

$$\begin{aligned} y(x) - \tilde{y}(x) &= y(x_{k-1}) - y_{k-1} + h_1(x, x_{k-1}) \Big(f(x_{k-1}, y(x_{k-1})) \\ &\quad - f(x_{k-1}, y_{k-1}) \Big) + h_2(x, x_{k-1}) E(x). \end{aligned}$$

Hence, using (7.5) and (7.6), we get

$$\alpha(x) = \alpha(x_{k-1}) + h_1(x, x_{k-1})\beta_1(x_{k-1}) + h_2(x, x_{k-1})E(x),$$

whereupon

$$\begin{aligned}
|\alpha(x)| &\leq |\alpha(x_{k-1})| + |h_1(x, x_{k-1})|L|\alpha(x_{k-1})| \\
&\quad + p|h_2(x, x_{k-1})| \\
&= (1 + L|h_1(x, x_{k-1})|)\,|\alpha(x_{k-1})| + |h_2(x, x_{k-1})|p \\
&\leq (1 + L|h_1(x, x_{k-1})|)\,|\alpha(x_{k-1})| + mp|h_1(x, x_{k-1})|.
\end{aligned} \tag{7.8}$$

If $L = 0$, then it follows that

$$|\alpha(x)| \leq |\alpha(x_{k-1})| + mph_1(x, x_{k-1}).$$

In particular,

$$|\alpha(x_k)| \leq |\alpha(x_{k-1})| + mph_1(x_k, x_{k-1}).$$

Therefore, we deduce

$$\begin{aligned}
|\alpha(x)| &\leq |\alpha(x_{k-2})| + mp(h_1(x_{k-1}, x_{k-2}) + h_1(x, x_{k-1})) \\
&= |\alpha(x_{k-2})| + mph_1(x, x_{k-2}) \\
&\leq \cdots \\
&\leq |\alpha(x_0)| + mph_1(x, x_0).
\end{aligned}$$

If $L > 0$, then we have

$$\begin{aligned}
|\alpha(x)| &\leq (1 + Lh_1(x, x_{k-1}))|\alpha(x_{k-1})| + mph_1(x, x_{k-1}) \\
&= (1 + Lh_1(x, x_{k-1}))|\alpha(x_{k-1})| + \frac{mp}{L}Lh_1(x, x_{k-1}) \\
&\quad + \frac{mp}{L} - \frac{mp}{L},
\end{aligned}$$

i.e.,

$$\begin{aligned}
|\alpha(x)| + \frac{mp}{L} &\leq (1 + Lh_1(x, x_{k-1}))|\alpha(x_{k-1})| + \frac{mp}{L}(1 + Lh_1(x, x_{k-1})) \\
&= (1 + Lh_1(x, x_{k-1}))\left(\frac{mp}{L} + |\alpha(x_{k-1})|\right)
\end{aligned}$$

$$\leq \; e_L(x, x_{k-1})\left(\frac{mp}{L} + |\alpha(x_{k-1})|\right).$$

In particular,

$$|\alpha(x_k)| + \frac{mp}{L} \leq e_L(x_k, x_{k-1})\left(\frac{mp}{L} + |\alpha(x_{k-1})|\right).$$

Hence, if $x \in [x_{k-1}, x_k]$, then we get

$$\begin{aligned}
|\alpha(x)| + \frac{mp}{L} &\leq e_L(x, x_{k-1})\left(\frac{mp}{L} + |\alpha(x_{k-1})|\right) \\
&\leq e_L(x, x_{k-1})e_L(x_{k-1}, x_{k-2})\left(\frac{mp}{L} + |\alpha(x_{k-2})|\right) \\
&\leq \cdots \\
&\leq e_L(x, x_0)\left(\frac{mp}{L} + |\alpha(x_0)|\right).
\end{aligned}$$

Combining the estimates found in the two cases and stating them formally, we have the following result.

Theorem 7.3.1 *Assuming that f satisfies a Lipschitz condition with a constant L, the global error satisfies the bound*

$$|y(x) - \tilde{y}(x)| \leq \begin{cases} |y(x_0) - \tilde{y}(x_0)| + mph_1(x, x_0) & \text{if} \quad L = 0 \\ e_{L,\beta}(x)\left(\frac{mp}{L} + |\alpha(x_0)|\right) - \frac{mp}{L} & \text{if} \quad L > 0. \end{cases} \tag{7.9}$$

Now we consider a sequence of approximations to $y(\overline{x})$. In each of these approximations, a computation using the Euler method is performed, starting from an approximation to $y(x_0)$, and taking a sequence of positive steps. Denote approximation number r by $\overline{y}_r$.

The only assumption we will make about $\overline{y_r}$, for each specific value of r, is that the initial error $y(x_0) - \overline{y_r}(x_0)$ is bounded by K_r and that the greatest step size is bounded by m_r. It is assumed that $K_r \to 0$ as $r \to \infty$. If $m_r \to 0$, then, by (7.9), we get that

$$|y(\overline{x}) - \overline{y_r}(\overline{x})| \to 0 \quad \text{as} \quad r \to \infty.$$

There are cases when m_r does not tend to zero as $r \to \infty$. For instance, when $I = \mathbb{R}$, $\beta(t) = 2t$, $t \in I$. We have $m_r \to 0$ as $r \to \infty$.

7.4 Advanced Practical Problems

Problem 7.4.1 *Let $I = \mathbb{R}$, $\beta(t) = 3t$, $t \in I$,*

$$a = x_0 = -3, \quad x_1 = -2, \quad x_2 = -\frac{1}{4}, \quad x_3 = 0, \quad x_4 = \frac{1}{6}, \quad x_5 = 2, \quad x_6 = 3.$$

Consider the IVP

$$\begin{aligned} D_\beta y(x) &= \frac{1+y(x)}{1+y(x)+(y(x))^2}, \quad x > -3, \\ y(-3) &= 1. \end{aligned}$$

Find the local truncation error.

Problem 7.4.2 *Let $I = \mathbb{R}$, $\beta(t) = 3t+4$, $t \in I$. Apply the Euler method for the following IVP*

$$\begin{aligned} D_\beta y(x) &= x^2 - (y(x))^2, \quad x > 1, \\ y(1) &= 1, \end{aligned}$$

where

$$a = x_0 = 1, \quad x_1 = 2, \quad x_2 = 4, \quad x_3 = 8, \quad x_4 = 16, \quad x_5 = 32.$$

Evaluate local and global truncation errors.

Problem 7.4.3 *Let $I = \mathbb{R}$, $\beta(t) = \frac{1}{2}t$, $t \in I$. Apply the Euler method for the following IVP*

$$\begin{aligned} D_\beta y(x) &= x + \frac{y(x)}{1+(y(x))^4}, \quad x > -\frac{7}{8}, \\ y\left(-\frac{7}{8}\right) &= \frac{1}{2}, \end{aligned}$$

where

$$a = x_0 - \frac{7}{8}, \quad x_1 = -\frac{3}{4}, \quad x_2 = 0, \quad x_3 = \frac{1}{6}, \quad x_4 = \frac{9}{2}, \quad x_5 = 5.$$

Evaluate the local and global truncation errors.

Problem 7.4.4 *Let $I = \mathbb{R}$, $\beta(t) = 2t+4$, $t \in I$. Rewrite the following IVP*

$$D_\beta y(x) + y(x) D_\beta y(x) + 4y(x) = x^2, \quad x > 0,$$

$$y(0) = D_\beta y(0) = 1,$$

as a first-order system and use the Euler method with

$$a = x_0 = 0, \quad x_1 = \frac{1}{6}, \quad x_2 = \frac{1}{2}, \quad x_3 = 2, \quad x_4 = \frac{8}{3}, \quad x_5 = \frac{10}{3}.$$

Evaluate the local and global truncation errors.

Problem 7.4.5 *Let $I = \mathbb{R}$, $\beta(t) = 3t$, $t \in I$. Rewrite the following IVP*

$$D^3_\beta y(x) + \sin_{1,\beta}(x) D^2_\beta y(x)$$

$$+xy(x)D_\beta y(x) + (y(x))^2 = x+1, \quad x > 1,$$

$$D^2_\beta y(1) = -1, \quad D_\beta y(1) = 0, \quad y(1) = 5,$$

as a first-order system and use the Euler method with

$$a = x_0 = 1, \quad x_1 = 9, \quad x_2 = 27, \quad x_3 = 81, \quad x_4 = 243.$$

Evaluate the local and global truncation errors.

Problem 7.4.6 *Let $I = \mathbb{R}$, $\beta(t) = 4t$, $t \in I$. Rewrite the following IVP*

$$D^4_\beta y(x) + \sin_{1,\beta}(x) D_\beta y(x) D^3_\beta y(x) + 2y(x) D^2_\beta y(x) + D_\beta y(x)$$

$$+(y(x))^2 = x^2 + x + 1 + e_{1,\beta}(x), \quad x > 1,$$

$$D^3_\beta y(1) = D^2_\beta y(1) = 2, \quad D_\beta y(1) = 1, \quad y(1) = -1,$$

as a first-order system and use the Euler method with

$$a = x_0 = 1, \quad x_1 = 4, \quad x_2 = 16, \quad x_3 = 64, \quad x_4 = 256.$$

Evaluate the local and global truncation errors.

7.5 Notes and References

In this chapter we investigate the Euler method for β-differential equations. We define local and global truncate errors and we evaluate them in both cases when $x_0 = s_0$ and $x_0 \neq s_0$.

8

The Order-Two Taylor Series Method-TS(2)

In this chapter we propose the Taylor series method for approximate solution of initial value problems associated with β-differential equations of first-order.

Let $I \subseteq \mathbb{R}$ and $\beta : I \to \mathbb{R}$ be first kind or second kind general quantum operator.

8.1 Analysing of the Method

Suppose that $t_0, t_f \in I$, $t_f < \infty$. Consider the initial value problem (IVP)

$$\begin{cases} D_\beta x(t) &= f(t,x(t)), \quad t \in [t_0,t_f], \\ x(t_0) &= x_0, \end{cases} \tag{8.1}$$

where $x_0 \in \mathbb{R}$. Throughout this chapter we assume the following conditions.

$$(H1)\begin{cases} |f(t,x)| \le A, \quad t \in I, x \in \mathbb{R}, \\ \text{there exist } D_{\beta t} f(t,x) \text{ and } \dfrac{\partial}{\partial x} f(t,x) \text{ such that} \\ |D_{\beta t} f(t,x)| \le A, \quad \left|\dfrac{\partial}{\partial x} f(t,x)\right| \le A, \quad t \in I, x \in \mathbb{R}. \end{cases}$$

$$(H2)\begin{cases} \text{If } g(t,x) = D_{\beta t} f(t,x) \\ + \left(\displaystyle\int_0^1 \frac{\partial}{\partial x} f(\beta(t), x + s(\beta(t)-t) f(t,x)) ds\right) f(t,x), \\ t \in I, x \in \mathbb{R}, \text{there exist } D_{\beta t} g(t,x) \text{ and } \dfrac{\partial}{\partial x} g(t,x) \text{ such that} \\ |D_{\beta t} g(t,x)| \le A, \quad \left|\dfrac{\partial}{\partial x} g(t,x)\right| \le A, \quad t \in I, x \in \mathbb{R}, \end{cases}$$

where $A > 0$ is a given constant.

Suppose that $r > 0$, $t, t+r \in [t_0, t_f]$, $(\beta^{-1})^2(t+r) \in [t, t_f]$. Then, by the Taylor formula of the second-order, we compute

$$\begin{aligned} x(t+r) &= x(t) + h_1(t+r,t) D_\beta x(t) + h_2(t+r,t) D_\beta^2 x(t) \\ &+ \int_t^{(\beta^{-1})^2(t+r)} h_2(t+r, \beta(\tau)) D_\beta^3 x(\tau) D_\beta \tau. \end{aligned}$$

DOI: 10.1201/9781003472131-8

Let

$$R_2(r) = \int_t^{(\beta^{-1})^2(t+r)} h_2(t+r,\beta(\tau))D_\beta^3 x(\tau)d_\beta\tau. \tag{8.2}$$

Then

$$x(t+r) = x(t) + h_1(t+r,t)D_\beta x(t) + h_2(t+r,t)D_\beta^2 x(t) + R_2(r).$$

Assume that $\{t_0 < t_1 < \ldots < t_{m+1} = t_f\}$ is a partition of the interval $[t_0,t_f]$ such that $t_{n+1} = t_n + r_{n+1} \in I$, $r_{n+1} > 0$, $n \in \{0,\ldots,m\}$. For example, when $I = \mathbb{R}$, $\beta(t) = 2t$, $t \in I$, and $[t_0,t_f] = [1,16]$, then $r_1 = 1$, $r_2 = 2$, $r_3 = 4$, $r_4 = 8$. Therefore,

$$\begin{aligned} x(t_{n+1}) &= x(t_n) + h_1(t_{n+1},t_n)D_\beta x(t_n) + h_2(t_{n+1},t_n)D_\beta^2 x(t_n) + R_2(r_{n+1}) \\ &= x(t_n) + r_{n+1}D_\beta x(t_n) + h_2(t_{n+1},t_n)D_\beta^2 x(t_n) + R_2(r_{n+1}). \end{aligned} \tag{8.3}$$

Neglecting the remainder term $R_2(r_{n+1})$ we get the formula

$$x(t_{n+1}) = x(t_n) + r_{n+1}D_\beta x(t_n) + h_2(t_{n+1},t_n)D_\beta^2 x(t_n). \tag{8.4}$$

Let $x_n = x(t_n)$, $D_\beta x_n = D_\beta x(t_n)$ and $D_\beta^2 x_n = D_\beta^2 x(t_n)$. Then the equation (8.4) can be written as

$$x_{n+1} = x_n + r_{n+1}D_\beta x_n + h_2(t_{n+1},t_n)D_\beta^2 x_n. \tag{8.5}$$

We shall refer to this relation as the order-two Taylor series method. The value $D_\beta x_n$ can be computed from the IVP (8.1) as

$$D_\beta x_n = f(t_n,x_n). \tag{8.6}$$

To determine $D_\beta^2 x_n$ we have to differentiate both sides of the equation in (8.1). Applying the chain rule, we get

$$\begin{aligned} D_\beta\left(f(t,x(t))\right) &= D_{\beta t}f(t,x(t)) \\ &+ \left(\int_0^1 \frac{\partial}{\partial x} f(\beta(t),x(t)+s(\beta(t)-t)D_\beta x(t))ds\right) D_\beta x(t), \end{aligned} \tag{8.7}$$

for $t \in I$, whereupon

$$\begin{aligned} D_\beta^2 x(t) &= D_{\beta t}f(t,x(t)) \\ &+ \left(\int_0^1 \frac{\partial}{\partial x} f(\beta(t),x(t)+s(\beta(t)-t)D_\beta x(t))ds\right) D_\beta x(t), \quad t \in I. \end{aligned}$$

Hence,

$$\begin{aligned} D_\beta^2 x_n &= D_{\beta t}f(t_n,x_n) \\ &+ \left(\int_0^1 \frac{\partial}{\partial x} f(\beta(t_n),x_n+s(\beta(t_n)-t_n)D_\beta x_n)ds\right) D_\beta x_n. \end{aligned}$$

Therefore, x_{n+1} can be determined by the formula (8.5).

Example 8.1.1 *Consider the IVP*

$$\begin{cases} D_\beta x(t) &= g(t) + \dfrac{1}{(x(t))^2+1}, \quad t \in [t_0, t_f], \\ x(t_0) &= x_0, \end{cases} \tag{8.8}$$

where $g : I \to \mathbb{R}$ is β-differentiable, $|g(t)| \le B$, $|D_\beta g(t)| \le B$ for some positive constant B and $x_0 \in \mathbb{R}$. Here

$$f(t,x(t)) = g(t) + \frac{1}{(x(t))^2+1}.$$

Then, if x is a solution of the IVP (8.8), *we have*

$$\begin{aligned} D_\beta^2 x(t) &= D_\beta g(t) \\ &\quad - \left(\int_0^1 \frac{2\left(x(t) + s(\beta(t)-t)D_\beta x(t)\right)}{\left(1 + \left(x(t) + s(\beta(t)-t)D_\beta x(t)\right)^2\right)^2} ds \right) D_\beta x(t) \\ &= D_\beta g(t) \\ &\quad - \left(\int_0^1 \frac{2\left(x(t) + s(\beta(t)-t)\left(g(t) + \frac{1}{(x(t))^2+1}\right)\right)}{\left(1 + \left(x(t) + s(\beta(t)-t)\left(g(t) + \frac{1}{(x(t))^2+1}\right)\right)^2\right)^2} ds \right) \\ &\quad \times \left(g(t) + \frac{1}{(x(t))^2+1} \right). \end{aligned}$$

8.2 Convergence of the Order-Two Taylor Series Method

In this section we give the convergence of the Taylor series method derived in the previous section. By the β-differential equation in the IVP (8.1) and the condition $(H1)$, we have

$$|D_\beta x(t)| \le A, \quad t \in [t_0, t_f]. \tag{8.9}$$

From (8.7) and the condition $(H2)$, we obtain

$$\begin{aligned} \left| D_\beta\left(f(t,x(t))\right) \right| &\le \left| D_{\beta t} f(t,x(t)) \right| \\ &\quad + \left(\int_0^1 \left| \frac{\partial}{\partial x} f(\beta(t), x(t) + s(\beta(t)-t)D_\beta x(t)) \right| ds \right) |D_\beta x(t)| \\ &\le A + A^2, \quad t \in [t_0, t_f]. \end{aligned} \tag{8.10}$$

Therefore,

$$|D_\beta^2 x(t)| \leq A + A^2, \quad t \in [t_0, t_f]. \tag{8.11}$$

On the other hand, by the properties of the monomial h_2, we have

$$\begin{aligned} h_2(t+r, \beta(\tau)) &\leq \frac{(t+r-\beta(\tau))^2}{2} \\ &\leq \frac{(t+r-t)^2}{2} \\ &= \frac{r^2}{2}, \quad \tau \in [t, (\beta^{-1})^2(t+r)], \\ & \quad t, t+r \in [t_0, t_f], \quad r > 0, \end{aligned} \tag{8.12}$$

and

$$\begin{aligned} h_1(t+r, \beta(\tau)) &\leq t+r-\beta(\tau) \\ &\leq t+r-t \\ &= r, \quad \tau \in [t, (\beta^{-1})^2(t+r)], \quad t, t+r \in [t_0, t_f], \quad r > 0. \end{aligned} \tag{8.13}$$

Applying again the chain rule, we obtain

$$\begin{aligned} D_\beta^3 x(t) &= D_{\beta t} g(t, x(t)) \\ &+ \left(\int_0^1 \frac{\partial}{\partial x} g(\beta(t), x(t) + s(\beta(t) - t) D_\beta x(t)) ds \right) D_\beta x(t), \end{aligned}$$

$$\begin{aligned} \left| D_\beta^3 x(t) \right| &\leq \left| D_{\beta t} g(t, x(t)) \right| \\ &+ \left(\int_0^1 \left| \frac{\partial}{\partial x} g(\beta(t), x(t) + s(\beta(t) - t) D_\beta x(t)) \right| ds \right) \left| D_\beta x(t) \right|, \\ &\leq A + A^2, \quad t \in [t_0, t_f]. \end{aligned}$$

Hence and (8.12), (8.13), we get

$$\begin{aligned} |R_2(r)| &= \left| \int_t^{(\beta^{-1})^2(t+r)} h_2(t+r, \beta(\tau)) D_\beta^3 x(\tau) d_\beta \tau \right| \\ &\leq \int_t^{(\beta^{-1})^2(t+r)} h_2(t+r, \beta(\tau)) \left| D_\beta^3 x(\tau) \right| d_\beta \tau \\ &\leq \frac{r^2}{2} (A + A^2)((\beta^{-1})^2(t+r) - t) \\ &\leq \frac{r^2}{2} (A + A^2)(t + r - t) \\ &\leq \frac{r^3}{2} (A + A^2), \quad t, t+r \in [t_0, t_f], \quad r > 0. \end{aligned}$$

On the other hand, by (8.11) and (8.13), we have

$$\begin{aligned}
|R_1(r)| &= \left| \int_t^{\beta^{-1}(t+r)} h_1(t+r,\beta(\tau)) D_\beta^2 x(\tau) d_\beta \tau \right| \\
&\leq \int_t^{\beta^{-1}(t+r)} h_1(t+r,\beta(\tau)) \left| D_\beta^2 x(\tau) \right| d_\beta \tau \\
&\leq r(A+A^2)(\beta^{-1}(t+r)-t) \\
&\leq r(A+A^2)(t+r-t) \\
&\leq r^2(A+A^2), \quad t, t+r \in [t_0,t_f], \quad r>0,
\end{aligned}$$

i.e.,

$$R_2(r) = O(r^3), \quad R_1(r) = O(r^2).$$

Now, we denote

$$e_n = x(t_n) - x_n.$$

For $t_{n+1} = t_n + r_{n+1}$, we have

$$x(t_{n+1}) = x(t_n) + r_{n+1} f(t_n, x(t_n)) + h_2(t_{n+1}, t_n) g(t_n, x(t_n)) + R_2(r_{n+1}),$$

and by (8.5),

$$x_{n+1} = x_n + r_{n+1} f(t_n, x_n) + h_2(t_{n+1}, t_n) g(t_n, x_n),$$

whereupon applying the mean value theorem in the classical case, we obtain

$$\begin{aligned}
x(t_{n+1}) - x_{n+1} &= x(t_n) - x_n + r_{n+1} \left(f(t_n, x(t_n)) - f(t_n, x_n) \right) \\
&+ h_2(t_{n+1}, t_n) \left(g(t_n, x(t_n)) - g(t_n, x_n) \right) + R_2(r_{n+1}) \\
&= x(t_n) - x_n + r_{n+1} \frac{\partial}{\partial x} f(t_n, \zeta) \left(x(t_n) - x_n \right) \\
&+ h_2(t_{n+1}, t_n) \frac{\partial}{\partial x} g(t_n, \eta) \left(x(t_n) - x_n \right) + R_2(r_{n+1}) \\
&= e_n + \left(r_{n+1} \frac{\partial}{\partial x} f(t_n, \zeta) + h_2(t_{n+1}, t_n) \frac{\partial}{\partial x} g(t_n, \eta) \right) e_n \\
&\quad + R_2(r_{n+1}),
\end{aligned}$$

where ζ and η are between $x(t_n)$ and x_n. Let

$$A_n = r_{n+1} \frac{\partial}{\partial x} f(t_n, \zeta) + h_2(t_{n+1}, t_n) \frac{\partial}{\partial x} g(t_n, \eta).$$

We have

$$\begin{aligned}
|A_n| &\leq r_{n+1} \left| \frac{\partial}{\partial x} f(t_n, \zeta) \right| + h_2(t_{n+1}, t_n) \left| \frac{\partial}{\partial x} g(t_n, \eta) \right| \\
&\leq A r_{n+1} + A \frac{r_{n+1}^2}{2} \\
&= r_{n+1} \left(1 + \frac{r_{n+1}}{2} \right) A.
\end{aligned}$$

Then,

$$\begin{aligned} e_0 &= 0, \\ e_1 &= R_2(r_1), \\ e_{n+1} &= (1+A_n)e_n + R_2(r_{n+1}), \quad n \in \mathbb{N}. \end{aligned}$$

In particular,

$$\begin{aligned} e_2 &= (1+A_1)e_1 + R_2(r_2), \\ e_3 &= (1+A_2)e_2 + R_2(r_3), \\ &= (1+A_2)((1+A_1)e_1 + R_2(r_2)) + R_2(r_3) \\ &= (1+A_2)(1+A_1)e_1 + (1+A_2)R_2(r_2) + R_2(r_3) \\ e_4 &= (1+A_3)e_3 + R_2(r_4), \\ &= (1+A_3)((1+A_2)(1+A_1)e_1 + (1+A_2)R_2(r_2) + R_2(r_3)) + R_2(r_4) \\ &= (1+A_3)(1+A_2)(1+A_1)e_1 + (1+A_3)(1+A_2)R_2(r_2) \\ &\quad +(1+A_3)R_2(r_3) + R_2(r_4), \end{aligned}$$

and so on. Let $r_{max} = \max\{r_1, \ldots, r_{m+1}\}$. Then, using that $0 < r_j \le t_f - t_0$, $j \in \{1, \ldots, m+1\}$, and $0 < t_f - t_0 < \infty$, we have that $R_2(r_j) = O(r_{max}^3)$, $j \in \{1, \ldots, m+1\}$. Since $0 < t_f - t_0 < \infty$ and $t_{j-1} + r_j \in [t_0, t_f]$, $j \in \{1, \ldots, m+1\}$, we have that there exists a constant $0 < B < \infty$ such that $mr_{max} \le B$. Then

$$\begin{aligned} |e_2| &\le \left(1 + r_{max}\left(1 + \frac{r_{max}}{2}\right)A\right) R_2(r_{max}) + R_2(r_{max}), \\ |e_3| &\le \left(1 + r_{max}\left(1 + \frac{r_{max}}{2}\right)A\right)^2 R_2(r_{max}) \\ &+ \left(1 + r_{max}\left(1 + \frac{r_{max}}{2}\right)A\right) R_2(r_{max}) + R_2(r_{max}), \\ &\cdots \\ |e_m| &\le \sum_{j=0}^{m-1} \left(1 + r_{max}\left(1 + \frac{r_{max}}{2}\right)A\right)^j R_2(r_{max}) \\ &\le \sum_{j=0}^{m-1} e^{r_{max}\left(1+\frac{r_{max}}{2}\right)Aj} R_2(r_{max}) \\ &\le m e^{mr_{max}\left(1+\frac{r_{max}}{2}\right)A} R_2(r_{max}) \\ &\le m e^{B\left(1+\frac{r_{max}}{2}\right)A} R_2(r_{max}) \\ &\le m e^{B\left(1+\frac{t_f-t_0}{2}\right)A} R_2(r_{max}) \\ &= m r_{max} r_{max}^2 e^{\left(1+\frac{t_f-t_0}{2}\right)A} C \\ &\le r_{max}^2 e^{B\left(1+\frac{t_f-t_0}{2}\right)A} BC. \end{aligned}$$

Since $t_f < \infty$, we conclude that

$$e_m = O(r_{max}^2)$$

that is, the order of convergence of the order-two Taylor series method is two.

8.3 Trapezoidal Rule

In this section we will introduce the trapezoidal rule for IVPs associated with first-order β-differential equations.

We start again with the Taylor series expansion of $D_\beta x$ which gives

$$\begin{aligned} D_\beta x(t+r) &= D_\beta x(t) + rD_\beta^2 x(t) + \int_t^{\beta^{-1}(t+r)} h_1(t+r,\beta(\tau))D_\beta^3 x(\tau)d_\beta\tau \\ &= D_\beta x(t) + rD_\beta^2 x(t) + R_1(r), \end{aligned}$$

whereupon

$$rD_\beta^2 x(t) = D_\beta x(t+r) - D_\beta x(t) - R_1(r), \quad t, t+r \in [t_0, t_f], \quad r > 0.$$

We substitute the last relation into the equation (8.3) and we find

$$\begin{aligned} x(t+r) &= x(t) + rD_\beta x(t) + h_2(t+r,t)D_\beta^2 x(t) + R_2(r) \\ &= x(t) + rD_\beta x(t) + \frac{h_2(t+r,t)}{r}\left(rD_\beta^2 x(t)\right) \\ &\quad + R_2(r) \\ &= x(t) + rD_\beta x(t) + \frac{h_2(t+r,t)}{r}\left(D_\beta x(t+r) - D_\beta x(t) - R_1(r)\right) \\ &\quad + R_2(r) \\ &= x(t) + \left(r - \frac{h_2(t+r,t)}{r}\right)D_\beta x(t) \\ &\quad + \frac{h_2(t+r,t)}{r}D_\beta x(t+r) \\ &\quad - \frac{h_2(t+r,t)}{r}R_1(r) + R_2(r) \\ &= x(t) + \left(r - \frac{h_2(t+r,t)}{r}\right)f(t,x(t)) \\ &\quad + \frac{h_2(t+r,t)}{r}f(t+r,x(t+r)) \\ &\quad - \frac{h_2(t+r,t)}{r}R_1(r) + R_2(r), \quad t, t+r \in [t_0, t_f], \quad r > 0. \end{aligned}$$

Evaluating this at $t = t_n$ and neglecting the remainder terms leads to the trapezoidal rule

$$x_{n+1} = x_n + \left(r_{n+1} - \frac{h_2(t_{n+1},t_n)}{r_{n+1}}\right)f(t_n,x_n) + \frac{h_2(t_{n+1},t_n)}{r_{n+1}}f(t_{n+1},x_{n+1}). \quad (8.14)$$

Definition 8.3.1 *The relation* (8.14) *will be called trapezoidal rule.*

Example 8.3.2 *Let* $I = \mathbb{R}$, $\beta(t) = t$, $t \in I$, *and* $t_{n+1} - t_n = r$ *be constant. Then*

$$h_2(t_{n+1},t_n) = \frac{1}{2}r^2,$$

and the trapezoidal rule takes the form

$$x_{n+1} = x_n + \frac{1}{2} r \left(f(t_n, x_n) + f(t_{n+1}, x_{n+1}) \right),$$

which is the classical trapezoidal rule.

Example 8.3.3 *Let $I = \mathbb{R}$, $\beta(t) = 2t$, $t \in I$. Then*

$$\begin{aligned} h_2(t_{n+1}, t_n) &= \frac{1}{3}(t_{n+1}^2 - t_n^2) \\ &= \frac{1}{3}(t_{n+1} - t_n)(t_{n+1} + t_n) \\ &= \frac{r_{n+1}}{3}(t_{n+1} + t_n), \end{aligned}$$

and the trapezoidal rule takes form

$$x_{n+1} = x_n + \left(h_{n+1} - \frac{1}{3}(t_{n+1} + t_n) \right) f(t_n, x_n) + \frac{1}{3}(t_{n+1} + t_n) f(t_{n+1}, x_{n+1}).$$

Exercise 8.3.4 *Let $I = \mathbb{R}$, $\beta(t) = 2t$, $t \in I$. Consider the IVP*

$$\begin{aligned} D_\beta y(x) &= 1 + y(x) + (y(x))^3, \quad x > 1, \\ y(1) &= 1 \end{aligned}$$

and assume that

$$a = x_0 = 1, \quad x_1 = 2, \quad x_2 = 4, \quad x_3 = 8, \quad x_4 = 16.$$

Write the trapezoidal rule.

8.4 Advanced Practical Problems

Problem 8.4.1 *Let $I = \mathbb{R}$, $\beta(t) = 2t$, $t \in I$. Consider the IVP*

$$\begin{aligned} D_\beta(x) &= 1 + \frac{1 + y(x)}{1 + (y(x))^2}, \quad x > 1, \\ y(1) &= 1 \end{aligned}$$

and assume that

$$a = x_0 = 1, \quad x_1 = 4, \quad x_2 = 16, \quad x_3 = 64, \quad x_4 = b = 216.$$

Write the trapezoidal rule.

Problem 8.4.2 *Let $I = \mathbb{R}$, $\beta(t) = \frac{2}{3}t + 1$, $t \in I$. Consider the IVP*

$$\begin{aligned} D_\beta y(x) &= (1+y(x))^3, \quad x > 0, \\ y(0) &= 0. \end{aligned}$$

Writhe the trapezoidal rule.

Problem 8.4.3 *Let $I = \mathbb{R}$, $\beta(t) = \frac{3}{4}t + 2$, $t \in I$. Apply TS(2) for the following IVP*

$$\begin{aligned} D_\beta y(x) &= \sin_{1,\beta}(x) + (y(x))^2, \quad x > 0, \\ y(0) &= 1, \end{aligned}$$

where

$$a = x_0 = 0, \quad x_1 = 2, \quad x_2 = 4, \quad x_3 = 5, \quad x_4 = b = 6.$$

Problem 8.4.4 *Let $I = \mathbb{R}$, $\beta(t) = \frac{1}{3}t + 4$, $t \in I$. Apply TS(2) for the following IVP*

$$\begin{aligned} D_\beta y(x) &= \frac{1}{1+(y(x))^2}, \quad x > 1, \\ y(1) &= 1, \end{aligned}$$

where

$$a = x_0 = 1, \quad x_1 = 2, \quad x_2 = 8, \quad x_3 = 16, \quad x_4 = b = 32.$$

Problem 8.4.5 *Let $I = \mathbb{R}$, $\beta(t) = \frac{5}{6}t + 2$, $t \in I$. Apply TS(2) for the following IVP*

$$\begin{aligned} D_\beta y(x) &= 1 + 3(y(x))^2, \quad x > 0, \\ y(0) &= 1, \end{aligned}$$

where

$$a = x_0 = 0, \quad x_1 = \frac{1}{2}, \quad x_2 = 1, \quad x_3 = 2, \quad x_4 = \frac{21}{8}, \quad x_5 = b = 3.$$

8.5 Notes and References

In this chapter, we introduce the order-two Taylor series method for first-order β-differential equations. We investigate its convergence. In the chapter, it is deducted a β-analogue of the trapezoidal rule.

9

The Order-p Taylor Series Method-TS(p)

In this chapter we derive the Taylor series method of order $p \geq 2$. Let $I \subseteq \mathbb{R}$ and β be first kind or second kind general quantum operator.

9.1 A Generalisation of the Chain Rule

To develop the Taylor series method, we need a chain rule for derivative of a composite function of $n+1$ variables. A chain rule for the case of a two variable function already exists and is known as the chain rule. The derivation of the method, requires a generalisation of this rule. Let $g : I \times \mathbb{R}^n \to \mathbb{R}$ be a given function. Then for the function $g(t, y_1, \ldots, y_n)$ we denote by $D_{\beta t} g(\cdot, y_1, \ldots, y_n)$ its delta derivative.

Theorem 9.1.1 *For some fixed $t_0 \in I$, let $y_j : I \to \mathbb{R}$, $j \in \{1, \ldots, n\}$, $f : I \times \mathbb{R}^n \to \mathbb{R}$ be continuous functions such that $f(\cdot, y_1(t_0), \ldots, y_n(t_0))$, and y_j, $j \in \{1, \ldots, n\}$, are differentiable at t_0. Let $U \subseteq I$ be a neighborhood of t_0 such that,*

1. *$f(t, \cdot, \ldots, \cdot)$ is continuously-differentiable for $t \in U \cup \{\beta(t_0)\}$,*
2. *$D_{\beta t} f(\cdot, y_1(\cdot), \ldots, y_n(\cdot))$ is continuous at t_0,*
3. $$\frac{\partial}{\partial y_j} f(\beta(t_0), y_1(\beta(t_0)), \ldots, y_{j-1}(\beta(t_0)), \cdot, y_{j+1}(t), \ldots, y_n(t))$$

   is continuous in the line segment

   $$\{y_j(t) + h(y_j(\beta(t_0)) - y_j(t)) \in \mathbb{R} : h \in [0,1]\},\ j \in \{1, \ldots, n\},$$

DOI: 10.1201/9781003472131-9

$\forall t \in U \cup \{t_0\}$,

4. $\dfrac{\partial}{\partial y_j} f$ *is continuous at* $(t_0, y_1(t_0), \ldots, y_n(t_0))$.

Then the composition function $F : I \to \mathbb{R}$, $F(t) = f(t, y_1(t), y_2(t), \ldots, y_n(t))$, *is differentiable at* t_0 *with derivative*

$$
\begin{aligned}
D_\beta F(t_0) &= D_{\beta t} f(t_0, y_1(t_0), y_2(t_0), \ldots, y_n(t_0)) \\
&+ \left(\int_0^1 \frac{\partial}{\partial y_1} f(\beta(t_0), y_1(t_0) + h(\beta(t_0) - t_0) D_\beta y_1(t_0), y_2(t_0), \ldots, y_n(t_0)) dh \right) D_\beta y_1(t_0) \\
&+ \left(\int_0^1 \frac{\partial}{\partial y_2} f(\beta(t_0), y_1(\beta(t_0)), y_2(t_0) + h(\beta(t_0) - t_0) D_\beta y_2(t_0), \ldots, y_n(t_0)) dh \right) D_\beta y_2(t_0) \\
&+ \cdots \\
&+ \left(\int_0^1 \frac{\partial}{\partial y_n} f(\beta(t_0), y_1(\beta(t_0)), y_2(\beta(t_0)), \ldots, y_{n-1}(\beta(t_0)), y_n(t_0) + h(\beta(t_0) - t_0) D_\beta y_n(t_0)) dh \right) . \\
&\times D_\beta y_n(t_0).
\end{aligned}
$$

Proof 9.1.2 *Let* $s \in (t_0 - \delta, t_0 + \delta) \cap I$, $s \neq \beta(t_0)$, *for* $\delta > 0$ *small enough, and* $s < \beta(t_0)$ *if* $\beta(t_0) > t_0$. *Then*

$$
\begin{aligned}
& F(\beta(t_0)) - F(s) \\
&= f(\beta(t_0), y_1(\beta(t_0)), y_2(\beta(t_0)), \ldots, y_n(\beta(t_0))) - f(s, y_1(s), y_2(s), \ldots, y_n(s)) \\
&= f(\beta(t_0), y_1(s), y_2(s), \ldots, y_n(s)) - f(s, y_1(s), y_2(s), \ldots, y_n(s)) \\
&\quad + f(\beta(t_0), y_1(\beta(t_0)), y_2(s), \ldots, y_n(s)) - f(\beta(t_0), y_1(s), y_2(s), \ldots, y_n(s)) \\
&\quad + f(\beta(t_0), y_1(\beta(t_0)), y_2(\beta(t_0)), \ldots, y_n(s)) \\
&\quad - f(\beta(t_0), y_1(\beta(t_0)), y_2(s), \ldots, y_n(s)) \\
&\quad + \cdots \\
&\quad + f(\beta(t_0), y_1(\beta(t_0)), y_2(\beta(t_0)), \ldots, y_n(\beta(t_0))) \\
&\quad - f(\beta(t_0), y_1(\beta(t_0)), y_2(\beta(t_0)), \ldots, y_n(s))
\end{aligned}
$$

Then, we have

$$
\begin{aligned}
&F(\beta(t_0)) - F(s)\\
&= f(\beta(t_0), y_1(s), y_2(s), \ldots, y_n(s)) - f(s, y_1(s), y_2(s), \ldots, y_n(s))\\
&\quad + \left(\int_0^1 \frac{\partial}{\partial y_1} f(\beta(t_0), y_1(s) + h(y_1(\beta(t_0)) - y_1(s)), y_2(s), \ldots, y_n(s))dh\right)\\
&\quad \times (y_1(\beta(t_0)) - y_1(s))\\
&\quad + \left(\int_0^1 \frac{\partial}{\partial y_2} f(\beta(t_0), y_1(\beta(t_0)), y_2(s) + h(y_2(\beta(t_0)) - y_2(s)), \ldots, y_n(s))dh\right)\\
&\quad \times (y_2(\beta(t_0)) - y_2(s))\\
&\quad + \cdots\\
&\quad + \left(\int_0^1 \frac{\partial}{\partial y_n} f(\beta(t_0), y_1(\beta(t_0)), y_2(\beta(t_0)), \ldots, y_n(s) + h(y_n(\beta(t_0)) - y_n(s)))dh\right)\\
&\quad \times (y_n(\beta(t_0)) - y_n(s)).
\end{aligned}
$$

If $\beta(t_0) > t_0$, *by the mean value theorem there exist* $\xi_1, \xi_2 \in [s, \beta(t_0)) = [s, t_0]$ *so that*

$$
\begin{aligned}
D_{\beta t} f(\xi_1, y_1(s), y_2(s), \ldots, y_n(s))(\beta(t_0) - s) &\leq f(\beta(t_0), y_1(s), y_2(s), \ldots, y_n(s))\\
- f(s, y_1(s), y_2(s), \ldots, y_n(s)) &\leq D_{\beta t} f(\xi_2, y_1(s), y_2(s), \ldots, y_n(s))(\beta(t_0) - s)
\end{aligned}
$$

and

$$
\begin{aligned}
D_{\beta t} f(t_0, y_1(t_0), y_2(t_0), \ldots, y_n(t_0)) &= \lim_{s \to t_0} D_{\beta t} f(\xi_1, y_1(s), y_2(s), \ldots, y_n(s))\\
&\leq \lim_{s \to t_0} \frac{1}{\beta(t_0) - s}\Big(f(\beta(t_0), y_1(s), y_2(s), \ldots, y_n(s))\\
&\qquad - f(s, y_1(s), y_2(s), \ldots, y_n(s))\Big)\\
&\leq \lim_{s \to t_0} D_{\beta t} f(\xi_2, y_1(s), y_2(s), \ldots, y_n(s))\\
&= D_{\beta t} f(t_0, y_1(t_0), y_2(t_0), \ldots, y_n(t_0)).
\end{aligned}
$$

If $t_0 = s_0$, *by the mean value theorem, there exist* ξ_1, ξ_2 *between* s *and* t_0 *so that*

$$
D_{\beta t} f(\xi_1, y_1(s), y_2(s), \ldots, y_n(s))(t_0 - s) \leq f(t_0, y_1(s), y_2(s), \ldots, y_n(s))
$$

$$-f(s,y_1(s),y_2(s),\dots,y_n(s)) \le D_{\beta t} f(\xi_2,y_1(s),y_2(s),\dots,y_n(s))(t_0-s).$$

Moreover,

$$\begin{aligned}
&\lim_{s\to t_0}\Bigg(\Bigg(\int_0^1 \frac{\partial}{\partial y_j} f(\beta(t_0),y_1(\beta(t_0)),\dots,y_{j-1}(\beta(t_0)),y_j(s)+h(y_j(\beta(t_0))-y_j(s)),\\
&\qquad y_{j+1}(s),\dots,y_n(s))dh\Bigg)\frac{y_j(\beta(t_0))-y_j(t_0)}{\beta(t_0)-s}\Bigg)\\
&= \lim_{s\to t_0}\Bigg(\int_0^1 \frac{\partial}{\partial y_j} f(\beta(t_0),y_1(\beta(t_0)),\dots,y_{j-1}(\beta(t_0)),y_j(s)\\
&\qquad +h(y_j(\beta(t_0))-y_j(s)),y_{j+1}(s),\dots,y_n(s))dh\Bigg)\lim_{s\to t_0}\frac{y_j(\beta(t_0))-y_j(t_0)}{\beta(t_0)-s}\\
&= \Bigg(\int_0^1 \frac{\partial}{\partial y_j} f(\beta(t_0),y_1(\beta(t_0)),\dots,y_{j-1}(\beta(t_0)),y_j(t_0)\\
&\qquad +h(y_j(\beta(t_0))-y_j(t_0)),y_{j+1}(t_0),\dots,y_n(t_0))dh\Bigg)D_\beta y_j(t_0)\\
&= \Bigg(\int_0^1 \frac{\partial}{\partial y_j} f(\beta(t_0),y_1(\beta(t_0)),\dots,y_{j-1}(\beta(t_0)),y_j(t_0)\\
&\qquad +h(\beta(t_0)-t_0)D_\beta y_j(t_0),y_{j+1}(t_0),\dots,y_n(t_0))dh\Bigg)\\
&\qquad \times D_\beta y_j(t_0), \quad j\in\{1,\dots,n\}.
\end{aligned}$$

Therefore

$$\begin{aligned}
&\lim_{s\to t_0}\frac{F(\beta(t_0))-F(s)}{\beta(t_0)-s}\\
&= \lim_{s\to t_0}\frac{f(\beta(t_0),y_1(s),y_2(s),\dots,y_n(s))-f(s,y_1(s),y_2(s),\dots,y_n(s))}{\beta(t_0)-s}\\
&\qquad +\lim_{s\to t_0}\Bigg(\Bigg(\int_0^1 \frac{\partial}{\partial y_1} f(\beta(t_0),y_1(s)\\
&\qquad +h(y_1(\beta(t_0))-y_1(s)),y_2(s),\dots,y_n(s))dh\Bigg)\\
&\qquad \times\frac{y_1(\beta(t_0))-y_1(s)}{\beta(t_0)-s}\Bigg)\\
&\qquad +\lim_{s\to t_0}\Bigg(\Bigg(\int_0^1 \frac{\partial}{\partial y_2} f(\beta(t_0),y_1(\beta(t_0)),y_2(s)
\end{aligned}$$

$$
\begin{aligned}
& +h(y_2(\beta(t_0)) - y_2(s)), \ldots, y_n(s))dh\Big) \\
& \times \frac{y_2(\beta(t_0)) - y_2(s)}{\beta(t_0) - s}\Bigg) \\
& + \cdots \\
& + \lim_{s \to t_0} \Bigg(\Bigg(\int_0^1 \frac{\partial}{\partial y_n} f(\beta(t_0), y_1(\beta(t_0)), y_2(\beta(t_0)), \ldots, y_n(s) \\
& +h(y_n(\beta(t_0)) - y_n(s)))dh\Bigg) \\
& \times \frac{y_n(\beta(t_0)) - y_n(s)}{\beta(t_0) - s}\Bigg) \\
= \; & D_{\beta t} f(t_0, y_1(t_0), y_2(t_0), \ldots, y_n(t_0)) \\
& + \Bigg(\int_0^1 \frac{\partial}{\partial y_1} f(\beta(t_0), y_1(t_0) \\
& +h(\beta(t_0) - t_0) D_\beta y_1(t_0), y_2(t_0), \ldots, y_n(t_0))dh\Bigg) D_\beta y_1(t_0) \\
& + \Bigg(\int_0^1 \frac{\partial}{\partial y_2} f(\beta(t_0), y_1(\beta(t_0)), y_2(t_0) \\
& +h(\beta(t_0) - t_0) D_\beta y_2(t_0), \ldots, y_n(t_0))dh\Bigg) D_\beta y_2(t_0) \\
& + \cdots \\
& + \Bigg(\int_0^1 \frac{\partial}{\partial y_n} f(\beta(t_0), y_1(\beta(t_0)), y_2(\beta(t_0)), \\
& \ldots, y_{n-1}(\beta(t_0)), y_n(t_0) + h(\beta(t_0) - t_0) D_\beta y_n(t_0))dh\Bigg) \\
& \times D_\beta y_n(t_0).
\end{aligned}
$$

This completes the proof.

9.2 Analysing of the Order-p Taylor Series Method

Suppose that $p \in \mathbb{N}$, $p \geq 2$, $t_0, t_f \in I$, $t_0 < t_f < \infty$, $r > 0$ be such that $t, t+r \in [t_0, t_f]$. Consider the initial value problem (IVP)

$$\left\{ \begin{array}{rcl} D_\beta x(t) & = & f(t, x(t)), \quad t \in [t_0, t_f], \\ x(t_0) & = & x_0, \end{array} \right. \tag{9.1}$$

where $x_0 \in \mathbb{R}$ is a given constant, the function f satisfies the following conditions

$$(\mathbf{H1}) \left\{ \begin{array}{l} |f(t,x)| \leq A, \quad t \in I, \quad x \in \mathbb{R}, \\ \text{there exists } g_k(t, x(t), \ldots, D_\beta^k x(t)) = D_\beta^k \left(f(t, x(t))\right), \\ k \in \{1, \ldots, p-1\}, \\ \text{such that } \left|\dfrac{\partial f}{\partial y}(t,z)\right| \leq A, \quad |D_{\beta t} g_k(t, y_1, \ldots, y_{k+1})| \leq A, \\ \text{and } \left|\dfrac{\partial}{\partial y_j} g_k(t, y_1, \ldots, y_{k+1})\right| \leq A, \quad j \in \{1, \ldots, k+1\}, \\ \text{for any } t \in I \text{ and for } z, y_j \in \mathbb{R}, j \in \{1, \ldots, p-1\}, \\ \text{where } pe^{t_f - t_0} A < 1 \text{ and } A > 0. \end{array} \right.$$

By the Taylor formula, we get

$$\begin{aligned} x(t+r) &= x(t) + h_1(t+r,t) D_\beta x(t) + h_2(t+r,t) D_\beta^2 x(t) \\ &\quad + \cdots + h_p(t+r,t) D_\beta^p x(t) \\ &\quad + \int_t^{(\beta^{-1})^p(t+r)} h_p(t+r, \beta(u)) D_\beta^{p+1} x(u) d_\beta u. \end{aligned}$$

Let

$$R_p(t) = \int_t^{(\beta^{-1})^p(t+r)} h_p(t+r, \beta(u)) D_\beta^{p+1} x(u) d_\beta u,$$

be the remainder term. Let also, $t_0 < t_1 < \ldots < t_{m+1} = t_f$ be a partition of the interval $[t_0, t_f]$ such that $t_{n+1} = t_n + r_{n+1} \in I$, $r_{n+1} > 0$, $n \in \{0, \ldots, m\}$. Then

$$\begin{aligned} x(t_{n+1}) &= x(t_n) + h_1(t_{n+1}, t_n) D_\beta x(t_n) + h_2(t_{n+1}, t_n) D_\beta^2 x(t_n) \\ &\quad + \cdots + h_p(t_{n+1}, t_n) D_\beta^p x(t_n) + R_p(t_{n+1}). \end{aligned}$$

Neglecting the remainder term $R_p(t)$, we obtain

$$x(t_{n+1}) = x(t_n) + h_1(t_{n+1}, t_n) D_\beta x(t_n) + h_2(t_{n+1}, t_n) D_\beta^2 x(t_n)$$

$$+\cdots+h_p(t_{n+1},t_n)D_\beta^p x(t_n).$$

Set

$$\begin{aligned} D_\beta^k x_n &= D_\beta^k x(t_n), \\ (D_\beta^k x_n)^\beta &= D_\beta^k x(\beta(t_n)), \quad k\in\{0,\ldots,p\}. \end{aligned}$$

Thus, we get

$$x_{n+1}=x_n+r_{n+1}D_\beta x_n+h_2(t_{n+1},t_n)D_\beta^2 x_n+\cdots+h_p(t_{n+1},t_n)D_\beta^p x_n. \tag{9.2}$$

To compute x_{n+1} we need to determine $D_\beta^q x_n$ for $q\in\{1,\ldots,p\}$. From the dynamic equation in the IVP (9.1) we determine $D_\beta x_n$ as

$$D_\beta x_n = f(t_n,x_n).$$

Now, we will determine $D_\beta^2 x_n$, ..., $D_\beta^p x_n$. By the general chain rule, we have

$$\begin{aligned} D_\beta(f(t,x(t))) &= D_{\beta t} f(t,x(t)) \\ &\quad+\left(\int_0^1 \frac{\partial}{\partial y_1} f(\beta(t),x(t)+h(\beta(t)-t)D_\beta x(t))dh\right) D_\beta x(t) \\ &= g_1(t,x(t),D_\beta x(t)) \end{aligned}$$

and for $q\in\{2,\ldots,p\}$, we compute

$$\begin{aligned} &D_\beta^q(f(t,x(t))) \\ &= D_\beta\left(g_{q-1}(t,x(t),D_\beta x(t),\ldots,D_\beta^{q-1}x(t))\right) \\ &= D_{\beta t} g_{q-1}(t,x(t),D_\beta x(t),\ldots,D_\beta^{q-1}x(t)) \\ &+\left(\int_0^1 \frac{\partial}{\partial y_1} g_{q-1}(\beta(t),x(t)+h(\beta(t)-t)D_\beta x(t),D_\beta x(t),\ldots,D_\beta^{q-1}x(t))dh\right) \\ &\times D_\beta x(t) \\ &+\left(\int_0^1 \frac{\partial}{\partial y_2} g_{q-1}(\beta(t),x(\beta(t)),D_\beta x(t)+h(\beta(t)-t)D_\beta^2 x(t),\ldots,D_\beta^{q-1}x(t))dh\right) \\ &\times D_\beta^2 x(t) \\ &+\cdots \\ &+\left(\int_0^1 \frac{\partial}{\partial y_q} g_{q-1}(\beta(t),x(\beta(t)),D_\beta x(\beta(t)),\ldots,D_\beta^{q-1}x(t)+h(\beta(t)-t)D_\beta^q x(t))dh\right) \\ &\times D_\beta^q x(t), \end{aligned}$$

for $t \in I$. Therefore, we have

$$
\begin{aligned}
D_\beta^2 x_n &= D_{\beta t} f(t_n, x_n) + \left(\int_0^1 \frac{\partial}{\partial y_1} f(\beta(t_n), x_n + h(\beta(t_n) - t_n) D_\beta x_n) dh \right) D_\beta x_n, \\
D_\beta^3 x_n &= D_{\beta t} g_1(t_n, x_n, D_\beta x_n) \\
&\quad + \left(\int_0^1 \frac{\partial}{\partial y_1} g_1(\beta(t_n), x_n + h(\beta(t_n) - t_n) D_\beta x_n, D_\beta x_n) dh \right) D_\beta x_n \\
&\quad + \left(\int_0^1 \frac{\partial}{\partial y_2} g_1(\beta(t_n), D_\beta x_n, D_\beta x_n + h(\beta(t_n) - t_n) D_\beta^2 x_n) dh \right) D_\beta^2 x_n \\
&\vdots \\
D_\beta^{p+1} x_n &= D_{\beta t} g_{p-1}(t_n, x_n, D_\beta x_n, \ldots, D_\beta^{p-1} x_n) \\
&\quad + \left(\int_0^1 \frac{\partial}{\partial y_1} g_{p-1}(\beta(t_n), x_n + h(\beta(t_n) - t_n) D_\beta x_n, D_\beta x_n, \ldots, D_\beta^{p-1} x_n) dh \right) D_\beta x_n \\
&\quad + \left(\int_0^1 \frac{\partial}{\partial y_2} g_{p-1}(\beta(t_n), D_\beta x_n, D_\beta x_n + h(\beta(t_n) - t_n) D_\beta^2 x_n, \ldots, D_\beta^{p-1} x_n) dh \right) D_\beta^2 x_n \\
&\quad + \cdots \\
&\quad + \left(\int_0^1 \frac{\partial}{\partial y_p} g_{p-1}(\beta(t_n), x_n^\beta, D_\beta x_n^\beta, \ldots, D_\beta^{p-1} x_n + h(\beta(t_n) - t_n) D_\beta^p x_n) dh \right) D_\beta^p x_n,
\end{aligned}
$$

from where we can find $D_\beta^2 x_n, \ldots, D_\beta^{p+1} x_n$.

9.3 Convergence and Error Analysis of the Taylor Series Method

Now, we will investigate the convergence of the Taylor series method of order p. The following estimate

$$0 \leq h_q(t,s) \leq \frac{(t-s)^q}{q!}, \quad t \geq s, \tag{9.3}$$

holds for all $q \in \mathbb{N}$.

By the condition $(H1)$ and the dynamic equation in (9.1), we find

$$|D_\beta x(t)| \leq A, \quad t \in [t_0, t_f].$$

Next, we estimate

$$\left| D_\beta^2 x(t) \right| \leq \left| D_{\beta t} f(t, x(t)) \right|$$

$$+\left(\int_0^1 \left|\frac{\partial}{\partial y_1} f(\beta(t), x(t)+h(\beta(t)-t)D_\beta x(t))\right| dh\right) |D_\beta x(t)|$$

$$\leq \quad A(1+A), \quad t \in [t_0, t_f],$$

and

$$\left|D_\beta^3 x(t)\right| \leq |D_{\beta t} g_1(t, x(t), D_\beta x(t))|$$

$$+\left(\int_0^1 \left|\frac{\partial}{\partial y_1} g_1(\beta(t), x(t)+h(\beta(t)-t)D_\beta x(t), D_\beta x(t))\right| dh\right) |D_\beta x(t)|$$

$$+\left(\int_0^1 \left|\frac{\partial}{\partial y_2} g_1(\beta(t), x(\beta(t)), D_\beta x(t)+h(\beta(t)-t)D_\beta^2 x(t))\right| dh\right) \left|D_\beta^2 x(t)\right|$$

$$\leq A+A^2+A(A+A^2)$$
$$= A(1+A)^2,$$

$$\left|D_\beta^4 x(t)\right| \leq \left|D_{\beta t} g_2(t, x(t), D_\beta x(t), D_\beta^2 x(t))\right|$$

$$+\left(\int_0^1 \left|\frac{\partial}{\partial y_1} g_2(\beta(t), x(t)+h(\beta(t)-t)D_\beta x(t), D_\beta x(t), D_\beta^2 x(t))\right| dh\right)$$

$$\times |D_\beta x(t)|$$

$$+\left(\int_0^1 \left|\frac{\partial}{\partial y_2} g_2(\beta(t), x(\beta(t)), D_\beta x(t)+h(\beta(t)-t)D_\beta^2 x(t), D_\beta^2 x(t))\right| dh\right)$$

$$\times \left|D_\beta^2 x(t)\right|$$

$$+\left(\int_0^1 \left|\frac{\partial}{\partial y_3} g_2(\beta(t), x(\beta(t)), D_\beta x(\beta(t)), D_\beta^2 x(t)+h(\beta(t)-t)D_\beta^3 x(t))\right| dh\right)$$

$$\times |D_\beta^3 x(t)|$$

$$\leq A+A^2+A(A+A^2)+A\left((A+A^2)+A(A+A^2)\right)$$
$$= A+A^2+2A(A+A^2)+A^2(A+A^2)$$
$$= A(1+A)^3,$$

$$\left|D_\beta^5 x(t)\right| \le \left|D_{\beta t} g_3(t,x(t),D_\beta x(t),D_\beta^2 x(t),D_\beta^3 x(t))\right|$$

$$+\left(\int_0^1 \left|\frac{\partial}{\partial y_1} g_3(\beta(t),x(t)+h(\beta(t)-t)D_\beta x(t),D_\beta x(t),D_\beta^2 x(t),D_\beta^3 x(t))\right| dh\right)$$

$$\times |D_\beta x(t)|$$

$$+\left(\int_0^1 \left|\frac{\partial}{\partial y_2} g_3(\beta(t),x(\beta(t)),D_\beta x(t)+h(\beta(t)-t)D_\beta^2 x(t),D_\beta^2 x(t),D_\beta^3 x(t))\right| dh\right)$$

$$\times \left|D_\beta^2 x(t)\right|$$

$$+\left(\int_0^1 \left|\frac{\partial}{\partial y_3} g_3(\beta(t),x(\beta(t)),D_\beta x(\beta(t)),D_\beta^2 x(t)+h(\beta(t)-t)D_\beta^3 x(t),D_\beta^3 x(t))\right| dh\right)$$

$$\times \left|D_\beta^3 x(t)\right|$$

$$+\left(\int_0^1 \left|\frac{\partial}{\partial y_4} g_3(\beta(t),x(\beta(t)),D_\beta x(\beta(t)),D_\beta^2 x(\beta(t)),D_\beta^3 x(t)+h(\beta(t)-t)D_\beta^4 x(t))\right| dh\right)$$

$$\times \left|D_\beta^4 x(t)\right|$$

$$\begin{aligned}
&\le A+A^2+A(A+A^2)+A(1+A)(A+A^2)+A(A+A^2)(1+A)^2\\
&= (A+A^2)(1+A+A+A^2+A+2A^2+A^3)\\
&= A(1+A)^4, \quad t\in[t_0,t_f],
\end{aligned}$$

so that we deduce

$$\begin{aligned}
\left|D_\beta^p x(t)\right| &\le A(1+A)^{p-1},\\
\left|D_\beta^{p+1} x(t)\right| &\le A(1+A)^p, \quad t\in[t_0,t_f].
\end{aligned}$$

Moreover, for the remainder terms

$$R_q(r)=\int_t^{(\beta^{-1})^q(t+r)} h_q(t+r,\beta(u))D_\beta^{q+1}x(u)d_\beta u, \quad q\in\{1,\ldots,p\}, \quad t\in[t_0,t_f],$$

employing the estimate (9.3) and the fact that $(\beta^{-1})^q(t+r)-t\le t+r-t=r$, we get

$$\begin{aligned}
\left|R_q(r)\right| &\le \int_t^{(\beta^{-1})^q(t+r)} h_q(t+r,\beta(u))\left|D_\beta^{q+1}x(u)\right| d_\beta u\\
&\le \frac{r^q}{q!}A(1+A)^q\left((\beta^{-1})^q(t+r)-t\right) \qquad (9.4)\\
&\le \frac{r^{q+1}}{q!}A(1+A)^q, \quad q\in\{1,\ldots,p\}.
\end{aligned}$$

Therefore

$$R_q(r) = O\left(r^{q+1}\right), \quad q \in \{1, \dots, p\}.$$

Denote

$$D^k_\beta e_n = D^k_\beta x(t_n) - D^k_\beta x_n, \quad k \in \{0, \dots, p-1\}.$$

Taking into account the fact that

$$\begin{aligned} D_\beta x_n &= f(t_n, x_n), \\ D_\beta x(t_n) &= f(t_n, x(t_n)), \\ D^q_\beta x_n &= g_{q-1}(t_n, x_n, D_\beta x_n, \dots, D^{q-1}_\beta x_n), \\ D^q_\beta x(t_n) &= g_{q-1}(t_n, x(t_n), D_\beta x(t_n), \dots, D^{q-1}_\beta x(t_n)), \quad q \in \{2, \dots, p\} \end{aligned}$$

we have

$$\begin{aligned} x(t_{n+1}) &= x(t_n) + r_{n+1} f(t_n, x(t_n)) + h_2(t_{n+1}, t_n) g_1(t_n, x(t_n), D_\beta x(t_n)) \\ &\quad + \dots + h_p(t_{n+1}, t_n) g_{p-1}(t_n, x(t_n), \dots, D^{p-1}_\beta x(t_n)) + R_p(r_{n+1}) \end{aligned}$$

and

$$\begin{aligned} x_{n+1} &= x_n + r_{n+1} f(t_n, x_n) + h_2(t_{n+1}, t_n) g_1(t_n, x_n, D_\beta x_n) \\ &\quad + \dots + h_p(t_{n+1}, t_n) g_{p-1}(t_n, x_n, \dots, D^{p-1}_\beta x_n). \end{aligned}$$

Then

$$\begin{aligned} x(t_{n+1}) - x_{n+1} = {} & (x(t_n) - x_n) + r_{n+1}(f(t_n, x(t_n)) - f(t_n, x_n)) \\ & + h_2(t_{n+1}, t_n) \left(g_1(t_n, x(t_n), D_\beta x(t_n)) - g_1(t_n, x_n, D_\beta x_n)\right) \\ & + \cdots \\ & + h_p(t_{n+1}, t_n) \left(g_{p-1}(t_n, x(t_n), \dots, D^{p-1}_\beta x(t_n)) - g_{p-1}(t_n, x_n, \dots, D^{p-1}_\beta x_n)\right) \\ & + R_p(r_{n+1}). \end{aligned}$$

Note that by the condition $(H1)$, the Mean value theorem for f and g_k implies that

$$f(t_n, x(t_n)) - f(t_n, x_n) = \frac{\partial f}{\partial y}(t_n, \xi^0_1)(x(t_n) - x_n) = \frac{\partial f}{\partial y}(t_n, \xi^0_1) e_n,$$

where ξ^0_1 is between $x(t_n)$ and x_n, and $\dfrac{\partial f}{\partial y}$ stands for the partial derivative with respect to the second variable. Also,

$$g_k(t_n, x(t_n), D_\beta x(t_n), \dots, D^k_\beta x(t_n)) - g_k(t_n, x_n, D_\beta x_n, \dots, D^k_\beta x_n)$$

$$
\begin{aligned}
= \; & g_k(t_n, x(t_n), D_\beta x(t_n), \ldots, D_\beta^k x(t_n)) - g_k(t_n, x_n, D_\beta x(t_n), \ldots, D_\beta^k x(t_n)) \\
& + g_k(t_n, x_n, D_\beta x(t_n), \ldots, D_\beta^k x(t_n)) - g_k(t_n, x_n, D_\beta x_n, \ldots, D_\beta^k x(t_n)) \\
& + \cdots \\
& + g_k(t_n, x_n, D_\beta x_n, \ldots, D_\beta^{k-1} x_n, D_\beta^k x(t_n)) - g_k(t_n, x_n, D_\beta x_n, \ldots, D_\beta^{k-1} x_n, D_\beta^k x_n) \\
= \; & \frac{\partial}{\partial y_1} g_k(t_n, \xi_1^k, D_\beta x(t_n), \ldots, D_\beta^k x(t_n)) e_n \\
& + \frac{\partial}{\partial y_2} g_k(t_n, x_n, \xi_2^k, \ldots, D_\beta^k x(t_n)) D_\beta e_n \\
& + \cdots \\
& + X \frac{\partial}{\partial y_{k+1}} g_k(t_n, x_n, \ldots, D_\beta^{k-1} x_n, \xi_{k+1}^k) D_\beta^k e_n, \quad k \in \{1, \ldots, p-1\},
\end{aligned}
$$

where ξ_j^k is between $D_\beta^{j-1} x(t_n)$ and $D_\beta^{j-1} x_n$, $j \in \{1, \ldots, k+1\}$ and the $\dfrac{\partial}{\partial y_j}$ denotes the partial derivative with respect to the $(j+1)$-st variable. Consequently,

$$
\begin{aligned}
e_{n+1} = \; & e_n + r_{n+1} \frac{\partial f}{\partial y_1}(t_n, \xi_1^0) e_n \\
& + h_2(t_{n+1}, t_n) \left(\frac{\partial g_1}{\partial y_1}(t_n, \xi_1^1, D_\beta x(t_n)) e_n + \frac{\partial g_1}{\partial y_2}(t_n, x_n, \xi_2^1) D_\beta e_n \right) \\
& + \cdots \\
& + h_p(t_{n+1}, t_n) \left(\frac{\partial g_{p-1}}{\partial y_1}(t_n, \xi_1^{p-1}, D_\beta x(t_n), \ldots, D_\beta^{p-1} x(t_n)) e_n \right. \\
& + \frac{\partial g_{p-1}}{\partial y_2}(t_n, x_n, \xi_2^{p-1}, \ldots, D_\beta^{p-1} x(t_n)) D_\beta e_n \\
& + \cdots \\
& \left. + \frac{\partial g_{p-1}}{\partial y_p}(t_n, x_n, D_\beta x_n, \ldots, \xi_p^{p-1}) D_\beta^{p-1} e_n \right) \\
& + R_p(r_{n+1}).
\end{aligned}
$$

Let $r_{max} = \max\{r_1, \ldots, r_{m+1}\}$. Since $t_f < \infty$, there is a constant $B > 0$ such that

$$
\frac{1}{p!} r_{max} A(1+A)^p \left(e^{r_{max}} A + 1\right) \leq B.
$$

Then

$$\begin{aligned}
|e_{n+1}| &\leq (1+h_1(t_{n+1},t_n)+\cdots+h_p(t_{n+1},t_n))A\left(|e_n|+|D_\beta e_n|+\cdots+\left|D_\beta^{p-1}e_n\right|\right)+|R_p(r_{n+1})| \\
&\leq \left(1+r_{max}+\frac{r_{max}^2}{2!}+\cdots+\frac{r_{max}^p}{p!}\right)A\left(|e_n|+|D_\beta e_n|+\cdots+|D_\beta^p e_n|\right)+|R_p(r_{n+1})| \\
&\leq e^{r_{max}}A\left(|e_n|+|D_\beta e_n|+\cdots+|D_\beta^{p-1}e_n|\right)+|R_p(r_{n+1})|
\end{aligned}$$

In a similar way, we make the following estimates.

$$\begin{aligned}
\left|D_\beta e_{n+1}\right| &\leq e^{r_{max}}A\left(\left|D_\beta e_n\right|+\cdots+\left|D_\beta^{p-1}e_n\right|\right)+|R_{p-1}(r_{n+1})|, \\
\left|D_\beta^2 e_{n+1}\right| &\leq e^{r_{max}}A\left(\left|D_\beta^2 e_n\right|+\cdots+\left|D_\beta^{p-1}e_n\right|\right)+|R_{p-2}(r_{n+1})|, \\
&\cdots \\
\left|D_\beta^{p-2}e_n\right| &\leq e^{r_{max}}A\left|D_\beta^{p-1}e_n\right|+|R_2(r_{n+1})|, \\
\left|D_\beta^{p-1}e_{n+1}\right| &\leq |R_1(r_{n+1})|.
\end{aligned}$$

Let

$$B_n = |e_n|+\left|D_\beta e_n\right|+\cdots+\left|D_\beta^{p-1}e_n\right|.$$

Then

$$B_{n+1} \leq pe^{r_{max}}AB_n+|R_1(r_{n+1})|+\cdots+|R_p(r_{n+1})|.$$

Observe that from (9.4), we get

$$\begin{aligned}
|R_1(r_{n+1})|+\cdots+|R_p(r_{n+1})| &\leq r_{max}^2A(1+A)+\frac{r_{max}^3}{2!}A(1+A)^2 \\
&+ \cdots+\frac{r_{max}^{p+1}}{p!}A(1+A)^p \\
&\leq r_{max}^2A(1+A)^p\left(1+\frac{r_{max}}{2!}+\cdots+\frac{r_{max}^{p-1}}{p!}\right) \\
&\leq r_{max}^2A(1+A)^p\left(1+r_{max}+\frac{r_{max}^2}{2!}+\cdots+\frac{r_{max}^{p-1}}{(p-1)!}\right) \\
&\leq r_{max}^2A(1+A)^pe^{r_{max}}.
\end{aligned}$$

Thus,

$$B_{n+1} \leq pe^{r_{max}}AB_n+r_{max}^2A(1+A)^pe^{r_{max}}$$

$$
\begin{aligned}
&\leq \ pe^{r_{max}}A\left(pe^{r_{max}}AB_{n-1}+r_{max}^2A(1+A)^pe^{r_{max}}\right)\\
&\quad \times + r_{max}^2A(1+A)^pe^{r_{max}}\\
&= \ \left(pe^{r_{max}}A\right)^2B_{n-1}+\left(pe^{r_{max}}A+1\right)r_{max}^2A(1+A)^pe^{r_{max}}\\
&\leq \ \cdots\\
&\leq \ \left(pe^{r_{max}}A\right)^{n+1}B_0\\
&\quad \times + \left(\left(pe^{r_{max}}A\right)^n+\cdots+pe^{r_{max}}A+1\right)r_{max}^2A(1+A)^pe^{r_{max}}\\
&\leq \ r_{max}^2A(1+A)^pe^{r_{max}}\sum_{j=0}^{\infty}\left(pe^{r_{max}}A\right)^j\\
&\leq \ r_{max}^2A(1+A)^pe^{t_f-t_0}\sum_{j=0}^{\infty}\left(pe^{t_f-t_0}A\right)^j\\
&= \ \frac{1}{1-pe^{t_f-t_0}A}r_{max}^2A(1+A)^pe^{t_f-t_0}.
\end{aligned}
$$

In the last inequality we have used the fact that $B_0 = 0$ and $r_{max} \leq t_f - t_0$. Consequently

$$
|e_n| + |D_\beta e_n| + \cdots + \left|D_\beta^{p-1}e_n\right| = O\left(r_{max}^2\right).
$$

9.4 The two-step Adams-Bashforth Method: AB(2)

In this section we consider the special case of the Taylor series method of order p, which in the case of $I = \mathbb{R}$, $\beta(t) = t$ reduces to the numerical method known as the two-step Adams-Bashforth method. We shall call this method the quantum two-step Adams-Bashforth method.

Consider again the IVP (9.1). Suppose that $r, l > 0$, $t, t+r, t-l \in [t_0, t_f]$, $(\beta^{-1})^2(t+r), \beta^{-1}(t-l) \in [t_0, t_f]$. Applying the Taylor formula of the second-order, we compute

$$
x(t+r) = x(t) + h_1(t+r,t)D_\beta x(t) + h_2(t+r,t)D_\beta^2 x(t) + R_2(r) \tag{9.5}
$$

and applying the Taylor formula of the first-order, we get

$$
D_\beta x(t-l) = D_\beta x(t) + h_1(t-l,t)D_\beta x(t) + R_1(l)
$$

$$= \quad D_\beta x(t) - l D_\beta^2 x(t) + R_1(l),$$

whereupon

$$D_\beta^2 x(t) = \frac{1}{l}\left(D_\beta x(t) - D_\beta x(t-l)\right) + \frac{1}{l} R_1(l).$$

We put this expression in (9.5) and we find

$$\begin{aligned} x(t+r) &= x(t) + r D_\beta x(t) \\ &\quad + \frac{h_2(t+r,t)}{l}\left(D_\beta x(t) - D_\beta x(t-l) + R_1(l)\right) + R_2(r) \\ &= x(t) + r f(t,x(t)) \\ &\quad + \frac{h_2(t+r,t)}{l}\left(f(t,x(t)) - f(t-l,x(t-l))\right) \\ &\quad + \frac{h_2(t+r,t)}{l} R_1(l) + R_2(r). \end{aligned} \tag{9.6}$$

Assume that $\{t_0 < t_1 < \ldots < t_{m+1} = t_f\}$ is a partition of the interval $[t_0, t_f]$ such that $t_{n+1} = t_n + r_{n+1} \in I$, $r_{n+1} > 0$, $n \in \{0, \ldots, m\}$. Take $t = t_n$, $r = r_{n+1}$, $l = r_n$ in (9.6) and we obtain

$$\begin{aligned} x(t_{n+1}) &= x(t_n) + r_{n+1} f(t_n, x(t_n)) \\ &\quad + \frac{h_2(t_{n+1},t_n)}{r_n}\left(f(t_n,x(t_n)) - f(t_{n-1},x(t_{n-1}))\right) \\ &\quad + \frac{h_2(t_{n+1},t_n)}{r_n} R_1(r_n) + R_2(r_{n+1}). \end{aligned}$$

Let $x_n = x(t_n)$, $f_n = f(t_n, x(t_n))$. Then, neglecting the remainder terms, we arrive at the two-step Adams-Bashforth method (AB(2) method).

$$x_{n+1} = x_n + r_{n+1} f_n + \frac{h_2(t_{n+1},t_n)}{r_n}(f_n - f_{n-1})$$

or

$$x_{n+1} = x_n + \left(r_{n+1} + \frac{h_2(t_{n+1},t_n)}{r_n}\right) f_n - \frac{h_2(t_{n+1},t_n)}{r_n} f_{n-1}. \tag{9.7}$$

Remark 9.4.1

1. *Note that the two-step Adams-Bashforth method* (9.7) *is of order* $(1 + O(r_n))O(r_{n+1}^2)$.

2. *If* $I = \mathbb{R}$ *and* $r_n = h$ *is constant, then we have* $h_2(t_{n+1}, t_n) = \frac{(t_{n+1} - t_n)^2}{2} = \frac{h^2}{2}$ *and hence,* (9.7) *takes the form*

$$x_{n+1} = x_n + \left(h + \frac{h}{2}\right) f_n - \frac{h}{2} f_{n-1} = x_n + \frac{3h}{2} f_n - \frac{h}{2} f_{n-1},$$

which is the classical two-step Adams-Bashforth method.

9.5 Notes and References

In this chapter we introduce the order-p Taylor series method. We investigate it for convergence and we give estimations of the errors. As application, we introduce two-step Adams-Bashforth method.

10

Linear Multistep Methods-LMMs

In the previous chapters it was shown the effectiveness of TS(p) methods. For order $p > 1$, the suffer a disadvantage in that they require the right-hand side of the β-differential equation to be β-differentiable a number of times. This often rules out their use in the real-world applications. The families of LMMs achieve higher order by exploiting x and $D_\beta x$, that were computed at the previous k-steps, are combined to generate an approximation of the next step. We begin with two-sstep methods to describe the strategy. Suppose that $I \subseteq \mathbb{R}$ and $\beta : I \to I$ is first kind or second kind general quantum operator. Let also, $a, b \in I$, $a < b$, and $x_j \in I$, $j \in \{0, 1, \ldots, m\}$, so that

$$a = x_0 < x_1 < \ldots < x_m = b.$$

Consider the IVP

$$\begin{aligned} D_\beta x(t) &= f(t, x(t)), \quad t \in [a, b], \\ x(a) &= x_0, \end{aligned} \tag{10.1}$$

where $x_0 \in \mathbb{R}$. Set $r_0 = r_{m+1} = 0$ and

$$x_j = x_{j-1} + r_j, \quad j \in \{1, \ldots, m\}, \quad r = \max_{j \in \{1, \ldots, m\}}.$$

10.1 Two-Step Methods

For a β-differentiable function z, we need to find

$$\alpha_{0j}, \quad \alpha_{1j}, \quad \alpha_{2j}, \quad \beta_{0j}, \quad \beta_{1j}, \quad \beta_{2j}, \quad \gamma_{0j}, \quad \gamma_{1j}, \quad \gamma_{2j}$$

so that

$$\begin{aligned} & z(t_j + r_{j+1} + r_{j+2}) + \alpha_{1j} z(t_j + r_{j-1}) + \alpha_{0j} z(t_j) \\ &= r_{j+2} \left(\beta_{2j} D_\beta z(t_j + r_{j+1} + r_{j+2}) + \beta_{1j} D_\beta z(t_j + r_{j+1}) + \beta_{0j} D_\beta z(t_j) \right) \\ &\quad + r_{j+1} \left(\gamma_{2j} D_\beta z(t_j + r_{j+1} + r_{j+2}) + \gamma_{1j} D_\beta z(t_j + r_{j+1}) + \gamma_{0j} D_\beta z(t_j) \right) \end{aligned}$$

DOI: 10.1201/9781003472131-10

$$+O(r^{p+1}), \quad j \in \{0, \ldots, m-2\},$$

where p might be specified in some cases or one might to try to make p as large as possible in others. We choose $z = x$, where x is a solution of the IVP (10.1) and dropping the $O(r^{p+1})$, we find

$$\begin{aligned} & x(t_j + r_{j+1} + r_{j+2}) + \alpha_{1j} x(t_j + r_{j+1}) + \alpha_{0j} x(t_j) \\ = & \; r_{j+2} \Big(\beta_{2j} f(t_j + r_{j+1} + r_{j+2}, x(t_j + r_{j+1} + r_{j+2})) \\ & + \beta_{1j} f(t_j + r_{j+1}, x(t_j + r_{j+1})) + \beta_{0j} f(t_j, x(t_j)) \Big) \\ & + r_{j+1} \Big(\gamma_{2j} f(t_j + r_{j+1} + r_{j+2}, x(t_j + r_{j+1} + r_{j+2})) \\ & + \gamma_{1j} f(t_j + r_{j+1}, x(t_j + r_{j+1})) + \gamma_{0j} f(t_j, x(t_j)) \Big), \end{aligned}$$

$j \in \{0, \ldots, m-2\}$. Set

$$x_j = x(t_j), \quad f_j = f(t_j, x(t_j)), \quad j \in \{0, 1, \ldots, m\}.$$

Then

$$\begin{aligned} x_{j+2} + \alpha_{1j} x_{j+1} + \alpha_{0j} x_j &= r_{j+2}(\beta_{2j} f_{j+2} + \beta_{1j} f_{j+1} + \beta_{0j} f_j) \\ & + r_{j+1}(\gamma_{2j} f_{j+2} + \gamma_{1j} f_{j+1} + \gamma_{0j} f_j), \quad j \in \{0, \ldots, m-2\}. \end{aligned} \tag{10.2}$$

Definition 10.1.1 *A two-step method is said to be explicit(of explicit type) if* $\beta_{2j} = \gamma_{2j} = 0$ *for any* $j \in \{0, \ldots, m-2\}$.

Definition 10.1.2 *A two-step method is said to be implicit if* $\beta_{2j} \neq 0$ *or* $\gamma_{2j} \neq 0$ *for some* $j \in \{0, \ldots, m-2\}$.

We may write (10.2) in the following form

$$\begin{aligned} x_{j+2} + \alpha_{1j} x_{j+1} + \alpha_{0j} x_j &= r_{j+2} \left(\beta_{2j} D_\beta x_{j+2} + \beta_{1j} D_\beta x_{j+1} + \beta_{0j} D_\beta x_j \right) \\ & + r_{j+1} \left(\gamma_{2j} D_\beta x_{j+2} + \gamma_{1j} D_\beta x_{j+1} + \gamma_{0j} D_\beta x_j \right), \end{aligned}$$

$j \in \{0, \ldots, m-2\}$.

10.2 Consistency of Two-Step Methods

Definition 10.2.1 *The linear difference operator associated with the two-step method* (10.2) *is defined for an arbitrary continuously-differentiable function z by*

$$\mathscr{L}_{r_{j+1},r_{j+2}}z(t) = z(t+r_{j+1}+r_{j+2})+\alpha_{1j}z(t+r_{j+1})+\alpha_{0j}z(t) \tag{10.3}$$

$$-r_{j+2}\left(\beta_{2j}D_\beta z(t+r_{j+1}+r_{j+2})+\beta_{1j}D_\beta z(t+t_{j+1})+\beta_{0j}D_\beta z(t)\right)$$

$$-r_{j+1}\left(\gamma_{2j}D_\beta z(t+r_{j+1}+r_{j+2})+\gamma_{1j}D_\beta z(t+r_{j+1})+\gamma_{0j}D_\beta z(t)\right),$$

$j\in\{0,\ldots,m-2\}$.

Obviously, for any $a,b\in\mathbb{R}$ and any continuously-differentiable functions z, w, we have

$$\mathscr{L}_{r_{j+1},r_{j+2}}(az(t)+bw(t)) = a\mathscr{L}_{r_{j+1},r_{j+2}}z(t)+b\mathscr{L}_{r_{j+1},r_{j+2}}w(t).$$

Definition 10.2.2 *A linear difference operator $\mathscr{L}_{r_{j+1},r_{j+2}}$ is said to be consistency of order p if*

$$\mathscr{L}_{r_{j+1},r_{j+2}}z(t) = O(r^{p+1})$$

with $p>0$ for any smooth function z.

Definition 10.2.3 *A two-step method is said to be consistent if its difference operator $\mathscr{L}_{r_{j+1},r_{j+2}}$ is consistent of order p for some $p>0$.*

Example 10.2.4 *Consider the Adams-Bashforth method. The associated linear difference operator is*

$$\mathscr{L}_{r_{n+1},r_{n=2}}z(t) = z(t+r_{n+1}+r_{n+2})-z(t+r_{n+1})$$

$$-\left(r_{n+2}+\frac{h_2(t+r_{n+1}+r_{n+2},t+r_{n+1})}{r_{n+1}}\right)D_\beta z(t+r_{n+1})$$

$$+\frac{h_2(t+r_{n+1}+r_{n+2},t+r_{n+1})}{r_{n+1}}D_\beta z(t).$$

By the Taylor expansion, we have

$$z(t+r_{n+1}+r_{n+2}) = z(t)+h_1(t+r_{n+1}+r_{n+2})D_\beta z(t)$$

$$+h_2(t+r_{n+1}+r_{n+2},t)D_\beta^2 z(t)+O(r^3),$$

$$z(t+r_{n+1}) = z(t)+h_1(t+r_{n+1},t)D_\beta z(t) + h_2(t+r_{n+1},t)D_\beta^2 z(t)+O(r^3),$$

$$D_\beta z(t+r_n) = D_\beta z(t)+h_1(t+r_n,t)D_\beta^2 z(t)+O(r^2).$$

Thus,

$$\begin{aligned}
\mathscr{L}_{r_{n+1},r_{n+2}} z(t) &= z(t)+h_1(t+r_{n+1}+r_{n+2},t)D_\beta z(t)\\
&\quad +h_2(t+r_{n+1}+r_{n+2},t)D_\beta^2 z(t)\\
&\quad -z(t)-h_1(t+r_{n+1},t)D_\beta z(t)-h_2(t+r_{n+1},t)D_\beta^2 z(t)\\
&\quad -\left(r_{n+2}+\frac{h_2(t+r_{n+1}+r_{n+2},t+r_{n+1})}{r_{n+1}}\right)\\
&\quad \times\left(D_\beta z(t)+h_1(t+r_n,t)D_\beta^2 z(t)+O(r^2)\right)\\
&\quad +\frac{h_2(t+r_{n+1}+r_{n+2},t+r_{n+1})}{r_{n+1}}D_\beta z(t)+O(r^3)\\
&= r_{n+2}D_\beta z(t)+\left(h_2(t+r_{n+1}+r_{n+2},t)-h_2(t+r_{n+1},t)\right)D_\beta^2 z(t)\\
&\quad -\left(r_{n+2}+\frac{h_2(t+r_{n+1}+r_{n+2},t+r_{n+1})}{r_{n+1}}\right)D_\beta z(t)\\
&\quad +h_1(t+r_n,t)\left(r_{n+2}+\frac{h_2(t+r_{n+1}+r_{n+2},t+r_{n+1})}{r_{n+1}}\right)D_\beta^2 z(t)\\
&\quad +\frac{h_2(t+r_{n+1}+r_{n+2},t+r_{n+1})}{r_{n+1}}D_\beta z(t)+O(r^2)+O(r^3)\\
&= \Big(h_2(t+r_{n+1}+r_{n+2},t)-h_2(t+r_{n+1},t)-r_{n+2}h_1(t+r_{n+1},t)\\
&\quad -h_1(t+r_n,t)\frac{h_2(t+r_{n+1}+r_{n+2},t+r_{n+1})}{r_{n+1}}\Big)D_\beta^2 z(t)\\
&\quad +O(r^2)+O(r^3).
\end{aligned}$$

Hence,

$$\left|\mathscr{L}_{r_{n+1},r_{n+2}} z(t)\right| \leq \Big(h_2(t+r_{n+1}+r_{n+2},t)+h_2(t+r_{n+1},t)$$

$$-r_{n+2}h_1(t+r_{n+1},t)$$

$$+h_1(t+r_n,t)\frac{h_2(t+r_{n+1}+r_{n+2},t+r_{n+1})}{r_{n+1}}\Bigg)D_\beta^2 z(t)$$

$$+O(r^2)+O(r^3)$$

$$\leq \left(\frac{(r_{n+1}+r_{n+2})^2}{2}+\frac{r_{n+1}^2}{2}+r_nr_{n+2}\right.$$

$$\left.+r_n\frac{(r_{n+2})^2}{r_{n+1}}\right)|D_\beta^2 z(t)|+O(r^2)+O(r^3)$$

and

$$\mathscr{L}_{r_{n+1},r_{n+2}}z(t)=O(r^2).$$

Therefore, the Adams-Bashforth method is consistent of order $p=1$.

Exercise 10.2.5 *Let*

$$x_{n+2}=x_{n+1}+r_{n+1}h_3(t_n+r_{n+1}+r_{n+2},t_n)f_{n+1}-2h_2(t_n+r_{n+1},t_n)f_n.$$

Find the associated linear difference operator and determine its order of consistency.

10.3 Construction of Two-Step Methods

For a general two-step method, given with (10.2), the associated linear difference operator is given by (10.3) and the right-hand side can be expanded with the help of the Taylor expansions

$$\begin{aligned}
z(t+r_{j+1}+r_{j+2}) &= z(t)+h_1(t+r_{j+1}+r_{j+2},t)D_\beta z(t)\\
&\quad +h_2(t+r_{j+1}+r_{j+2},t)D_\beta^2 z(t)\\
&\quad +h_3(t+r_{j+1}+r_{j+2},t)D_\beta^3 z(t)+\cdots,\\
z(t+r_{j+1}) &= z(t)+h_1(t+r_{j+1},t)D_\beta z(t)+h_2(t+r_{j+1},t)D_\beta^2 z(t)\\
&\quad +h_3(t+r_{j+1},t)D_\beta^3 z(t)+\cdots,
\end{aligned}$$

$$\begin{aligned} D_\beta z(t+r_{j+1}+r_{j+2}) &= D_\beta z(t)+h_1(t+r_{j+1}+r_{j+2},t)D_\beta^2 z(t) \\ &\quad +h_2(t+r_{j+1}+r_{j+2},t)D_\beta^3 z(t) \\ &\quad +h_3(t+r_{j+1}+r_{j+2},t)D_\beta^4 z(t)+\cdots, \\ D_\beta z(t+r_{j+1}) &= D_\beta z(t)+h_1(t+r_{j+1},t)D_\beta^2 z(t) \\ &\quad +h_2(t+r_{j+1},t)D_\beta^3 z(t) \\ &\quad +h_3(t+r_{j+1},t)D_\beta^4 z(t)+\cdots. \end{aligned}$$

To precise number of terms that have to be retained depends on either the order required or the maximum order possible with the "template" used some coefficients in the two-step method may be set zero in order to achieve a method with particular pattern of terms.

Firstly, we focuss our attention on the case $p=1$. We have

$$\begin{aligned} &\mathscr{L}_{r_{j+1},r_{j+2}} z(t) = z(t)+h_1(t+r_{j+1}+r_{j+2},t)D_\beta z(t)+O(r^2) \\ &\quad +\alpha_{1j}\left(z(t)+h_1(t+r_{j+1},t)D_\beta z(t)+O(r^2)\right)+\alpha_{0j}z(t) \\ &\quad -r_{j+2}\left(\beta_{2j}\left(D_\beta z(t)+O(r)\right)+\beta_{1j}\left(D_\beta z(t)+O(r)\right)+\beta_{0j}D_\beta z(t)\right) \\ &\quad -r_{j+1}\left(\gamma_{2j}\left(D_\beta z(t)+O(r)\right)+\gamma_{1j}\left(D_\beta z(t)+O(r)\right)+\gamma_{0j}D_\beta z(t)\right) \\ &= \left(1+\alpha_{1j}+\alpha_{0j}\right)z(t) \\ &\quad +\Big(h_1(t+r_{j+1}+r_{j+2},t)+\alpha_{1j}h_1(t+r_{j+1},t)-r_{j+2}(\beta_{0j}+\beta_{1j}+\beta_{2j}) \\ &\quad -r_{j+1}(\gamma_{2j}+\gamma_{1j}+\gamma_{0j})\Big)D_\beta z(t)+O(r^2) \\ &= \left(1+\alpha_{1j}+\alpha_{0j}\right)z(t) \\ &\quad +\Big(r_{j+1}+r_{j+2}+\alpha_{1j}r_{j+1}-r_{j+2}(\beta_{0j}+\beta_{1j}+\beta_{2j}) \\ &\quad -r_{j+1}(\gamma_{0j}+\gamma_{1j}+\gamma_{2j})\Big)D_\beta z(t)+O(r^2) \end{aligned}$$

$$\begin{aligned} &= (1+\alpha_{1j}+\alpha_{0j})\, z(t) \\ &\quad + \Big(r_{j+1}(1+\alpha_{1j}-\gamma_{0j}-\gamma_{1j}-\gamma_{2j}) \\ &\quad + r_{j+2}(1-\beta_{0j}-\beta_{1j}-\beta_{2j}) \Big) D_\beta z(t) + O(r^2). \end{aligned}$$

Therefore, $\mathscr{L}_{r_{j+1},r_{j+2}} z(t) = O(r^2)$ or the two-step method is consistent of order $p = 1$ if

$$\begin{aligned} 1+\alpha_{0j}+\alpha_{1j} &= 0 \\ \beta_{0j}+\beta_{1j}+\beta_{2j} &= 1 \\ \gamma_{0j}+\gamma_{1j}+\gamma_{2j} &= 1+\alpha_{1j}. \end{aligned} \tag{10.4}$$

Definition 10.3.1 *The first, second and third characteristic polynomials of the two-step method* (10.2) *are defined to be*

$$\begin{aligned} \beta_{1j}^{-1}(q) &= q^2+\alpha_{1j}q+\alpha_{0j}, \\ \beta_{2j}^{-1}(q) &= \beta_{2j}q^2+\beta_{1j}q+\beta_{0j}, \\ \beta_{3j}^{-1}(q) &= \gamma_{2j}q^2+\gamma_{1j}q+\gamma_{0j}, \quad q \in \mathbb{R}, \end{aligned}$$

respectively.

Example 10.3.2 *Consider the following two-step method*

$$x_{j+2}-x_{j+1}+x_j = r_{j+2}(f_{j+2}-f_{j+1})+r_{j+1}(f_{j+2}+f_{j+1}+f_j).$$

The first, second and third characteristic polynomials are as follows.

$$\begin{aligned} \beta_{1j}^{-1}(q) &= q^2-q-1, \\ \beta_{2j}^{-1}(q) &= q^2-q, \\ \beta_{3j}^{-1}(q) &= q^2+q+1. \end{aligned}$$

Exercise 10.3.3 *Write the first, second and third characteristic polynomials of the following two-step method.*

$$x_{j+2}+3x_{j+1}-4x_j = r_{j+2}(f_{j+2}-f_{j+1}-f_j)+r_{j+1}(f_{j+2}-f_j).$$

Theorem 10.3.4 *The two-step method* (10.2) *is consistent with the dynamic equation* (10.1) *if and only if*

$$\begin{aligned} (\beta_{ij}^{-1})'(1) &= 0 \\ \frac{1}{2}(\beta_{1j}^{-1})''(1) &= \beta_{2j}^{-1}(1) \\ (\beta_{ij}^{-1})'\left(\frac{1}{2}\right) &= \beta_{3j}^{-1}(1). \end{aligned} \tag{10.5}$$

Proof 10.3.5 *Note that*

$$\begin{aligned} \beta_{1j}^{-1}(1) &= 1+\alpha_{0j}+\alpha_{1j}, \\ (\beta_{1j}^{-1})'(q) &= 2q+\alpha_{1j}, \\ (\beta_{1j}^{-1})'\left(\frac{1}{2}\right) &= 1+\alpha_{1j}, \\ (\beta_{1j}^{-1})''(q) &= 2, \\ \frac{1}{2}(\beta_{1}^{-1})'(1) &= 1, \\ \beta_{2j}^{-1}(1) &= \beta_{0j}+\beta_{1j}+\beta_{2j}, \\ \beta_{3j}^{-1}(1) &= \gamma_{0j}+\gamma_{1j}+\gamma_{2j}. \end{aligned}$$

1. *Suppose that* (10.2) *is consistent with the dynamic equation* (10.1). *Then* (10.4) *hold and hence,* (10.5) *hold.*
2. *Suppose that* (10.5) *hold. Then* (10.4) *hold. This completes the proof.*

Example 10.3.6 *Consider the two-step method*

$$\begin{aligned} x_{j+2}-x_j &= r_{j+2}(f_{j+2}-f_{j+1}+f_j) \\ &\quad +r_{j+1}(-3f_{j+2}+3f_{j+1}+2f_j). \end{aligned}$$

Here

$$\begin{aligned} \alpha_{0j} &= -1, \quad \alpha_{1j}=0, \\ \beta_{0j} &= 1, \quad \beta_{1j}=-1, \quad \beta_{2j}=1, \end{aligned}$$

$$\gamma_{0j} = 2, \quad \gamma_{1j} = 3, \quad \gamma_{2j} = -3.$$

Then

$$\begin{aligned}
1+\alpha_{0j}+\alpha_{1j} &= 1+(-1)+0\\
&= 0,\\
\beta_{0j}+\beta_{1j}+\beta_{2j} &= 1+(-1)+1\\
&= 1,\\
\gamma_{0j}+\gamma_{1j}+\gamma_{2j} &= 2+3+(-3)\\
&= 2\\
&= 1+\alpha_{1j}.
\end{aligned}$$

Therefore, (10.4) *hold and the considered two-step method is consistent.*

Exercise 10.3.7 *Prove that the two-step method*

$$\begin{aligned}
x_{j+2}+3x_{j+1}-4x_j &= r_{j+2}(-4f_{j+2}+3f_{j+1}+2f_j)\\
&\quad +r_{j+1}(3f_{j+2}+2f_{j+1}-f_j)
\end{aligned}$$

is consistent.

Now we will consider the general case $p \geq 1$. We have

$$\begin{aligned}
&\mathscr{L}_{r_{j+1},r_{j+2}}z(t) = z(t)+h_1(t+r_{j+1}+r_{j+2},t)D_\beta z(t)\\
&\quad +h_2(t+r_{j+1}+r_{j+2},t)D_\beta^2 z(t)\\
&\quad +\cdots+h_{p+1}(t+r_{j+1}+r_{j+2},t)D_\beta^{p+1}D_\beta^{p+1}z(t)+O(r^{p+2})\\
&\quad +\alpha_{1j}\Big(z(t)+h_1(t+r_{j+1},t)D_\beta z(t)+h_2(t+r_{j+1},t)D_\beta^2 z(t)+\cdots\\
&\quad +h_{p+1}(t+r_{j+1},t)D_\beta^{p+1}z(t)+O(r^{1+p})\Big)\\
&\quad +\alpha_{0j}z(t)
\end{aligned}$$

$$
\begin{aligned}
&-r_{j+2}\beta_{2j}\Big(D_\beta z(t)+h_1(t+r_{j+1}+r_{j+2},t)D_\beta^2 z(t)\\
&+h_2(t+r_{j+1}+r_{j+2},t)D_\beta^3 z(t)\\
&+\cdots+h_p(t+r_{j+1}+r_{j+2},t)D_\beta^{p+1}z(t)+O(r^{p+2})\Big)\\
&-r_{j+2}\beta_{1j}\Big(D_\beta z(t)+h_1(t+r_{j+1},t)D_\beta^2 z(t)\\
&+h_2(t+r_{j+1},t)D_\beta^3 z(t)\\
&+\cdots+h_p(t+r_{j+1},t)D_\beta^{p+1}z(t)+O(r^{p+2})\Big)\\
&-r_{j+2}\beta_{0j}D_\beta z(t)\\
&-r_{j+2}\gamma_{2j}\Big(D_\beta z(t)+h_1(t+r_{j+1}+r_{j+2},t)D_\beta^2 z(t)\\
&+h_2(t+r_{j+1}+r_{j+2},t)D_\beta^3 z(t)\\
&+\cdots+h_p(t+r_{j+1}+r_{j+2},t)D_\beta^{p+1}z(t)+O(r^{p+2})\Big)\\
&-r_{j+2}\gamma_{1j}\Big(D_\beta z(t)+h_1(t+r_{j+1},t)D_\beta^2 z(t)\\
&+h_2(t+r_{j+1},t)D_\beta^3 z(t)\\
&+\cdots+h_p(t+r_{j+1},t)D_\beta^{p+1}z(t)+O(r^{p+2})\Big)\\
=\;&(1+\alpha_{0j}+\alpha_{1j})z(t)\\
&+\Big(h_1(t+r_{j+1}+r_{j+2},t)+\alpha_{1j}h_1(t+r_{j+1},t)-r_{j+2}\beta_{2j}\\
&-r_{j+2}\beta_{1j}-r_{j+2}\beta_{0j}
\end{aligned}
$$

$$-r_{j+1}\gamma_{2j} - r_{j+1}\gamma_{1j} - r_{j+1}\gamma_{0j}\Big) D_\beta z(t)$$

$$+\Big(h_2(t+r_{j+1}+r_{j+2},t) + \alpha_{1j}h_2(t+r_{j+1},t)$$

$$-r_{j+2}\beta_{2j}h_1(t+r_{j+1}+r_{j+2},t)$$

$$-r_{j+2}\beta_{1j}h_1(t+r_{j+1},t) - r_{j+1}\gamma_{2j}h_1(t+r_{j+1}+r_{j+2},t)$$

$$-r_{j+1}\gamma_{1j}h_1(t+r_{j+1},t)\Big) D_\beta^2 z(t)$$

$$+\Big(h_{p+1}(t+r_{j+1}+r_{j+2},t) + \alpha_{1j}h_{p+1}(t+r_{j+1},t)$$

$$-r_{j+2}\beta_{2j}h_p(t+r_{j+1}+r_{j+2},t)$$

$$-r_{j+2}\beta_{1j}h_p(t+r_{j+1},t) - r_{j+1}\gamma_{2j}h_p(t+r_{j+1}+r_{j+2},t)$$

$$-r_{j+1}\gamma_{1j}h_p(t+r_{j+1},t)\Big) D_\beta^{p+1} z(t) + O(r^{p+2}).$$

If the coefficients of $h_1, \ldots, h_p$ are equal to zero, then

$$\mathscr{L}_{r_{j+1},r_{j+2}} z(t) = O(r^{p+1}),$$

i.e., the two-step method is consistent of order p.

10.4 k-Step Methods

The most general method takes the form

$$\begin{aligned}
& x_{j+k} + \alpha_{k-1j}x_{j+k-1} + \alpha_{k-2j}x_{j+k-2} + \cdots + \alpha_{0j}x_j \\
= \; & r_{j+k}\left(\beta_{kkj}f_{j+k} + \beta_{kk-1j}f_{j+k-1} + \cdots + \beta_{k0j}f_j\right) \\
& + r_{j+k-1}\left(\beta_{k-1kj}f_{j+k} + \beta_{k-1k-1j}f_{j+k-1} + \cdots + \beta_{k-10j}f_j\right) \\
& + \cdots \\
& + r_{j+1}\left(\beta_{1kj}f_{j+k} + \beta_{1k-1j}f_{j+k-1} + \cdots + \beta_{10j}f_j\right).
\end{aligned} \tag{10.6}$$

Definition 10.4.1 *The first, second, third and son on $k+1$-th characteristic polynomials of the k-step method are defined as follows.*

$$\beta_{1j}^{-1}(q) = q^k + \alpha_{k-1j}q^{k-1} + \alpha_{k-2j}q^{k-2} + \cdots + \alpha_{0j},$$

$$\beta_{2j}^{-1}(q) = \beta_{kkj}q^k + \beta_{kk-1j}q^{k-1} + \beta_{kk-2j}q^{k-2} + \cdots + \beta_{k0j},$$

$$\vdots$$

$$\beta_{k+1j}^{-1}(q) = \beta_{1kj}q^k + \beta_{1k-1j}q^{k-1} + \beta_{1k-2j}q^{k-2} + \cdots + \beta_{10j},$$

respectively.

Example 10.4.2 *Consider the four-step method*

$$x_{j+4} - x_{j+3} - x_{j+2} + 2x_{j+1} - 3x_j = r_{j+4}(f_{j+4} - f_{j+3} - f_{j+2})$$

$$+r_{j+3}(f_{j+4} + f_j)$$

$$+r_{j+2}(2f_{j+4} - 3f_{j+3} - f_{j+2} - f_{j+1} - f_j)$$

$$+r_{j+1}(f_{j+4} + f_{j+3} + f_{j+2} + f_{j+1} + f_j).$$

Then the first, second, third, fourth and fifth characteristic polynomials are as follows.

$$\beta_{1j}^{-1}(q) = q^4 - q^3 - q^2 + 2q - 3,$$

$$\beta_{2j}^{-1}(q) = q^4 - q^3 - q^2,$$

$$\beta_{3j}^{-1}(q) = -q^4 + 1,$$

$$\beta_{4j}^{-1}(q) = 2q^4 - 3q^3 - q^2 - q - 1,$$

$$\beta_{5j}^{-1}(q) = q^4 + q^3 + q^2 + q + 1.$$

Exercise 10.4.3 *Write the first, second, third, fourth, fifth and sixth characteristic polynomials of the following five-step method*

$$\begin{aligned} & x_{j+5} - x_{j+4} - 2x_{j+3} + 3x_{j+2} - 4x_{j+1} - x_j \\ = \quad & r_{j+5}(f_{j+5} = f_{j+4} - 2f_{j+3} - 3f_{j+2} - f_{j+1} - f_j) \\ & + r_{j+4}(f_{j+1} - f_j) \\ & + r_{j+3}(2f_{j+5} + f_{j+4} - f_{j+3} - 2f_{j+2} - 3f_{j+1} - 4f_j) \\ & + r_{j+2}(f_{j+5} + 4f_{j+4} - 3f_{j+1} - f_j) \\ & + r_{j+1}(f_{j+4} + f_{j+3} - 2f_{j+1} - 4f_j). \end{aligned}$$

10.5 Consistency of k-Set Methods

Definition 10.5.1 *The associated linear difference operator of the k-step method* (10.6) *is defined as follows.*

$$\begin{aligned} \mathscr{L}_{r_{j+1}, r_{j+2}, \ldots, r_{j+k}} z(t) \quad = \quad & z(t + r_{j+1} + r_{j+2} + \cdots + r_{j+k}) \\ & + \alpha_{k-1j} z(t + r_{j+1} + r_{j+2} + \cdots + r_{j+k-1}) \\ & + \alpha_{k-2j} z(t + r_{j+1} + r_{j+2} + \cdots + r_{j+k-2}) \\ & + \cdots \\ & + \alpha_{0j} z(t) \\ & - r_{j+k}\Big(\beta_{kkj} D_\beta z(t + r_{j+1} + r_{j+2} + \cdots + r_{j+k}) \\ & + \beta_{kk-1j} D_\beta z(t + r_{j+1} + r_{j+2} + \cdots + r_{j+k-1}) \\ & + \cdots \end{aligned}$$

$$+\beta_{k0j}D_{\beta}z(t)\Big)$$

$$-r_{j+k-1}\Big(\beta_{k-1kj}D_{\beta}z(t+r_{j+1}+r_{j+2}+\cdots+r_{j+k})$$

$$+\beta_{k-1k-1j}D_{\beta}z(t+r_{j+1}+r_{j+2}+\cdots+r_{j+k-1})$$

$$+\cdots$$

$$+\beta_{k-10j}D_{\beta}z(t)\Big)$$

$$-\cdots$$

$$-r_{j+1}\Big(\beta_{1kj}D_{\beta}z(t+r_{j+1}+r_{j+2}+\cdots+r_{j+k})$$

$$+\beta_{1k-1j}D_{\beta}z(t+r_{j+1}+r_{j+2}+\cdots+rf_{j+k-1})$$

$$+\cdots$$

$$+\beta_{10j}D_{\beta}z(t)\Big).$$

Definition 10.5.2 *A linear difference operator $\mathcal{L}_{r_{j+1},r_{j+2},\ldots,r_{j+k}}$ is said to be consistent of order p if*

$$\mathcal{L}_{r_{j+1},r_{j+2},\ldots,r_{j+k}}z(t) = O(r^{p+1})$$

with $p > 0$ for any smooth function z.

Definition 10.5.3 *A k-step method is said to be consistent if its linear difference operator $\mathcal{L}_{r_{j+1},r_{j+2},\ldots,r_{j+k}}$ is consistent of order p for some $p > 0$.*

Taylor expanding the right-hand side of $\mathcal{L}_{r_{j+1},r_{j+2},\ldots,r_{j+k}}$ and collecting the terms in $h_l(\cdot,\cdot)$ of z, $D_{\beta}z$, …, $D_{\beta}^{p}z$ and choosing these terms to be equal to zero, we get that this method has order p. We will illustrate this for the case $p = 1$. We have

$$\begin{aligned} z(t+r_{j+1}+r_{j+2}+\cdots+r_{j+k}) &= z(t) \\ &\quad +(r_{j+1}+r_{j+2}+\cdots+r_{j+k})D_{\beta}z(t) \end{aligned}$$

$$
\begin{aligned}
& +O(r^2), \\
z(t+r_{j+1}+r_{j+2}+\cdots+r_{j+k-1}) &= z(t)+(r_{j+1}+r_{j+2}+\cdots \\
& +r_{j+k-1})D_\beta z(t)+O(r^2), \\
z(t+r_{j+1}+r_{j+2}+\cdots+r_{j+k-2}) &= z(t)+(r_{j+1}+r_{j+2}+\cdots \\
& +r_{j+k-2})D_\beta z(t)+O(r^2), \\
& \vdots \\
z(t+r_{j+1}) &= z(t)+r_{j+1}D_\beta z(t)+O(r^2).
\end{aligned}
$$

Then

$$
\begin{aligned}
\mathscr{L}_{r_{j+1},r_{j+2},\ldots,r_{j+k}}z(t) &= z(t)+(r_{j+1}+r_{j+2}+\cdots+r_{j+k})D_\beta z(t)+O(r^2) \\
& +\alpha_{k-1j}\left(z(t)+(r_{j+1}+r_{j+2}+\cdots+r_{j+k-1})D_\beta z(t)+O(r^2)\right) \\
& +\alpha_{k-2j}\left(z(t)+(r_{j+1}+r_{j+2}+\cdots+r_{j+k-2})D_\beta z(t)+O(r^2)\right) \\
& +\cdots \\
& +\alpha_{0j}z(t) \\
& -r_{j+k}\Big(\beta_{kkj}(D_\beta z(t)+O(r))+\beta_{kk-1j}(D_\beta z(t)+O(r)) \\
& +\beta_{k0j}D_\beta z(t)\Big) \\
& -r_{j+k-1}\Big(\beta_{k-1kj}(D_\beta z(t)+O(r))+\beta_{k-1k-1j}(D_\beta z(t)+O(r)) \\
& +\beta_{k-10j}D_\beta z(t)\Big) \\
& -r_{j+k-2}\Big(\beta_{k-2kj}(D_\beta z(t)+O(r))+\beta_{k-2k-1j}(D_\beta z(t)+O(r))
\end{aligned}
$$

$$
\begin{aligned}
& +\beta_{k-20j} D_\beta z(t)\Big) \\
& +\cdots \\
& -r_{j+1}\Big(\beta_{1kj}(D_\beta z(t)+O(r))+\beta_{1k-1j}(D_\beta z(t)+O(r)) \\
& +\beta_{10j} D_\beta z(t)\Big) \\
= \; & (1+\alpha_{k+-1j}+\alpha_{k-2j}+\cdots+\alpha_{kj})z(t) \\
& +r_{j+1}(1+\alpha_{k-1j}+\alpha_{k-2j}+\cdots+\alpha_{1j} \\
& -\beta_{1kj}-\beta_{1k-1j}-\cdots-\beta_{10j}\Big) D_\beta z(t) \\
& +r_{j+2}(1+\alpha_{k-1j}+\alpha_{k-2j}+\cdots+\alpha_{2j} \\
& -\beta_{2kj}-\beta_{2k-1j}-\cdots-\beta_{20j}\Big) D_\beta z(t) \\
& +r_{j+3}(1+\alpha_{k-1j}+\alpha_{k-2j}+\cdots+\alpha_{3j} \\
& -\beta_{3kj}-\beta_{3k-1j}-\cdots-\beta_{30j}\Big) D_\beta z(t) \\
& +\cdots \\
& +r_{j+k}(1-\beta_{kkj}-\beta_{kk-1j}-\cdots-\beta_{k0j}) D_\beta z(t) \\
& +O(r^2).
\end{aligned}
$$

Thus,

$$
\mathscr{L}_{r_{j+1},r_{j+2},\ldots,r_{j+k}} z(t) = O(r^2)
$$

for any smooth function z if and only if

$$\begin{aligned}
1+\alpha_{k-1j}+\alpha_{k-2j}+\cdots+\alpha_{0j} &= 0, \\
1+\alpha_{k-1j}+\alpha_{k-2j}+\cdots+\alpha_{1j} &= \beta_{1kj}+\beta_{1k-1j}+\cdots+\beta_{10j}, \\
1+\alpha_{k-1j}+\alpha_{k-2j}+\cdots+\alpha_{2j} &= \beta_{2kj}+\beta_{2k-1j}+\cdots+\beta_{20j}, \\
&\vdots \\
1+\alpha_{k-1j} &= beta_{k-1kj}+\beta_{k-1k-1j}+\cdots+\beta_{k-10j} \\
1 &= beta_{kkj}+\beta_{kk-1j}+\cdots+\beta_{k0j}.
\end{aligned} \tag{10.7}$$

Example 10.5.4 *Consider the four-step method*

$$\begin{aligned}
& x_{j+4}-7x_{j+3}+3x_{j+2}+2x_{j+1}+x_j \\
&= r_{j+4}\left(-6f_{j+4}+3f_{j+3}+f_{j+2}+2f_{j+1}+f_j\right) \\
&\quad +r_{j+3}\left(-10f_{j+4}+3f_{j+3}+2f_{j+2}-f_{j+1}\right) \\
&\quad +r_{j+2}\left(-8f_{j+4}-f_{j+3}+3f_{j+2}+2f_{j+1}+f_j\right) \\
&\quad +r_{j+1}\left(-4f_{j+4}+f_{j+3}+f_{j+2}+f_{j+1}\right).
\end{aligned}$$

Here

$$\begin{aligned}
\alpha_{0j} &= 1, \quad \alpha_{1j}=2, \quad \alpha_{2j}=3, \quad \alpha_{3j}=-7, \\
\beta_{40j} &= 1, \quad \beta_{41j}=2, \quad \beta_{42j}=1, \quad \beta_{43j}=3, \quad \beta_{44j}=-6, \\
\beta_{30j} &= 0, \quad \beta_{31j}=-1, \quad \beta_{32j}=2, \quad \beta_{33j}=3, \quad \beta_{34j}=-10, \\
\beta_{20j} &= 1, \quad \beta_{21j}=2, \quad \beta_{22j}=3, \quad \beta_{23j}=-1, \quad \beta_{24j}=-8, \\
\beta_{10j} &= 0, \quad \beta_{11j}=1, \quad \beta_{12j}=1, \quad \beta_{13j}=1, \quad \beta_{14j}=-4.
\end{aligned}$$

Hence,

$$1+\alpha_{0j}+\alpha_{1j}+\alpha_{2j}+\alpha_{3j}+\alpha_{4j} = 1+1+2+3-7$$

$$
\begin{aligned}
&= 0,\\
1+\alpha_{1j}+\alpha_{2j}+\alpha_{3j} &= 1+2+3-7\\
&= -1,\\
\beta_{10j}+\beta_{11j}+\beta_{12j}+\beta_{13j}+\beta_{14j} &= 0+1+1+1-4\\
&= -1,\\
1+\alpha_{2j}+\alpha_{3j} &= 1+3-7\\
&= -3,\\
\beta_{20j}+\beta_{21j}+\beta_{22j}+\beta_{23j}+\beta_{24j} &= 1+2+3-1-8\\
&= -3,\\
1+\alpha_{3j} &= 1-7\\
&= -6,\\
\beta_{30j}+\beta_{31j}+\beta_{32j}+\beta_{33j}+\beta_{34j} &= 0-1+2+3-10\\
&= -6,\\
\beta_{40j}+\beta_{41j}+\beta_{42j}+\beta_{43j}+\beta_{44j} &= 1+2+1+3-6\\
&= 1.
\end{aligned}
$$

Therefore, the considered four-step method is consistent of order $p=1$.

Exercise 10.5.5 *Check if the following two-step method satisfies the conditions* (10.7).

$$
\begin{aligned}
x_{j+2}-4x_{j+1}+3x_j &= r_{j+2}(-3f_{j+2}+3f_{j+1}+f_j)\\
&\quad +r_{j+1}(-6f_{j+2}+2f_{j+1}+f_j).
\end{aligned}
$$

10.6 Advanced Practical Problems

Problem 10.6.1 *Let*

$$x_{n+2} = 3x_{n+1} + 2r_{n+1}h_2(t_n + r_{n+1} + r_{n+2}, t_n)f_{n+1}$$
$$-2h_3(t_n + r_{n+1}, t_n)f_n.$$

Write the associated linear difference operator and determine its order of consistency.

Problem 10.6.2 *Write the first, second and third characteristic polynomials of the following two-step method.*

$$x_{j+2} + 4x_j = r_{j+2}(f_{j+2} + 3f_j) + r_{j+1}(f_{j+2} + 4f_{j+1} + 4f_j).$$

Problem 10.6.3 *Prove that the two-step method*

$$x_{j+2} - x_j = r_{j+2}f_{j+2} + r_{j+1}(6f_{j+2} - 2f_{j+1} - 3f_j)$$

is consistent.

Problem 10.6.4 *Prove that the two-step method*

$$x_{j+2} + x_{j+1} - 2x_j = r_{j+2}(f_{j+2} + 3f_{j+1} - 4f_j)$$
$$+r_{j+1}(5f_{j+2} - 2f_{j+1} - f_j)$$

is consistent.

Problem 10.6.5 *Write the first, second, third and fourth characteristic polynomials of the following three-step method.*

$$x_{j+3} - 2x_{j+2} + x_{j+1} - 3x_j = r_{j+3}(f_{j+3} + 2f_{j+2} - 4f_{j+1} - 3f_j)$$
$$+r_{j+2}(-f_{j+3} - f_{j+2} - f_j)$$
$$+r_{j+1}(2f_{j+3} - f_{j+2} + f_{j+1} - f_j).$$

Problem 10.6.6 *Check if the following five-step method satisfies the conditions* (10.7).

$$x_{j+5} - 3x_{j+4} + x_{j+3} - x_{j+2} + x_{j+1} - x_j$$

$$
\begin{aligned}
= \; & r_{j+5}\left(f_{j+5} - f_{j+4} + f_{j+3} - f_{j+2} + f_{j+1} - f_j\right) \\
& + r_{j+4}\left(2f_{j+5} + f_{j+4} - f_{j+3} - f_{j+2} - f_{j+1} - f_j\right) \\
& + r_{j+3}\left(-f_{j+5} - 2f_{j+4} + 2f_{j+3} + f_{j+2} + f_{j+1} - f_j\right) \\
& + r_{j+2}\left(f_{j+5} - 7f_{j+4} + 6f_{j+3} - f_{j+2} + 3f_{j+1} - 4f_j\right) \\
& + r_{j+1}\left(7f_{j+5} - 10f_{j+1} + 4f_j\right).
\end{aligned}
$$

10.7 Notes and References

In this chapter, we introduce linear two- and k-steps methods. We investigate these methods for consistency and we give their constructions. The results in the chapter are provided with suitable examples.

11

Runge-Kutta Methods-RMMs

Suppose that $I \subseteq \mathbb{R}$ and $\beta : I \to I$ is first kind or second kind general quantum operator. Let $a, b \in I$, $a < b$, $m \in \mathbb{N}$, and

$$a = t_0 < t_1 < \ldots < t_m = b,$$

and $(t_{j-1}, t_j) \cap I \neq \emptyset$, $j \in \{1, \ldots, m\}$. Denote

$$t_{j+1} = t_j + r_{j+1}, \quad j \in \{0, 1, \ldots, m-1\}, \quad r = \max_{j \in \{1, \ldots, m\}} r_j.$$

Consider the following IVP

$$\begin{aligned} D_\beta x(t) &= f(t, x(t)), \quad t \in [a, b], \\ x(t_0) &= x_0, \end{aligned} \tag{11.1}$$

where $x_0 \in \mathbb{R}$ and

$$(H1) \begin{cases} |f(t,x)| \leq A, \quad t \in I, x \in \mathbb{R}, \\ \text{there exist } D_\beta f(t,x) \text{ and } \dfrac{\partial}{\partial x} f(t,x) \text{ such that} \\ |D_\beta f(t,x)| \leq A, \quad \left| \dfrac{\partial}{\partial x} f(t,x) \right| \leq A, \quad t \in I, x \in \mathbb{R}, \end{cases}$$

for some positive constant A.

11.1 One-Stage Methods

Define

$$x_{n+1} = x_n + b_1 r_{n+1} f_n,$$

where $x_n = x(t_n)$, $f_n = f(t_n, x_n)$, $b_1 \in \mathbb{R}$ will be determined below. By the chain rule, we have

$$D_\beta(f(t, x(t))) = D_{\beta t} f(t, x(t))$$

DOI: 10.1201/9781003472131-11

$$+\left(\int_0^1 \frac{\partial}{\partial x} f(\beta(t), x(t) + s(\beta(t) - t) f(t, x(t))) ds\right) f(t, x(t)),$$

$t \in [a, b]$, where x is a solution of the IVP (11.1). Set

$$\begin{aligned} g(t, x(t)) &= D_{\beta t} f(t, x(t)) \\ &\quad + \left(\int_0^1 \frac{\partial}{\partial x} f(\beta(t), x(t) + s(\beta(t) - t) f(t, x(t))) ds\right) f(t, x(t)), \end{aligned}$$

$t \in [a, b]$. Then

$$D_\beta^2 x(t) = g(t, x(t)), \quad t \in [a, b].$$

Now, applying the Taylor formula, we arrive at

$$\begin{aligned} x(t_{n+1}) &= x(t_n) + r_{n+1} D_\beta x(t_n) + h_2(t_{n+1}, t_n) D_\beta^2 x(t_n) + O(r^3) \\ &= x(t_n) + r_{n+1} f(t_n, x(t_n)) + h_2(t_{n+1}, t_n) g(t_n, x(t_n)) + O(r^3) \\ &= x_n + r_{n+1} f_n + h_2(t_{n+1}, t_n) g(t_n, x_n) + O(r^3). \end{aligned}$$

Hence,

$$\begin{aligned} x(t_{n+1}) - x_{n+1} &= x_n + r_{n+1} f_n + h_2(t_{n+1}, t_n) g(t_n, x_n) + O(r^3) \\ &\quad - x_n - b_1 r_{n+1} f_n \\ &= r_{n+1}(1 - b_1) f_n + h_2(t_{n+1}, t_n) g(t_n, x_n) + O(r^3). \end{aligned}$$

Therefore, the method will be consistent of order $p = 1$ if we choose $b_1 = 1$. Thus, for $b_1 = 1$, we get the Euler method. Note that the Euler method is the only first-order one-stage explicit RKM.

11.2 Two-Stage Methods

In this section, in addition of $(H1)$, suppose that

$$(H2) \quad \frac{\partial f}{\partial x}(t, x) \neq 0, \quad (t, x) \in [a, b] \times \mathbb{R}.$$

Let $t_n + r_{n+1}^1 = t_{n+1}^1 \in I$, $t_n < t_n + r_{n+1}^1 < t_n + r_{n+1}$,

$$x_n = x(t_n),$$

$$\begin{aligned}
k_1 &= f(t_n, x_n), \\
k_2 &= a_1 f\left(t_{n+1}^1, x_n\right) + a_2 f\left(t_n, x_n + \frac{h_2\left(t_{n+1}^1, t_n\right)}{r_{n+1}^1} \cdot \frac{g(t_n, x_n)}{f_x(t_n, x_n)}\right), \\
x_{n+1} &= x_n + r_{n+1}^1 (b_1 k_1 + b_2 k_2),
\end{aligned}$$

where $a_1, a_2, b_1, b_2 \in \mathbb{R}$ will be determined below. We have

$$\begin{aligned}
x\left(t_{n+1}^1\right) &= x\left(t_n + r_{n+1}^1\right) \\
&= x(t_n) + r_{n+1}^1 D_\beta x(t_n) + h_2\left(t_{n+1}^1, t_n\right) g(t_n, x_n) + O(r^3) \\
&= x_n + r_{n+1}^1 f(t_n, x_n) + h_2\left(t_{n+1}^1, t_n\right) g(t_n, x_n) + O(r^3), \\
f\left(t_{n+1}^1, x_n\right) &= f(t_n, x_n) + r_{n+1}^1 D_{\beta t} f(t_n, x_n) + O(r^3),
\end{aligned}$$

and

$$\begin{aligned}
& f\left(t_n, x_n + \frac{h_2\left(t_{n+1}^1, t_n\right)}{r_{n+1}^1} \cdot \frac{g(t_n, x_n)}{f_x(t_n, x_n)}\right) \\
&= f(t_n, x_n) + \frac{h_2\left(t_{n+1}^1, t_n\right)}{r_{n+1}^1} \cdot \frac{g(t_n, x_n)}{f_x(t_n, x_n)} f_x(t_n, x_n) + O(r^2) \\
&= f(t_n, x_n) + g(t_n, x_n) \frac{h_2\left(t_{n+1}^1, t_n\right)}{r_{n+1}^1} + O(r^2),
\end{aligned}$$

and

$$\begin{aligned}
k_2 &= a_1 f\left(t_{n+1}^1, x_n\right) + a_2 f\left(t_n, x_n + \frac{h_2\left(t_{n+1}^1, t_n\right)}{r_{n+1}^1} \cdot \frac{g(t_n, x_n)}{f_x(t_n, x_n)}\right) \\
&= a_1 \left(f(t_n, x_n) + r_{n+1}^1 D_{\beta t} f(t_n, x_n) + O(r^2)\right) \\
&\quad + a_2 \left(f(t_n, x_n) + g(t_n, x_n) \frac{h_2\left(t_{n+1}^1, t_n\right)}{r_{n+1}^1} + O(r^2)\right) \\
&= (a_1 + a_2) f(t_n, x_n) + a_1 r_{n+1}^1 D_{\beta t} f(t_n, x_n) \\
&\quad + a_2 g(t_n, x_n) \frac{h_2\left(t_{n+1}^1, t_n\right)}{r_{n+1}^1} + O(r^2),
\end{aligned}$$

and

$$\begin{aligned} b_1k_1+b_2k_2 &= (b_1+b_2(a_1+a_2))\,f(t_n,x_n) \\ &\quad +a_1b_2r_{n+1}^1 D_{\beta t}f(t_n,x_n) \\ &\quad +a_2b_2g(t_n,x_n)\frac{h_2\left(t_{n+1}^1,t_n\right)}{r_{n+1}^1}+O(r^2), \end{aligned}$$

and

$$\begin{aligned} x_{n+1} &= x_n+(b_1+b_2(a_1+a_2))\,r_{n+1}^1 f(t_n,x_n) \\ &\quad +a_1b_2\left(r_{n+1}^1\right)^2 D_{\beta t}f(t_n,x_n) \\ &\quad +a_2b_2g(t_n,x_n)h_2\left(t_{n+1}^1,t_n\right)+O(r^3). \end{aligned}$$

Hence,

$$\begin{aligned} x\left(t_{n+1}^1\right)-x_{n+1} &= x_n+r_{n+1}^1 f(t_n,x_n)+h_2\left(t_{n+1}^1,t_n\right)g(t_n,x_n) \\ &\quad -x_n-(b_1+b_2(a_1+a_2))\,r_{n+1}^1 f(t_n,x_n) \\ &\quad -a_1b_2\left(r_{n+1}^1\right)^2 D_{\beta t}f(t_n,x_n) \\ &\quad -a_2b_2g(t_n,x_n)h_2\left(t_{n+1}^1,t_n\right)+O(r^3) \\ &= (1-(b_1+b_2(a_1+a_2)))\,r_{n+1}^1 f(t_n,x_n) \\ &\quad -a_1b_2\left(r_{n+1}^1\right)^2 D_{\beta t}f(t_n,x_n) \\ &\quad +(1-a_2b_2)g(t_n,x_n)h_2\left(t_{n+1}^1,t_n\right)+O(r^3). \end{aligned}$$

Therefore

1. if

$$\begin{cases} b_1+b_2(a_1+a_2)=1 \\ a_1b_2\neq 0 \end{cases} \quad \text{or} \quad \begin{cases} b_1+b_2(a_1+a_2)=1 \\ a_2b_2\neq 1, \end{cases}$$

then the method is consistent of order $p=1$.

2. if
$$\begin{cases} b_1 + b_2(a_1 + a_2) = 1 \\ a_1 b_2 = 0 \\ a_2 b_2 = 1, \end{cases}$$
which is equivalent to
$$\begin{cases} a_1 = b_1 = 0 \\ a_2 b_2 = 1, \end{cases}$$
then the method is consistent of order $p = 2$.

11.3 Three-Stage Methods

Suppose $(H1)$ and $(H2)$ and $t_{n+1}^2 = t_n + r_{n+1}^1 \in I$, $t_n < t_n + r_{n+1}^1 < t_n + r_{n+1}$. Let

$$\begin{aligned} k_1 &= f(t_n, x_n), \\ k_2 &= a_{11} f(t_{n+1}^1, x_n) + a_{12} f\left(t_n, x_n + \frac{D_{\beta t} f(t_n, x_n)}{f_x(t_n, x_n)} r_{n+1}^1\right), \\ k_3 &= a_{21} f(t_{n+1}^1, x_n) + a_{22} f\left(t_n, x_n + \frac{h_2\left(t_{n+1}^1, t_n\right) g(t_n, x_n)}{f_x(t_n, x_n) r_{n+1}^1}\right), \\ x_{n+1} &= x_n + r_{n+1}^1 (b_1 k_1 + b_2 k_2 + b_3 k_3), \end{aligned}$$

where $a_{11}, a_{12}, a_{21}, a_{22}, b_1, b_2, b_3 \in \mathbb{R}$ will be determined below. We have

$$\begin{aligned} x(t_{n+1}^1) &= x_n + r_{n+1}^1 f(t_n, x_n) + h_2\left(t_{n+1}^1, t_n\right) g(t_n, x_n) + O(r^3), \\ f\left(t_n + r_{n+1}^1, x_n\right) &= f(t_n, x_n) + r_{n+1}^1 D_{\beta t} f(t_n, x_n) + O(r^2), \\ f\left(t_n, x_n + \frac{D_{\beta t} f(t_n, x_n)}{f_x(t_n, x_n)} r_{n+1}^1\right) &= f(t_n, x_n) \\ &\quad + \frac{D_{\beta t} f(t_n, x_n)}{f_x(t_n, x_n)} r_{n+1}^1 f_x(t_n, x_n) + O(r^2), \\ f\left(t_n, x_n + \frac{h_2\left(t_{n+1}^1, t_n\right) g(t_n, x_n)}{f_x(t_n, x_n)}\right) &= f(t_n, x_n) \end{aligned}$$

$$+\frac{h_2\left(t_{n+1}^1,t_n\right)g(t_n,x_n)}{f_x(t_n,x_n)r_{n+1}^1}f_x(t_n,x_n)+O(r^2)$$

$$= \quad f(t_n,x_n)$$

$$+\frac{h_2\left(t_{n+1}^1,t_n\right)g(t_n,x_n)}{r_{n+1}^1}+O(r^2).$$

Thus,

$$\begin{aligned}
k_2 &= a_{11}f(t_{n+1}^1,x_n)+a_{12}f\left(t_n,x_n+\frac{D_{\beta t}f(t_n,x_n)}{f_x(t_n,x_n)}r_{n+1}^1\right)\\
&= a_{11}f(t_n,x_n)+a_{11}r_{n+1}^1D_{\beta t}f(t_n,x_n)\\
&\quad +a_{12}f(t_n,x_n)+a_{12}\frac{h_2\left(t_{n+1}^1,t_n\right)g(t_n,x_n)}{r_{n+1}^1}+O(r^2)\\
&= (a_{11}+a_{12})f(t_n,x_n)+a_{11}r_{n+1}^1D_{\beta t}f(t_n,x_n)\\
&\quad +a_{12}\frac{h_2\left(t_{n+1}^1,t_n\right)g(t_n,x_n)}{r_{n+1}^1}+O(r^2),\\
b_2k_2 &= b_2(a_{11}+a_{12})f(t_n,x_n)+a_{11}r_{n+1}^1D_{\beta t}f(t_n,x_n)\\
&\quad +b_2a_{12}\frac{h_2\left(t_{n+1}^1,t_n\right)g(t_n,x_n)}{r_{n+1}^1}+O(r^2),\\
k_3 &= a_{21}f(t_{n+1}^1,x_n)+a_{22}f\left(t_n,x_n+\frac{D_{\beta t}f(t_n,x_n)}{f_x(t_n,x_n)}r_{n+1}^1\right)\\
&= a_{21}f(t_n,x_n)+a_{21}r_{n+1}^1D_{\beta t}f(t_n,x_n)\\
&\quad +a_{22}f(t_n,x_n)+a_{12}\frac{h_2\left(t_{n+1}^1,t_n\right)g(t_n,x_n)}{r_{n+1}^1}+O(r^2)\\
&= (a_{21}+a_{22})f(t_n,x_n)+a_{21}r_{n+1}^1D_{\beta t}f(t_n,x_n)\\
&\quad +a_{22}\frac{h_2\left(t_{n+1}^1,t_n\right)g(t_n,x_n)}{r_{n+1}^1}+O(r^2),\\
b_3k_3 &= b_3(a_{21}+a_{22})f(t_n,x_n)+a_{21}r_{n+1}^1D_{\beta t}f(t_n,x_n)
\end{aligned}$$

$$+b_3a_{22}\frac{h_2\left(t_{n+1}^1,t_n\right)g(t_n,x_n)}{r_{n+1}^1}+O(r^2),$$

and

$$\begin{aligned}b_1k_1+b_2k_2+b_3k_3 &= (b_1+b_2(a_{11}+a_{12})+b_3(a_{21}+a_{22}))f(t_n,x_n)\\ &+(b_2a_{11}+b_3a_{21})r_{n+1}^1D_{\beta t}f(t_n,x_n)\\ &+(b_2a_{12}+b_3a_{22})\frac{h_2\left(t_{n+1}^1,t_n\right)g(t_n,x_n)}{r_{n+1}^1}+O(r^2),\end{aligned}$$

and

$$\begin{aligned}r_{n+1}^1\left(b_1k_1+b_2k_2+b_3k_3\right) &= (b_1+b_2(a_{11}+a_{12})+b_3(a_{21}+a_{22}))\\ &\times r_{n+1}^1f(t_n,x_n)\\ &+(b_2a_{11}+b_3a_{21})\left(r_{n+1}^1\right)^2D_{\beta t}f(t_n,x_n)\\ &+(b_2a_{12}+b_3a_{22})h_2\left(t_{n+1}^1,t_n\right)g(t_n,x_n)\\ &+O(r^3),\end{aligned}$$

and

$$\begin{aligned}x_{n+1} &= x_n+(b_1+b_2(a_{11}+a_{12})+b_3(a_{21}+a_{22}))r_{n+1}^1f(t_n,x_n)\\ &+(b_2a_{11}+b_3a_{21})\left(r_{n+1}^1\right)^2D_{\beta t}f(t_n,x_n)\\ &+(b_2a_{12}+b_3a_{22})h_2\left(t_{n+1}^1,t_n\right)g(t_n,x_n)+O(r^3).\end{aligned}$$

Hence,

$$\begin{aligned}x(t_{n+1}^1)-x_{n+1} &= x_n+r_{n+1}^1f(t_n,x_n)+h_2\left(t_{n+1}^1,t_n\right)g(t_n,x_n)\\ &-x_n-(b_1+b_2(a_{11}+a_{12})+b_3(a_{21}+a_{22}))r_{n+1}^1f(t_n,x_n)\\ &-(b_2a_{11}+b_3a_{21})\left(r_{n+1}^1\right)^2D_{\beta t}f(t_n,x_n)\end{aligned}$$

$$-(b_2a_{12}+b_3a_{22})h_2\left(t^1_{n+1},t_n\right)g(t_n,x_n)+O(r^3)$$

$$= \quad (1-(b_1+b_2(a_{11}+a_{12})+b_3(a_{21}+a_{22})))\,r^1_{n+1}f(t_n,x_n)$$

$$-(b_2a_{11}+b_3a_{21})\left(r^1_{n+1}\right)^2 D_{\beta t}f(t_n,x_n)$$

$$+(1-(b_2a_{12}+b_3a_{22}))\,h_2\left(t^1_{n+1},t_n\right)g(t_n,x_n)+O(r^3).$$

Therefore

1. if
$$\begin{cases} b_1+b_2(a_{11}+a_{12})+b_3(a_{21}+a_{22})=1 \\ b_2a_{11}+b_3a_{21}\neq 0 \end{cases}$$
or
$$\begin{cases} b_1+b_2(a_{11}+a_{12})+b_3(a_{21}+a_{22})=1 \\ b_2a_{12}+b_3a_{22}\neq 1, \end{cases}$$
then the three-stage method is consistent of order $p=1$.
2. if
$$\begin{cases} b_1+b_2(a_{11}+a_{12})+b_3(a_{21}+a_{22})=1 \\ b_2a_{11}+b_3a_{21}=0 \\ b_2a_{12}+b_3a_{22}=1, \end{cases}$$
then the three-stage method is consistent of order $p=2$.

11.4 s-Stage Methods

Suppose $(H1)$ and $(H2)$ and $t^2_{n+1}=t_n+r^1_{n+1}\in I$, $t_n<t_n+r^1_{n+1}<t_n+r_{n+1}$. The general s-stage RK method can be written in the form

$$x_{n+1}=x_n+r^1_{n+1}\sum_{j=1}^{s}b_jk_j,$$

where

$$k_1 \quad = \quad f(t_n,x_n),$$

$$k_j \quad = \quad k_j\left(f,D_{\beta t}f,f_x\right), \quad j\in\{2,\ldots,s\},$$

are determined so that the method to be consistent of order $p > 0$. To determine the coefficients b_j, $j \in \{1, \ldots, s\}$, we use the Taylor expansions of f, x, k_j, $j \in \{1, \ldots, s\}$, of order $p > 0$ so that the considered s-stage method to be consistent of order 1 or 2 or so on or p.

11.5 Notes and References

In this chapter, we analyse quantum analogues of the Runge-Kutta method. We consider one-, two-, three- and s-stages methods.

12

The Adomian Polynomials Method

In this chapter we derive a quantum analogue of the Adomian polynomial method and present its application to a β-differential equations of arbitrary order with a nonlinear term. We will use some properties of β-Laplace transform. For the definition of the β-Laplace transform and some of its properties we refer the reader to the book [5].

Let $I \subseteq \mathbb{R}$ and $\beta : I \to I$ be first kind or second kind general quantum operator.

12.1 Analysing of the Method

Denote the set consisting of all possible strings of length n, containing exactly k times β and $n-k$ times D_β operators by $S_k^{(n)}$. Then, we have

$$h_n(t,s)h_m(t,s) = \sum_{l=m}^{m+n} \left(\sum_{\Lambda_{l,m} \in S_m^{(l)}} h_n^{\Lambda_{l,m}}(s,s) \right) h_l(t,s)$$

for every $t,s \in I$. For $s \in I$, $l,m,n \in \mathbb{N}_0$, set

$$A_{l,m,n,s} = \sum_{\Lambda_{l,m} \in S_m^{(l)}} h_n^{\Lambda_{l,m}}(s,s)$$

and for any $m,n \in \mathbb{N}_0$, we have

$$h_n(t,s)h_m(t,s) = \sum_{l=m}^{m+n} A_{l,m,n,s} h_l(t,s). \tag{12.1}$$

For $n \in \mathbb{N}_0$, $t,s \in I$, define the polynomials

$$H_n^1(t,s) = (h_1(t,s))^n, \quad t,s \in I.$$

Note that

$$H_n^1(t,s)H_m^1(t,s) = H_{n+m}^1(t,s), \quad t,s \in I.$$

DOI: 10.1201/9781003472131-12

Note also that

$$H_1^1(t,s) = h_1(t,s), \tag{12.2}$$

and by (12.1), we get

$$\begin{aligned} H_2^1(t,s) &= h_1(t,s)h_1(t,s) \\ &= \sum_{l=1}^{2} A_{l,1,1,s}h_l(t,s) \\ &= A_{1,1,1,s}h_1(t,s) + A_{2,1,1,s}h_2(t,s) \\ &= A_{1,1,1,s}H_1^1(t,s) + A_{2,1,1,s}h_2(t,s), \end{aligned}$$

whereupon

$$h_2(t,s) = -\frac{A_{1,1,1,s}}{A_{2,1,1,s}}H_1^1(t,s) + \frac{1}{A_{2,1,1,s}}H_2^1(t,s),$$

and so on. Below we denote by B_i^j, $i, j \in \mathbb{N}$, the constants for which

$$H_n^1(t,s) = B_1^n h_1(t,s) + B_2^n h_2(t,s) + \cdots + B_n^n h_n(t,s), \quad t, s \in I. \tag{12.3}$$

Example 12.1.1 *Let* $\alpha \in \mathbb{R}$. *Then*

$$\begin{aligned} e_{\alpha,\beta}(t) &= 1 + \alpha h_1(t,s) + \alpha^2 h_2(t,s) + \cdots \\ &= 1 + \alpha H_1^1(t,s) \\ &\quad + \alpha^2\left(-\frac{A_{1,1,1,s}}{A_{2,1,1,s}}H_1^1(t,s) + \frac{1}{A_{2,1,1,s}}H_2^1(t,s)\right) + \cdots \\ &= 1 + \left(\alpha - \alpha^2\frac{A_{1,1,1,s}}{A_{2,1,1,s}} + \cdots\right)H_1^1(t,s) \\ &\quad + \left(\frac{\alpha^2}{A_{2,1,1,s}} + \cdots\right)H_2^1(t,s) + \cdots. \end{aligned}$$

Suppose that $u : I \to \mathbb{R}$ is a given function which has a convergent series expansion of the form

$$u = \sum_{j=0}^{\infty} u_j. \tag{12.4}$$

Suppose also that $f : \mathbb{R} \to \mathbb{R}$ is a given analytic function such that

$$f(u) = \sum_{n=0}^{\infty} A_n(u_0, u_1, \ldots, u_n), \tag{12.5}$$

where A_n, $n \in \mathbb{N}_0$, are given by

$$\begin{aligned} A_0 &= f(u_0) \\ A_n &= \sum_{\nu=1}^{n} c(\nu,n) f^{(\nu)}(u_0), \quad n \in \mathbb{N}. \end{aligned}$$

Here the functions $c(\nu,n)$ denote the sum of products of ν components u_j of u given in (12.4), whose subscripts sum up to n, divided by the factorial of the number of repeated subscripts, i.e.,

$$\begin{aligned} A_0 &= f(u_0), \\ A_1 &= c(1,1)f'(u_0) \\ &= u_1 f'(u_0), \\ A_2 &= c(1,2)f'(u_0) + c(2,2)f''(u_0) \\ &= u_2 f'(u_0) + \frac{u_1^2}{2!} f''(u_0), \\ A_3 &= c(1,3)f'(u_0) + c(2,3)f''(u_0) + c(3,3)f'''(u_0) \\ &= u_3 f'(u_0) + u_1 u_2 f''(u_0) + \frac{u_1^3}{3!} f'''(u_0), \\ A_4 &= c(1,4)f'(u_0) + c(2,4)f''(u_0) + c(3,4)f'''(u_0) \\ &\quad + c(4,4) f^{(4)}(u_0) \\ &= u_4 f'(u_0) + \left(u_1 u_3 + \frac{u_2^2}{2} \right) f''(u_0) + \frac{u_1^2 u_2}{2} f'''(u_0) \\ &\quad + \frac{u_1^4}{4!} f^{(4)}(u_0) \end{aligned}$$

and so on. Suppose now that u is given by the convergent series

$$u = \sum_{n=0}^{\infty} c_n H_n^1(x,x_0). \tag{12.6}$$

We wish to find the respected transformed series for $f(u)$. From (12.4), we have

$$u = \sum_{n=0}^{\infty} u_n = \sum_{n=0}^{\infty} c_n H_n^1(x,x_0),$$

and hence,

$$u_n = c_n H_n^1(x, x_0) \quad n \in \mathbb{N}_0.$$

Thus,

$$\begin{aligned} f(u) &= \sum_{n=0}^{\infty} A_n(u_0, u_1, \ldots, u_n) \\ &= f\left(\sum_{n=0}^{\infty} c_n H_n^1(x, x_0)\right) \\ &= \sum_{n=0}^{\infty} A^n(c_0, c_1, \ldots, c_n) H_n^1(x, x_0). \end{aligned}$$

Hence,

$$A_n(u_0, u_1, \ldots, u_n) = A^n(c_0, c_1, \ldots, c_n) H_n^1(x, x_0).$$

For $n = 0$, we have

$$\begin{aligned} u_0 &= c_0 H_0^1(x, x_0). \\ &= c_0. \end{aligned}$$

Thus,

$$\begin{aligned} A_0(u_0) &= A^0(c_0) H_0^1(x, x_0) \\ &= A^0(c_0). \end{aligned}$$

For $n = 1$, we find

$$\begin{aligned} A_1(u_0, u_1) &= u_1 f'(u_0) \\ &= A^1(c_0, c_1) H_1^1(x, x_0) \end{aligned}$$

or

$$c_1 H_1^1(x, x_0) f'(u_0) = A^1(c_0, c_1) H_1^1(x, x_0),$$

whereupon

$$\begin{aligned} A^1(c_0, c_1) &= c_1 f'(u_0) \\ &= c_1 f'(c_0) \\ &= A_1(c_0, c_1). \end{aligned}$$

For $n = 2$, we have

$$A_2(u_0, u_1, u_2) = A^2(c_0, c_1, c_2)H_2^1(x, x_0)$$

or

$$u_2 f'(u_0) + \frac{u_1^2}{2} f''(u_0) = A^2(c_0, c_1, c_2)H_2^1(x, x_0).$$

Then

$$c_2 H_2^1(x, x_0) f'(c_0) + \frac{c_1^2 (H_1^1(x, x_0))^2}{2} f''(c_0) = A^2(c_0, c_1, c_2)H_2^1(x, x_0),$$

or

$$\left(c_2 f'(c_0) + \frac{c_1^2}{2} f''(c_0) \right) H_2^1(x, x_0) = A^2(c_0, c_1, c_2)H_2^1(x, x_0),$$

whereupon

$$\begin{aligned} A^2(c_0, c_1, c_2) &= c_2 f'(c_0) + \frac{c_1^2}{2} f''(c_0) \\ &= A_2(c_0, c_1, c_2). \end{aligned}$$

For $n = 3$, we find

$$\begin{aligned} u_3 f'(u_0) + u_1 u_2 f''(u_0) + \frac{u_1^3}{3!} f'''(u_0) &= A_3(u_0, u_1, u_2, u_3) \\ &= A^3(c_0, c_1, c_2, c_3)H_3^1(x, x_0) \end{aligned}$$

or

$$\begin{aligned} &c_3 H_3^1(x, x_0) f'(c_0) + c_1 c_2 H_3^1(x, x_0) f''(x_0) + \frac{c_1^3}{3!} f'''(c_0) H_3^1(x, x_0) \\ &= A^3(c_0, c_1, c_2, c_3)H_3^1(x, x_0), \end{aligned}$$

whereupon

$$\begin{aligned} c_3 f'(c_0) + c_1 c_2 f''(x_0) + \frac{c_1^3}{3!} f'''(c_0) &= A^3(c_0, c_1, c_2, c_3) \\ &= A_3(c_0, c_1, c_2, c_3), \end{aligned}$$

and so on. Therefore, we get the following result.

Theorem 12.1.2 *Let $u : I \to \mathbb{R}$ be a function with a convergent expansion given in* (12.6). *Let $f : \mathbb{R} \to \mathbb{R}$ be an analytic function having the form* (12.5). *Then*

$$f(u) = f\left(\sum_{n=0}^{\infty} c_n H_n^1(x, x_0) \right) = \sum_{n=0}^{\infty} A_n(c_0, c_1, \ldots, c_n) H_n^1(x, x_0).$$

Example 12.1.3 *For $\alpha = 1$, consider $u = e_\alpha(x,x_0)$ and $f(u) = u^2$. Using Example 12.1.1, we have*

$$e_\alpha(x,x_0) = \sum_{m=0}^{\infty} c_m H_m^1(x,x_0)$$

where

$$\begin{aligned} c_0 &= 1, \\ c_1 &= \alpha - \alpha^2 \frac{A_{1,1,1,s}}{A_{2,1,1,s}} + \cdots +, \\ c_3 &= \frac{\alpha^2}{A_{2,1,1,s}} + \cdots, \\ &\vdots \end{aligned}$$

Note that

$$(e_\alpha(x,x_0))^2 = c_0^2 + 2c_0c_1H_1^1(x,x_0) + \cdots. \tag{12.7}$$

On the other hand, by Theorem 12.1.2, we obtain

$$(e_\alpha(x,x_0))^2 = \sum_{m=0}^{\infty} A_m H_m^1(x,x_0)$$

and

$$\begin{aligned} A_0(u_0) &= A_0(c_0) \\ &= 1 \\ &= c_0^2, \\ A_1(u_0,u_1) &= c_1 f'(c_0) \\ &= 2c_0c_1 \end{aligned}$$

and so on, i.e., we get (12.7).

In what follows, we present the Adomian polynomials method for a dynamic equation of arbitrary order on ageneral time scale I. With $\mathscr{L}$ we will denote the Laplace transform on I. Suppose that $t_0 \in I$. Consider the initial value problem (IVP)

$$\begin{aligned} &D_\beta^n y + a_1 D_\beta^{n-1} y + \cdots + a_n y = f(y), \quad t > t_0, \\ &y(t_0) = y_0, \quad D_\beta y(t_0) = y_1, \quad \ldots, \quad D_\beta^{n_1} y(t_0) = y_{n-1}, \end{aligned} \tag{12.8}$$

where $a_i \in \mathbb{R}$, $i \in \{1,\ldots,n\}$, $y_i \in \mathbb{R}$, $i \in \{0,\ldots,n-1\}$, are given constants, $f:\mathbb{R}\to\mathbb{R}$ is an analytic function. We will search a solution of the IVP (12.8), in the form

$$y(t)=\sum_{j=0}^{\infty} c_j H_j^1(t,t_0), \quad t\ge t_0.$$

Assume that

$$f(y)=\sum_{j=0}^{\infty} A_j(c_0,\ldots,c_j)H_j^1(t,t_0), \quad t\ge t_0.$$

Using that

$$\mathscr{L}_\beta\left(h_k(t,t_0)\right)(z)=\frac{1}{z^{k+1}}, \quad k\in\mathbb{N}_0,$$

we get

$$\begin{aligned}
\mathscr{L}_\beta\left(H_0^1(t,t_0)\right)(z) &= \frac{1}{z},\\
\mathscr{L}_\beta\left(H_j^1(t,t_0)\right)(z) &= \sum_{k=1}^{j} B_k^j \mathscr{L}_\beta(h_k(t,t_0))(z)\\
&= \sum_{k=1}^{j} B_k^j \frac{1}{z^{k+1}}, \quad j\in\mathbb{N}.
\end{aligned}$$

Let $Y(z)=\mathscr{L}_\beta(y(t))(z)$. We take the β-Laplace transform of both sides of the β-differential equation in (12.8) and using the initial conditions we obtain

$$z^nY(z)-\sum_{l=0}^{n-1} z^l y_{n-1-l}+a_1z^{n-1}Y(z)-a_1\sum_{l=0}^{n-2} z^l y_{n-2-l}$$
$$+\cdots+a_nY(z)=\sum_{j=0}^{\infty}\left(A_j(c_0,\ldots,c_j)\sum_{k=1}^{j} B_k^j\frac{1}{z^{k+1}}\right)$$

or

$$\begin{aligned}
\left(z^n+a_1z^{n-1}+\cdots+a_n\right)Y(z) &= \sum_{l=0}^{n-1} z^l y_{n-1-l}+a_1\sum_{l=0}^{n-2} z^l y_{n-2-l}\\
&\quad +\cdots+a_{n-1}y_0+A_0(c_0)\frac{1}{z}\\
&\quad +\sum_{j=1}^{\infty}\left(A_j(c_0,\ldots,c_j)\sum_{k=1}^{j} B_k^j\frac{1}{z^{k+1}}\right),
\end{aligned}$$

From this equation we get

$$Y(z) = \frac{1}{z^n+a_1z^{n-1}+\cdots+a_n}\left(\sum_{l=0}^{n-1} z^l y_{n-1-l}+a_1\sum_{l=0}^{n-2} z^l y_{n-2-l}\right.$$

$$+\cdots+a_{n-1}y_0+A_0(c_0)\frac{1}{z}$$

$$+\sum_{j=1}^{\infty}\left(A_j(c_0,\ldots,c_j)\sum_{k=1}^{j}B_k^j\frac{1}{z^{k+1}}\right)\Bigg).$$

Consequently,

$$\begin{aligned} y(t) &= \mathscr{L}_\beta^{-1}\left(\frac{1}{z^n+a_1z^{n-1}+\cdots+a_n}\left(\sum_{l=0}^{n-1}z^l y_{n-1-l}+a_1\sum_{l=0}^{n-2}z^l y_{n-2-l}\right.\right. \\ &\quad +\cdots+a_{n-1}y_0+A_0(c_0)\frac{1}{z} \\ &\quad \left.\left.+\sum_{j=1}^{\infty}\left(A_j(c_0,\ldots,c_j)\sum_{k=1}^{j}B_k^j\frac{1}{z^{k+1}}\right)\right)\right)(t) \end{aligned}$$

or by the linearity of the inverse β-Laplace transform,

$$\begin{aligned} y(t) &= \sum_{l=0}^{n-1}y_{n-1-l}\mathscr{L}_\beta^{-1}\left(\frac{z^l}{z^n+a_1z^{n-1}+\cdots+a_n}\right)(t) \\ &\quad +a_1\sum_{l=0}^{n-2}y_{n-2-l}\mathscr{L}_\beta^{-1}\left(\frac{z^l}{z^n+a_1z^{n-1}+\cdots+a_n}\right)(t) \\ &\quad +\cdots \\ &\quad +a_{n-1}y_0\mathscr{L}_\beta^{-1}\left(\frac{1}{z^n+a_1z^{n-1}+\cdots+a_n}\right)(t) \\ &\quad +A_0(c_0)\mathscr{L}_\beta^{-1}\left(\frac{1}{z^{n+1}+a_1z^n+\cdots+a_nz}\right)(t) \\ &\quad +\sum_{j=1}^{\infty} \\ &\quad \times\left(A_j(c_0,\ldots,c_j)\sum_{k=1}^{j}B_k^j\mathscr{L}_\beta^{-1}\left(\frac{1}{z^{n+k+1}+a_1z^{n+k}+\cdots+a_nz^{k+1}}\right)(t)\right), \end{aligned}$$

$t\geq t_0$.

After computing the inverse β-Laplace transform of the right-hand-side, we equate the coefficients of the functions $h_k(t,t_0)$ on both sides. In general, this results in a nonlinear system for the constants c_k, $k\in\mathbb{N}_0$.

12.2 Examples of First-Order Nonlinear β-Differential Equations

As a particular case, we consider an IVP associated with a first-order β-differential equation of the form

$$D_\beta y = f(y), \quad t > t_0, \quad y(t_0) = 0, \tag{12.9}$$

where $f : \mathbb{R} \to \mathbb{R}$ is an analytic function. We propose a solution of the IVP (12.9), in the form

$$y(t) = \sum_{j=0}^{\infty} c_j H_j^1(t,t_0), \quad t \geq t_0.$$

Like in the general case, we suppose that

$$f(y) = \sum_{j=0}^{\infty} A_j(c_0,\ldots,c_j) H_j^1(t,t_0), \quad t \geq t_0.$$

On the other hand, by (12.3) we have

$$y(t) = c_0 + \sum_{j=1}^{\infty} \sum_{k=1}^{j} c_j B_k^j h_k(t,t_0), \quad t \geq t_0. \tag{12.10}$$

and

$$f(y) = A_0(c_0) + \sum_{j=1}^{\infty} \sum_{k=1}^{j} A_j(c_0,\ldots,c_j) B_k^j h_k(t,t_0), \quad t \geq t_0. \tag{12.11}$$

Let

$$\mathcal{L}_\beta\left(y(t)\right)(z) = Y(z).$$

Then, we have

$$\mathcal{L}_\beta\left(D_\beta y(t)\right)(z) = zY(z) - y(t_0) = zY(z).$$

Taking the β-Laplace transform of both sides of the β-differential equation (12.9) we obtain

$$\begin{aligned} zY(z) &= \mathcal{L}_\beta\left(A_0(c_0) + \sum_{j=1}^{\infty} \sum_{k=1}^{j} A_j(c_0,\ldots,c_j) B_k^j h_k(t,t_0)\right)(z) \\ &= A_0(c_0)\frac{1}{z} + \sum_{j=1}^{\infty} \sum_{k=1}^{j} A_j(c_0,\ldots,c_j) B_k^j \frac{1}{z^{k+1}}. \end{aligned}$$

Then, we obtain

$$Y(z) = A_0(c_0)\frac{1}{z^2} + \sum_{j=1}^{\infty} \sum_{k=1}^{j} A_j(c_0,\ldots,c_j) B_k^j \frac{1}{z^{k+2}}.$$

Now, by taking inverse β-Laplace transform of both sides, we get

$$y(t) = A_0(c_0)h_1(t,t_0) + \sum_{j=1}^{\infty}\sum_{k=1}^{j} A_j(c_0,\dots,c_j)B_k^j h_{k+1}(t,t_0).$$

Employing (12.10), we have

$$\begin{aligned} c_0 + \sum_{j=1}^{\infty}\sum_{k=1}^{j} c_j B_k^j h_k(t,t_0) &= A_0(c_0)h_1(t,t_0) \\ &\quad + \sum_{j=1}^{\infty}\sum_{k=1}^{j} A_j(c_0,\dots,c_j)B_k^j h_{k+1}(t,t_0). \end{aligned}$$

In order to equate the coefficients of the time scale monomials $h_k(t,t_0)$ on both sides, we reorder the sums as follows.

$$\begin{aligned} & c_0 + \left(\sum_{j=1}^{\infty} c_j B_1^j\right) h_1(t,t_0) + \sum_{k=2}^{\infty}\left(\sum_{j=k}^{\infty} c_j B_k^j\right) h_k(t,t_0) \\ &= A_0(c_0)h_1(t,t_0) + \sum_{k=2}^{\infty}\sum_{j=k-1}^{\infty} A_j(c_0,\dots,c_j)B_{k-1}^j h_k(t,t_0). \end{aligned}$$

This results in the following nonlinear system for determining the constants c_j, $j = 0,1,\dots$.

$$\begin{aligned} c_0 &= 0, \\ \sum_{j=1}^{\infty} c_j B_1^j &= A_0(c_0) = f(0) \\ \sum_{j=k}^{\infty} c_j B_k^j &= \sum_{j=k-1}^{\infty} A_j(c_0,\dots,c_j)B_{k-1}^j, \quad k \geq 2. \end{aligned} \tag{12.12}$$

Notice that the system is infinite and nonlinear in its unknowns. However, the nonlinearity is of polynomial type. This is a results of the nonlinear structure of the function f.

Remark 12.2.1 *If $I = \mathbb{R}$, $\beta(t) = t$, $t \in I$, we have $H_1^k(t,t_0) = h_k(t,t_0) = \dfrac{(t-t_0)^k}{k!}$ for $k \in \mathbb{N}$ and hence, $B_k^j = k!\delta_{k,j}$ for $k \in \mathbb{N}$ and $j = 1,\dots k$. In this case, the system* (12.12) *becomes:*

$$\begin{aligned} c_0 &= 0, \\ k!c_k &= (k-1)!A_{k-1}(c_0,\dots,c_{k-1}), \quad k = 1,2,3,\dots. \\ \text{or simply } c_k &= \frac{1}{k}A_{k-1}(c_0,\dots,c_{k-1}) \quad k = 1,2,3,\dots. \end{aligned} \tag{12.13}$$

12.3 Numerical Examples

Example 12.3.1 *As a first example we consider an IVP associated with a linear β-differential equation of first-order of the form*

$$D_\beta y(t) = ay(t) + b, \quad y(0) = 0, \tag{12.14}$$

where a, b are real constants. Assume that

$$y(t) = \sum_{j=0}^{\infty} c_j H_j^1(t,0), \quad t \geq 0,$$

where c_j, $j = 0, 1, \ldots$ are the coefficients to be determined. By Theorem 12.1.2, we have

$$f(y) = ay(t) + b = \sum_{j=0}^{\infty} A_j(c_0, \ldots, c_j) H_j^1(t,0), \quad t \geq 0,$$

where

$$\begin{aligned}
A_0 &= f(c_0) \\
&= ac_0 + b \\
A_1 &= c_1 f'(c_0) \\
&= ac_1 \\
A_2 &= c_2 f'(c_0) + \frac{c_1^2}{2!} f''(c_0) \\
&= ac_2 \\
A_3 &= c_3 f'(c_0) + c_1 c_2 f''(c_0) + \frac{c_1^3}{3!} f'''(c_0) \\
&= ac_3 \\
A_4 &= c_4 f'(c_0) + \left(c_1 c_3 + \frac{c_2^2}{2}\right) f''(c_0) + \frac{c_1^2 c_2}{2} f'''(c_0) + \frac{c_1^4}{4!} f^{(4)}(c_0) \\
&= ac_4 \\
&\cdots \\
A_n &= ac_n
\end{aligned}$$

since $f'(c_0) = a$ and $f^{(k)}(c_0) = 0$ for $k \geq 2$. Therefore, the system (12.12) *for this example takes the form*

$$\begin{aligned}
c_0 &= 0, \\
\sum_{j=1}^{\infty} c_j B_1^j &= c_0 \\
\sum_{j=k}^{\infty} c_j B_k^j &= \sum_{j=k-1}^{\infty} ac_{j-1} B_{k-1}^j, \quad k \in \mathbb{N}, k \geq 2.
\end{aligned} \tag{12.15}$$

This is an infinite linear system having the following triangular form

$$\begin{array}{rcl} c_0 & = & 0 \\ c_1B_1^1+c_2B_1^2+c_3B_1^3+\cdots & = & b \\ c_2B_2^2+c_3B_2^3+c_4B_2^4+\cdots & = & a(c_1B_1^1+c_2B_1^2+c_3B_1^3+\cdots)=ab \\ c_3B_3^3+c_4B_3^4+c_5B_3^5+\cdots & = & a(c_2B_2^2+c_3B_2^3+c_4B_2^4+\cdots)=a^2b \\ \cdots & & \\ c_nB_n^n+c_{n+1}B_n^{n+1}+\cdots & = & a(c_{n-1}B_{n-1}^{n-1}+c_nB_{n-1}^n+\cdots)=a^{n-1}b \\ & \cdots & \end{array}$$

In the next two examples we take f to be a nonlinear function.

Example 12.3.2 *Consider the initial value problem associated with the first-order nonlinear β- equation of the form*

$$D_\beta y(t)=e^{y(t)}, \quad t\geq 0, \quad y(0)=0, \tag{12.16}$$

where $e^{y(t)}$ is the exponential function on the set of real numbers. Assume that the solution has the series representation

$$y(t)=\sum_{j=0}^{\infty} c_jH_j^1(t,0), \quad t\geq 0,$$

where c_j, $j\in\mathbb{N}_0$ are the coefficients to be determined. By Theorem 12.1.2 we have

$$f(y)=e^{y(t)}=\sum_{j=0}^{\infty} A_j(c_0,\ldots,c_j)H_j^1(t,0), \quad t\geq 0,$$

where

$$\begin{array}{rcl} A_0 & = & f(c_0) \\ & = & e^{c_0} \\ A_1 & = & c_1f'(c_0) \\ & = & c_1e^{c_0} \\ A_2 & = & c_2f'(c_0)+\frac{c_1^2}{2!}f''(c_0) \\ & = & \left(c_2+\frac{c_1^2}{2!}\right)e^{c_0} \\ A_3 & = & c_3f'(c_0)+c_1c_2f''(c_0)+\frac{c_1^3}{3!}f'''(c_0) \\ & = & \left(c_3+c_1c_2+\frac{c_1^3}{3!}\right)e^{c_0} \\ A_4 & = & c_4f'(c_0)+\left(c_1c_3+\frac{c_2^2}{2}\right)f''(c_0)+\frac{c_1^2c_2}{2}f'''(c_0)+\frac{c_1^4}{4!}f^{(4)}(c_0) \\ & = & \left(c_4+c_1c_3+\frac{c_2^2}{2}+\frac{c_1^2c_2}{2}+\frac{c_1^4}{4!}\right)e^{c_0} \\ \cdots & & \end{array} \tag{12.17}$$

The infinite nonlinear system (12.12) *for this example has the form*

$$\begin{aligned} c_0 &= 0, \\ \sum_{j=1}^{\infty} c_j B_1^j &= A_0(c_0) \\ \sum_{j=k}^{\infty} c_j B_k^j &= \sum_{j=k-1}^{\infty} A_j B_{k-1}^j, \quad k \geq 2, \end{aligned} \tag{12.18}$$

or, more explicitly,

$$\begin{aligned} c_0 &= 0, \\ c_1 B_1^1 + c_2 B_1^2 + c_3 B_1^3 + \cdots &= 1 \\ c_2 B_2^2 + c_3 B_2^3 + c_4 B_2^4 + \cdots &= c_1 B_1^1 + \left(c_2 + \frac{c_1^2}{2!}\right) B_1^2 + \cdots \\ c_3 B_3^3 + c_4 B_3^4 + c_5 B_3^5 + \cdots &= \left(c_2 + \frac{c_1^2}{2!}\right) B_2^2 + \cdots \\ \cdots \end{aligned}$$

Solving this nonlinear system one can approximately obtain c_i, $i \in \mathbb{N}$, and hence, the approximate solution of the initial value problem which is

$$y(t) = c_1 H_1^1(t,0) + c_2 H_2^1(t,0) + c_3 H_3^1(t,0) + \cdots \tag{12.19}$$

Example 12.3.3 *In the last example we consider the initial value problem associated with the first-order nonlinear dynamic equation of the form*

$$D_\beta y(t) = y^2 + 1, \quad y(0) = 0. \tag{12.20}$$

Assume that

$$y(t) = \sum_{j=0}^{\infty} c_j H_j^1(t,0), \quad t \geq 0,$$

where the coefficients c_j, $j \in \mathbb{N}$ will be determined from the nonlinear system (12.12). *Let*

$$f(y) = y^2 + 1 = \sum_{j=0}^{\infty} A_j(c_0, \ldots, c_j) H_j^1(t,0), \quad t \geq 0,$$

where

$$\begin{aligned} A_0 &= f(c_0) \\ A_1 &= c_1 f'(c_0) \\ A_2 &= c_2 f'(c_0) + \frac{c_1^2}{2!} f''(c_0) \\ A_3 &= c_3 f'(c_0) + c_1 c_2 f''(c_0) + \frac{c_1^3}{3!} f'''(c_0) \\ A_4 &= c_4 f'(c_0) + \left(c_1 c_3 + \frac{c_2^2}{2}\right) f''(c_0) + \frac{c_1^2 c_2}{2} f'''(c_0) + \frac{c_1^4}{4!} f^{(4)}(c_0) \\ &\cdots \end{aligned} \tag{12.21}$$

Since $f'(c_0) = 2c_0$, $f''(c_0) = 2$ and $f^{(m)}(c_0) = 0$ for $m \geq 3$, then we obtain

$$\begin{array}{rcl} A_0 & = & c_0^2 + 1 \\ A_1 & = & 2c_0c_1 \\ A_2 & = & 2c_0c_2 + c_1^2 \\ A_3 & = & 2c_0c_3 + 2c_1c_2 \\ A_4 & = & 2c_0c_4 + 2c_1c_3 + 2c_2^2 \\ & \cdots & \end{array} \tag{12.22}$$

The nonlinear infinite system (12.12) *becomes*

$$\begin{array}{rcl} c_0 & = & 0, \\ \sum\limits_{j=1}^{\infty} c_j B_1^j & = & A_0(c_0) \\ \sum\limits_{j=k}^{\infty} c_j B_k^j & = & \sum\limits_{j=k-1}^{\infty} A_j(c_0, \ldots, c_j) B_{k-1}^j, \quad k \geq 2 \end{array} \tag{12.23}$$

If, in particular, the time scale under consideration is $I = \mathbb{Z}$, then

$$h_0(t,0) = 1, \quad h_1(t,0) = t, \quad h_k(t,0) = \frac{t(t-1)\ldots(t-k+1)}{k!}, k = 2, 3, \cdots,$$

and hence, we compute

$$\begin{array}{rcl} H_1^1(t,0) & = & t = h_1(t,0) \\ H_2^1(t,0) & = & t^2 = 2h_2(t,0) + h_1(t,0) \\ H_3^1(t,0) & = & t^3 = 6h_3(t,0) + 6h_2(t,0) + h_1(t,0) \\ H_4^1(t,0) & = & t^4 = 24h_4(t,0) + 36h_3(t,0) + 14h_2(t,0) + h_1(t,0) \\ & \cdots & \end{array}$$

Then, the system (12.23) *turns into*

$$\begin{array}{rcl} c_0 & = & 0, \\ c_1 + c_2 + c_3 + c_4 + \cdots & = & 1 \\ 2c_2 + 6c_3 + 14c_4 + \cdots & = & c_1^2 + 2c_1c_2 + (2c_1c_3 + 2c_2^2) + \cdots \\ 6c_3 + 36c_4 + \cdots & = & 2c_1^2 + 12c_1c_2 + 14(2c_1c_3 + 2c_2^2) + \cdots \\ 24c_4 + \cdots & = & 6c_1c_2 + 36(2c_1c_3 + 2c_2^2) + \cdots \\ & \cdots & \end{array} \tag{12.24}$$

12.4 Notes and References

In this chapter we derive a quantum analogue of the Adomian polynomial method and present its application to a
beta-differential equations of arbitrary order with a nonlinear term. The results are provided with numerical examples for linear and nonlinear β-differential equations.

Bibliography

[1] A. Hamza, A-S. Sarhan, K. Shehata and K. Aldwoah. A general quantum calculus, Advances in Difference Equations, (2015) 2015:182.

[2] A. Hamza, A-S. Sarhan and E. Shehata. Exponential, trigonometric and hyperbolic functions associated with a general quantum difference operator, Advances in Dynamical Systems and Applications, Vol. 12, No. 1, pp. 25–37, 2017.

[3] V. Kac and P. Cheung. Quantum Calculus, Springer, 2002.

[4] E. Shehata, N. Faried and R. Zafarani. A general quantum Laplace transform, Advances in Difference Equations, (2020), 2020;613.

[5] S. Georgiev and S. Tikare. Generalized Quantum Calculus with Applications, Elsevier, 2024.

Index